HISTORY
SCIENCE ·
TECHNOL
REPRINT SERIES

Other titles in the History of Science and Technology Reprint Series:

A Brief History of Science by A. Rupert Hall and Marie Boas Hall
The Discovery of Our Galaxy by Charles A. Whitney
From Watt to Clausius: The Rise of Thermodynamics in the Early Industrial Age by D. S. L. Cardwell
The Gene: A Critical History by Elof Axel Carlson
A History of Biology to About the Year 1900: A General Introduction to the Study of Living Things by Charles Singer
A History of Cytology by Arthur Hughes
The Major Achievements of Science: The Development of Science from Ancient Times to the Present by A. E. E. McKenzie
Mind and Matter: Man's Changing Concepts of the Material World by Cecil J. Schneer
A Short History of Genetics: The Development of Some of the Main Lines of Thought: 1864–1939 by L. C. Dunn

A History of Western Technology

Friedrich Klemm

Translated by
Dorothea Waley Singer

IOWA STATE UNIVERSITY PRESS ■ Ames

The late **Friedrich Klemm** was librarian of the Deutsches Museum in Munich, Germany.

Originally published in German, *Technik: Eine Geschichte ihrer Probleme,* by Verlag Karl Alber © 1954

Original English translation published by George Allen and Unwin Ltd, Great Britain © 1959

First paperback edition published by Massachusetts Institute of Technology Press © 1964

This edition published in North America, © 1991, by Iowa State University Press, Ames, Iowa 50010

⊗ Text reprinted without correction on acid-free paper in the United States of America

History of Science and Technology Reprint Series
First printing, 1991

Library of Congress Cataloging-in-Publication Data

Klemm, Friedrich
[Technik. English]
A history of Western technology / Friedrich Klemm ; translated by Dorothea Waley Singer.
p. cm.—(History of science and technology reprint series)
Translation of: Technik : eine Geschichte ihrer Probleme.
Reprint. Originally published: London : Allen and Unwin, c1959.
ISBN 0-8138-0499-X
1. Technology—History. I. Title. II. Series.
T15.K5713 1991
609—dc20 91-15169

PREFATORY NOTE

THIS book derives from a series of lectures delivered by the author on the history of the exact sciences and technology. In it, attention is directed less to individual aspects of technology than to its problems in general. Historical details have therefore been omitted, and an effort has been made, by extensive use of contemporary documents, to reveal the forces which guided the development of technical advance in this or that direction. These formative influences may come from within technology itself or from outside. The book reveals how those influences which determine the intellectual climate of a period have their effect on technology and, on the other hand, how technology influences culture in general.

The documents which are quoted begin with Graeco-Roman times. Although, in many respects, it would have been desirable to include the technological activities of the pre-Greek cultures, the author decided they must be omitted except for a few indications considering that the trickle of written original matter from these epochs is very thin and that only those who understand ancient oriental languages are in a position to draw an unbiased picture of their technological development.

Quotations from foreign writers are given in English. Where classical translations into English exist already, these have been employed (whether contemporary with the original writer or not); but otherwise writers have been translated into modern English, whatever their period. To show important technical developments more clearly, a number of contemporary illustrations have been reproduced.

The quotations are drawn from a wide circle of sources. Alongside the statements of technologists are writings by philosophers, churchmen, naturalists, national economists, statesmen and poets. This must clearly indicate what varied spiritual forces have influenced technological development.

The author's main object is to contribute towards a wide popular understanding of technological history. In recent times especially, a number of notable works on this subject have appeared, but this book may claim something of a special character. It aims, as has already been emphasized, at relating historical problems to technological history derived from original sources.

A special word of thanks is due to Mr F. W. Kent of University College, London, for his invaluable help in the translation, especially of passages containing technical descriptions. The author has contributed some new material to the translation which was not in the original.

D. W. S.

CONTENTS

CONTENTS

LIST OF FIGURES

LIST OF FIGURES

PLATES

PART I

GRAECO-ROMAN ANTIQUITY

1. Old Roman water supply channels, carried on aqueducts. Painting by Zeno Diemer in the German Museum, Munich
The road is the Via Latina

2. Hinged rudder at stern. Earliest illustration. Bronze, Winchester Cathedral, 1180

Bellfounders, 1505. Miniature from the Cracow Behaim Codex

PART I

INTRODUCTION

WHEN man appeared at the dawn of history, he already had at his disposal a considerable treasury of technical instruments.

Excavation has revealed from the Palaeolithic or Old Stone Age the axe, the flint scraper, the spear, bow and arrow, gimlet, oil-lamp and all sorts of bone instruments. Among the cave drawings of the later Palaeolithic period in southern France and in Spain we even find pictures of what might be called, as J. E. Lips[1] has remarked, man's first machine. For here are portrayed big game such as bison, mammoth and reindeer, together with traps for them. The animal on entering the trap releases a lever mechanism which causes a number of transversely piled, heavy tree trunks to fall on him and bury him (Fig. 1). We still have to discover how it was possible to fell and to pile the great trees needed for mammoth traps. At that period, from 12,000 to 20,000 years ago, man lived in a world of magical identity between objects and their images, between matter and innate forces. Thus a representation of an animal imprisoned in a trap or transfixed by a spear was regarded as equivalent to actual possession of the animal. The actual snaring or slaying of the wild animal in open country was in some sort merely the completion of the magically achieved seizure. We are dealing with the hunting witchcraft and with the magical technique of a period when symbol and object were still regarded as one, as were also this world and the next. Man himself was not yet differentiated from Nature.

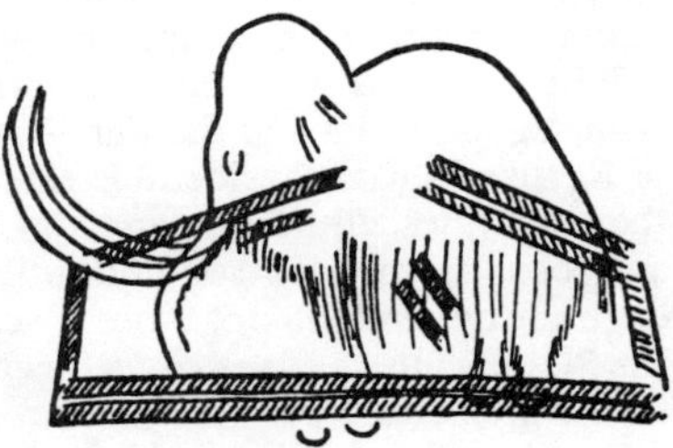

FIG. 1. *Mammoth in Gravity Trap (?)*
Cave drawing at Font-de-Gaume
(Late Palaeolithic)

In the succeeding periods, the Middle and the New Stone Age, belief in this identification of object with image or with its powers, of the visible with the spirit world, died out. An object was no longer identified with its powers; rather it was regarded as possessed by demoniac powers. Human self-consciousness began to develop; Man became increasingly distinguished from Nature and thereby he confronted Nature with increased power. To be sure, technical achievement at that period had to be linked with a ritual which was itself rooted in the effort to placate and to control the demoniac world within and surrounding the objects of our own world.

The late Stone Age saw the step forward toward agriculture and also toward animal husbandry. In the Neolithic Age a peasantry pursuing those activities appeard. There appeared the sharpened stone axe, the flint sickle, the sharp-edged chopper, saws, bone needles, bow string drills the spinning spindle, the loom, pottery, the hand grinding-mill, and finally the plough and the use of draught animals. Yet the employment of these instruments still remained rooted in ritual. Thus the plough that penetrated Mother Earth was at the same time a tool and a symbol of generation and fertility.

About 2000 B.C. the Stone Age implements of stone yielded slowly, in Europe, to instruments cast in copper and in bronze. About 1000 B.C. copper and bronze were increasingly replaced by iron which could be wrought in fire to the shape desired.

History opens with the civilizations of the great Empires of Western Asia, Egypt, India and the Far East. The technology of Mesopotamia and of Egypt was to a considerable extent determined by the three great rivers, the Euphrates, the Tigris and the Nile. Irrigation, the building of dykes and canals, the control and utilization of flood waters, were major engineering undertakings on a scale that necessitated organization by the State. Thus in place of technical crafts on a more or less family scale in the prehistoric period, there arose after the unification of the various territories of Egypt, or as Hans Freyer strikingly expresses it, when these had fought out their differences, a technique organized by the State, in close alliance with the priesthood. The great technical achievements of Egypt that included the building of the Pyramids and the preparation, conveyance and erection of obelisks. Such immense works could be undertaken only by a government with absolute powers which also supervised the organization of religion. Vast armies of workmen were at the disposal of the State and their disposition and provisioning required long term plans (Plate 1). Their tools and primitive cranes were extremely simple. Of the other technical accomplishments of the ancient Orient during the period roughly 4000–12000 B.C., a number of which influenced Graeco-Roman civilization, we mention only the wheel, bellows and tongs, all from the western Asiatic cultures; and from Egypt, mining and working of various metals, the sailing ship, papyrus, brewing of beer, tanning, glass-making, parchment, the steelyard, and the potter's wheel.

THE SUBORDINATE POSITION OF TECHNOLOGY

The great cultural achievement of ancient Greece was undoubtedly the development of a scientific sense. The Greek was in fact the first Man of

Theory. His life was devoted to and in a higher sense formed by, scientific understanding. Technology generally ranked in the Greek world below pure science. Platonic realism, in particular, regarded as 'Reality' the distant and changeless realm of Ideas rather than the objects of our mundane sphere, which were considered as but shadows and therefore of lesser rank. For this reason, experiment was of little import among the Greeks. All the more important was Geometry, dealing with conceptions appertaining to the world of Ideas. If Greece could match her development of mathematicized Statics with no corresponding science of Dynamics, that is to say no theory of Motion, the reason must, as Dingler[2] has shown, be sought in the conception of the changelessness and immovability of the Idea, the Form. The ancient world, with a purely static conception of Form, could not realize that Motion must itself be conceived as an Idea, a Form.

As already noted, the Greek world attained a real understanding of Statics just because the mathematical concept was regarded as the formative principle of the objects of our world. But the Greek was reluctant to take the step from theory to practical application.[3] A freeman devoted himself to the State, to pure science, to art. Practical production was in general the business of the foreigner and of slaves. The latter were very numerous in Greece, especially from the Hellenistic period.

Witness of the estimation in which manual and technical work was held is borne by Plato himself (*circ.* 380 B.C.) as follows:

If a man who is afflicted by great and incurable bodily diseases is in no way benefited by having been saved from drowning, much less he who has great and incurable diseases, not of the body, but of the soul, which is the more valuable part of him; neither is life worth having nor of any profit to the bad man, whether he be delivered from the sea, or the law-courts, or any other devourer; . . . such a one had better not live, for he cannot live well.

And this is the reason why the pilot, although he is our saviour, is not usually conceited, any more than the engineer, who is not at all behind either the general, or the pilot, or any one else, in his saving power, for he sometimes saves whole cities. Is there any comparison between him and the pleader? And if he were to talk, Kallikles, in your grandiose style, he would bury you under a mountain of words, declaring and insisting that we ought all of us to be engine-makers, and that no other profession is worth thinking about; he would have plenty to say. Nevertheless you despise him and his art, and sneeringly call him an engine-maker, and you will not allow your daughter to marry his son, or marry your son to his daughter. And yet, on your principle, what justice or reason is there in your refusal? What right have you to despise the engine-maker, and the others whom I was just now mentioning? I know that you will say, 'I am better, and better born'. But if the better is not what I say, and virtue consists only in a man saving himself and his, whatever may be his character, then your censure of the engine-maker, and of the physician, and of the other arts of salvation, is ridiculous. O my friend! I want you to see that the noble and the good may possibly be something different from saving and being saved.[1]

Now of artisans, let the regulations be as follows: In the first place, let no citizen or servant of a citizen be occupied in handicraft arts; for he who is to secure and preserve the public order of the State, has an art which requires much study and many kinds of knowledge, and does not admit of being made a secondary occupation; and hardly any human being is capable of pursuing two professions or two arts rightly, or of practising one art himself, and superintending someone else who is practising another. . . . Let every man in the state have one art, and get his living by that. Let the wardens of the city labour to maintain this law, and if any citizen incline to any other art rather than the study of virtue, let them punish him with disgrace and infamy, until they bring him back into his own right course; and if any stranger profess two arts, let them chastise him with bonds and money penalties, and expulsion from the state, until they compel him to be one only and not many.[2]

The following well-known passage from Plutarch concerning Archimides (who died 212 B.C.) is significant testimony to the extreme value ascribed to pure theory as against its practical application to which recourse would be had only from dire necessity and which was widely regarded as unworthy of an author's attention:

Marcellus proceeded to attack the city by land and sea, Appius leading up the land forces, and he himself having a fleet of sixty quinqueremes filled with all sorts of arms and missiles. Moreover, he had erected an engine of artillery on a huge platform supported by eight galleys fastened together, and with this sailed up to the city wall, confidently relying on the extent and splendour of his equipment and his own great fame. But all this proved to be of no account in the eyes of Archimedes and in comparison with the engines of Archimedes. To these he had by no means devoted himself as work worthy of his serious effort, but most of them were mere accessories of a geometry practised for amusement, since in bygone days Hiero the king had eagerly desired and at last persuaded him to turn his art somewhat from abstract notions to material things, and by applying his philosophy somehow to the needs which make themselves felt, to render it more evident to the common mind.

For the art of mechanics, now so celebrated and admired, was first originated by Eudoxus and Archytas, who embellished geometry with its subtleties, and gave to problems incapable of proof by word and diagram, a support derived from mechanical illustrations that were patent to the senses. For instance, in solving the problem of finding two mean proportional lines, a necessary requisite for many geometrical figures, both mathematicians had recourse to mechanical arrangements, adapting to their purposes certain intermediate portions of curved lines and sections. But Plato was incensed at this, and inveighed against them as corrupters and destroyers of the pure excellence of geometry, which thus turned her back upon the incorporeal things of abstract thought and descended to the things of sense, making use, moreover, of objects which required much mean and manual labour. For this reason mechanics was made entirely distinct from geometry, and being for a long time ignored by philosophers, came to be regarded as one of the military arts.

And yet even Archimedes, who was a kinsman and friend of King Hiero, wrote to him that with any given force it was possible to move any given weight; and emboldened, as we are told, by the strength of his demonstration, he declared that, if there were another world, and he could go to it, he could move this. Hiero was astonished, and begged him to put his proposition into execution, and show him some great weight moved by a slight force. Archimedes therefore fixed upon a three-masted merchantman of the royal fleet, which had been dragged ashore by the great labours of many men, and after putting on board many passengers and the customary freight, he seated himself at a distance from her, and without any great effort, but quietly setting in motion with his hand a system of compound pulleys, drew her towards him smoothly and evenly, as though she were gliding through the water. Amazed at this, then, and comprehending the power of his art, the king persuaded Archimedes to prepare for him offensive and defensive engines to be used in every kind of siege warfare. These he had never used himself, because he spent the greater part of his life in freedom from war and amid the festal rites of peace; but at the present time his apparatus stood the Syracusans in good stead, and, with the apparatus, its fabricator.

And yet Archimedes possessed such a lofty spirit, so profound a soul, and such a wealth of scientific theory, that although his inventions had won for him a name and fame for superhuman sagacity, he would not consent to leave behind him any treatise on this subject, but regarding the work of an engineer and every art that ministers to the needs of life as ignoble and vulgar, he devoted his earnest efforts only to those studies the subtlety and charm of which are not affected by the claims of necessity. These studies, he thought, are not to be compared with any others; in them the subject matter vies with the demonstration, the former supplying grandeur and beauty, the latter precision and surpassing power.

. . . .

But what most of all afflicted Marcellus was the death of Archimedes. For it chanced that he was by himself, working out some problem with the aid of a diagram, and having fixed his thoughts and his eyes as well upon the matter of his study, he was not aware of the incursion of the Romans or of the capture of the city. Suddenly a soldier came upon him and ordered him to go with him to Marcellus. This Archimedes refused to do until he had worked out his problem and established his demonstration, whereupon the soldier flew into a passion, drew his sword, and despatched him. However, it is generally agreed that Marcellus was afflicted at his death, and turned away from his slayer as from a polluted person, and sought out the kindred of Archimedes and paid them honour.[3]

Finally, we will cite from Imperial Rome two letters of Seneca (d. A.D. 65) to Lucilius. They betray a similar contempt for all manual labour which is executed with bowed body and lowered eyes:

Posidonius says there are four kinds of arts—the workaday-mercenary, the scenic, the pupillary, and the liberal. The workaday are those of the artisan, manual crafts engaged in catering for life and making no pretence of any aesthetic or moral ideals. The scenic are those which aim at the diversion of eye and ear. Among their professors you may count the engineers who invent automatically rising platforms, stages that noiselessly sprout upwards story upon story, and various other surprises—solid floors that yawn, great chasms mysteriously closing, tall structures slowly telescoped. It is by these devices that the uneducated, who wonder at anything sudden because they don't know its cause, are dazzled. The pupillary (which aren't altogether unlike the really liberal) are those arts which the Greeks call ἐγκυκλίους—encyclic—and our own writers liberal. But the only liberal arts, or rather, to speak more truly, arts of liberty, are those which concern themselves with virtue.[4]

Further, I dispute the statement that it was the philosopher who discovered iron and copper mines, when the earth, white-hot from a forest-fire, spewed up molten veins of surface-ore: these things are found by the persons who are interested in them. Even the question whether the hammer or the pincers came into use first doesn't seem to me so subtle as it does to Posidonius. Both were invented by some one whose intellect was nimble and keen, not massive and sublime: so was everything else the quest of which involves bowed shoulders and earthward gaze.

. . . .

All those handicrafts which fill the city either with the solicitations of the hawker or the clatter of the factory, are about the business of the body, which was once a slave with slave's allowance, but is now the master whose every esteemed command is executed. Thus on either side we see the cloth-mill and the foundry, here the perfumer's distillery, there the academy of effeminacy in the shape of dancing and deportment or the more sentimental forms of music. For the natural standard, which defines a man's desires by the satisfaction of his needs, has disappeared.

. . . .

Some things we know to have appeared only within our own memory; the use, for example, of glass windows which let in the full brilliance of day through a transparent pane, or the substructures of our baths and the pipes let into their walls to distribute heat and preserve an equal warmth above and below. Need I enlarge on the marbles with which our temples and houses gleam? On the columns, vast polished cylinders of stone, on which we support colonnades and halls in which whole nations could find room? Or on the shorthand which catches even the quickest speech, the hand keeping pace with the tongue? All these are the inventions of the meanest slaves. Philosophy sits more loftily enthroned: she doesn't train the hand, but is instructress of the spirit.

. . . .

She doesn't contrive arms and fortifications and munitions of war: she is the friend of peace, and summons the human race to brotherhood. No, she's not, I say, an artisan producing tools for the mere everyday necessities.[5]

. . . .

FIG. 2. *Greek Smithy.* 500 B.C.
Picture from an Attic amphora of 500 B.C. Left, the master smith; centre, the smith's mate; right, spectators.—From: Blümner, *Technologie und Terminologie der Gewerbe und Künste bei Griechen und Römern*

Handworkers were, however, not so universally despised as might be concluded from the evidence of certain authors of antiquity. In the Homeric age, even those of noble birth were by no means ashamed to perform manual labour, and in Athens the number of manual workers in the popular Assembly was relatively quite high.[4]

TECHNOLOGY OUTWITS NATURE

Archimedes created mathematical mechanics, embodied in two works on the Statics of solid and of liquid bodies respectively. He published, as we saw above, no work concerning the application of the new knowledge. He developed the principles of his mechanics by a process of pure deduction in the true Euclidean sense from a few simple observations and definitions. Mathematics was the principal interest. There is a work, written some hundred years before Archimedes, and formerly ascribed to Aristotle which is also concerned with Mechanics; the pseudo-Aristotelian *Mechanical Problems* dating probably from the time of about 280 B.C. mentions numerous applications of Mechanics, although not for their own sake. The concern is rather with dialectical discussion and solution of the so-called *Aporisms*, that is to say problems—in this case of practical Mechanics—which present special contradictions and difficulties. There is an *Aporie* or absurdity when a small force is said to move a heavy weight. It is noteworthy that the author regards such a technical procedure as *against Nature*. Technology is *Machination*, an artful method, the word derived from the Greek 'mechanáomai' 'I contrive a deception'. Thus what was at stake in mechanical technology was to outwit Nature by solving the apparent contradictions, and overcoming the difficulties. The author of these *Mechanical Problems* was by no means mathematicizing in the manner of Archimedes. He turned to concrete events in the realm of mechanical technique, though chiefly for the purpose of demonstrating by his dialectical skill that here in fact were no contradictions. Most of the mechanical devices considered bring the author back to the lever, and he demonstrates that the absurdities which he cites of the apparently miraculous and self-contradictory moving of a heavy body by a small force is explained by the dialectically contradictory miraculous properties of the circle! For with the lever, both force and burden move through an arc of a circle. The circumference appertains indeed to the essential nature of the lever. In the lever, the remarkable nature of the circle reappears whereby opposed qualities are miraculously joined in unity. Here then the significance of the *Philosophy of Nature* occupies the foreground. With Archimedes it is Mathematics, but here it is Philosophy which relegates practical Mechanics to the background. Despite the different point of view, we have in each case an expression of typical Greek mentality.

We will now give a few passages from the pseudo-Aristotelian *Mechanical Problems*:

Our wonder is excited, firstly, by phenomena which occur in accordance with nature but of which we do not know the cause, and secondly by those which are produced by art despite Nature, for the benefit of mankind. Nature often operates contrary to human expediency; for she always has her own methods. When, therefore, we have to do something contrary to Nature, the difficulty of it causes us perplexity, and art has to be called to our aid. The kind of art which helps us in such perplexities we call Mechanical Skill. The words of the poet Antiphon are quite true:

'Mastered by Nature, we o'ercome by Art.'

Instances of this are those cases in which the lesser prevails over the greater, and where small forces move great weights—in fact practically all those problems which we call Mechanical Problems. They are not quite identical nor yet entirely unconnected with Natural Problems. They have something in common both with mathematical and with natural speculations; for theory is determined by mathematics and practical problems by physics.

Among questions of a mechanical kind are included those which are connected with the lever. It seems strange that a great weight can be moved with but little force, and even when the addition of more weight is involved; for the very same weight, which one cannot move at all without a lever, one can move quite easily with it, in spite of the additional weight of the lever.

The original cause of all such phenomena is the circle. It is quite natural that this should be so; for there is nothing strange in a lesser marvel being caused by a greater marvel, and it is a very great marvel that contraries should be present together, and the circle is made up of contraries. For to begin with, it is formed by motion and rest,* things which are by nature opposed to one another. Hence in examining the circle we need not be much astonished at the contradictions which occur in connexion with it. Firstly, in the line which encloses the circle, being without breadth, two contraries somehow appear, namely, the concave and the convex. These are as much opposed to one another as the great and the small; the mean being in the latter case the equal, in the former the straight. Therefore just as, if they are to change into one another, the greater and smaller must become equal before they can pass into the other extreme; so a line must become straight in passing from convex into concave, or on the other hand from concave into convex and curved. This, then, is one peculiarity of the circle.

Another peculiarity of the circle is that it moves in two contrary directions at the same time; for it moves simultaneously to a forward and a backward position. Such, too, is the nature of the radius which describes a circle. For its extremity comes back again to the same position from which it starts; for, when it moves continuously, its last position is a return to its original position, in such a way that it has clearly undergone a change from that position.

Therefore, as has already been remarked, there is nothing strange in the circle being the origin of any and every marvel. The phenomena observed in the balance can be referred to the circle, and those observed in the lever to the balance; while practically all the other phenomena of mechanical motion are connected with the lever.[6]

Why is it that those rowers who are amidships move the ship most? Is it because the oar acts as a lever and the fulcrum is the thole-pin; for it remains in the same place; and the weight is the sea which the oar displaces; and the power that moves the lever is the rower. The further he who moves a weight is from the fulcrum, the greater is the weight which he moves; for then the radius becomes

* i.e. by the motion of a line round a fixed point.

greater, and the thole-pin acting as the fulcrum is the centre. Now amidships there is more of the oar inside the ship than elsewhere; for there the ship is widest, so that on both sides a longer portion of the oar can be inside the two walls of the vessel. The ship then moves because as the blade presses against the sea, the handle of the oar, which is inside the ship, advances forward, and the ship, being firmly attached to the thole-pin, advances with it in the same direction as the handle of the oar. For where the blade displaces most water, there necessarily must the ship be propelled most; and it displaces most water where the handle is furthest from the thole-pin. This is why the rowers who are amidships move the ship most; for it is in the middle of the ship that the length of the oar from the thole-pin inside the ship is greatest.[7]

Why is it that a missile travels further from a sling than from the hand, although the thrower moves it further with the hand alone than when using a sling? Because in the latter case he moves two weights, that of the sling and the missile, while in the former case he moves only the missile. Because he who casts the missile does so when it is already in motion in the sling (for he swings it round many times before he lets it go), whereas when cast from the hand it starts from a state of rest? In using a sling the hand becomes the centre and the sling the radius, and the longer the radius is the more quickly it

FIG. 3. *Roman hoisting gear for columns.*
Driven by Treadmill. Relief in Amphitheatre in Capua.—From: H. Blümner, *Technologie und Terminologie der Gewerbe und Künste bei Griechen und Römern*

moves, and so a cast from the hand is short as compared with a cast from a sling?[8]

The example of the oar is evidently intended to show that it is solely by reason of leverage that with a stroke of equal force, a longer oar moving through the same angle will propel a ship at greater speed than will an oar that is shorter, notwithstanding an equal ratio in each case between the load and the leverage.[5] In that matter this chapter is, like the treatment of many of the thirty-five problems in the work, very far from clear. The ambiguities were recognized by Renaissance writers who nevertheless valued highly both the *Mechanics* of Archimedes and the pseudo-Aristotelian *Mechanical Problems.*[6]

SPECIALIZATION AND THE DIVISION OF LABOUR

Already in pre-Hellenistic Greece, besides home industries and small scale land cultivation, there existed, at least in the larger towns, an advanced manufacturing technology. We hear of large workshops in which actual division of labour was undoubtedly practised for the execution of different manual tasks.[7] Machinery here played almost no part. Again, we find in Athens at the end of the fifth century B.C. twenty slaves occupied in bedstead manufacture. A workshop with thirty-two slaves produced knives. The well-known shield factory of Lysias employed 120 slaves. Odysseus still boasted of having constructed his own bridal couch; but a high-class Athenian of the period of Pericles would obtain such a commodity from a furniture dealer. Workshops depending on slave labour increased during the Hellenistic period; and in the Roman empire there were large workshops producing cheap goods in quantity.[8] But the large workshop did not eliminate the home industry of antiquity, that is to say the home production of the most important goods; and machinery was hardly ever introduced, for the supply of slave labour rendered it unnecessary to introduce costly labour-saving machinery. Slave labour was indeed not cheap; but the slave could always be sold, if necessary, and therefore constituted mobile capital. In late Roman times, as Rostovtzeff endeavours to demonstrate, general social and economic conditions in any case probably did not permit initially expensive mechanized industry.

For evidence of specialization and division of labour in ancient times, we will cite Xenophon (about 370 B.C.).

For in small towns the same workman makes chairs and doors and ploughs and tables, and often this same artisan builds houses, and even so he is thankful if he can only find employment enough to support him. And it is, of course, impossible for a man of many trades to be proficient in all of them. In large cities, on the other hand, inasmuch as many people have demands to make upon each branch of industry, one trade alone, and very often even less than a whole trade, is enough to support a man: one man, for instance, makes shoes for men, and another for women; and there are places even where one man earns a living by only stitching shoes, another by cutting them out, another by sewing the uppers together, while

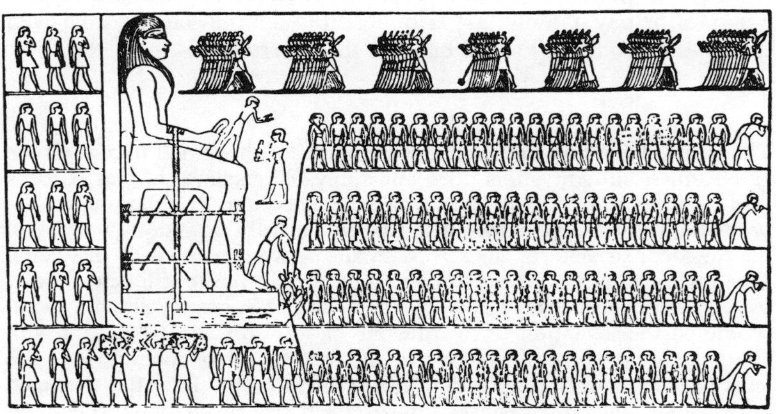

FIG. 4. *Slave labour in antiquity.*
Transport of an alabaster colossus, 7 m. high, on skids, by 172 slaves. Relief from El Berscheh (Egypt). Middle Kingdom, 11/12th Dynasty, 2000 B.C.

there is another who performs none of these operations but only assembles the parts. It follows, therefore, as a matter of course, that he who devotes himself to a very highly specialized line of work is bound to do it in the best possible manner.

Exactly the same thing holds true also in reference to the kitchen: in any establishment where one and the same man arranges the dining couches, lays the table, bakes the bread, prepares now one sort of dish and now another, he must necessarily have things go as they may; but where it is all one man can do to stew meats and another to roast them, for one man to boil fish and another to bake them, for another to make bread and not every sort at that, but where it suffices if he makes one kind that has a high reputation—everything that is prepared in such a kitchen will, I think, necessarily be worked out with superior excellence.[9]

As an example of division of labour with the different workshops separated by some 438 miles of sea, we will cite the *Natural History* of Pliny (d. A.D. 97):

Aegina specialized in producing only the upper parts of chandeliers, and similarly Taranto made only the stems, and consequently credit for manufacture is, in the matter of these articles, shared between these two localities. Nor are people ashamed to buy these at a price equal to the pay of a military tribune, although they clearly take even their name from the lighted candles they carry.[10]

MAKING ARTILLERY

Though we have said that machinery played a rather small part in Greek and Roman antiquity, this does not apply to machines of war. In this field Greece utilized both practical experience and calculation for the production of large machines. The efficient catapult of antiquity relied on the torsional strength of thick ropes made of women's hair or of sinews. This torsional artillery was probably invented in the fourth century B.C. by the adherents of the elder Dionysius, tyrant of Syracuse, who in general promoted the invention of instruments of war to sustain his fight with Carthage. Be it noted, however, that the first use of torsional power goes back to the torsion spring traps of the prehistoric age. But after the reign of Alexander the Great marked improvement was effected in artillery and in other implements of war by the application of scientific knowledge. In Alexandria in the second or third century B.C. by the use of experiment and theory, a formula was discovered establishing the relation between calibre and the length of the bolt to be fired or between the calibre and the weight of the stone missile. Thus the name calibre was given to the diameter of the so-called straining-hole through which the bundles of sinew were drawn. (Fig. 5.) The dimensions of each of the other parts were stated in multiples of the calibre. The method used was that which in eighteenth and nineteenth century machine construction was called the proportional coefficient. The material for these weapons was wood, strengthened where necessary with iron bands. Such a torsional missile thrower may be regarded as a rationally designed and constructed machine as regards both material and function. Regarding the construction of Greek artillery, which was adopted by the Romans, we will quote Philo of Byzantium (*circ.* 200 B.C.), Vitruvius (first century B.C.) and Hero of Alexandria (first century A.D.):

I think, Ariston, that you are not unaware that this art of constructing artillery appears to most men somewhat hard to understand

and to judge. Thus many who have attempted to build several pieces of equally large scale artillery, and have used similar woods and constructed the same parts of iron, introducing moreover no variation in the weight, have produced some weapons of long range and great penetration and others that have proved less successful; and when asked the reason for this, they can give no reply. Very apposite therefore appears to me the remark of the sculptor Polykleitos that success depends on a number of proportional relationships, so that a small deviation may decide the issue. Similarly in this art, the work depends on the correct ranging of many numerical relationships, so that a small change in an individual part may lead to a gross error in the production. Therefore in my opinion the disposition of a successful missile throwing weapon must be copied with the greatest care, especially if either a larger or smaller size is to be adopted. I hope that those who use what I have written will be in no uncertainty on this point. . . . Later workers, partly by recognition of the mistakes of their predecessors, partly by observation of modern experiments, have now established the principle and theory of the construction of a missile thrower on a sound basis; I mean the diameter (caliber) of the circle which holds the torsion strands. Alexandrian craftsmen have lately attained success in this matter, because they have been provided with ample means by monarchs who love renown and the arts. For it has become abundantly clear that not everything can be achieved by theory and the methods of mechanics, but much may be discovered by experiment. . . . It is, however, the aim of the art of constructing artillery to enable the missile to be thrown far and with greater penetrative force and it is for this purpose that most of the experiments and researches have been undertaken. I will only tell you of my own experience, gained through meeting many of these craftsmen in Alexandria, and through acquaintance with numerous engineers in Rhodes with whom I saw many well proven missile throwing weapons. . . .[11]

All the dimensions of the machines as designed are given from the proposed length of the arrow which the machine is to let fly. The ninth part of this gives the size of the opening in the frame. Through these openings twisted cords are stretched,* which are to hold back the arms of the catapults themselves.

The height and breadth of the frame are fixed by the size of the holes. The cross-pieces at the top and bottom of the frame are called

* The cords are wound as tightly as possible round the nuts above and below the cross-pieces. The arm passes between the cords and still further stretches them.

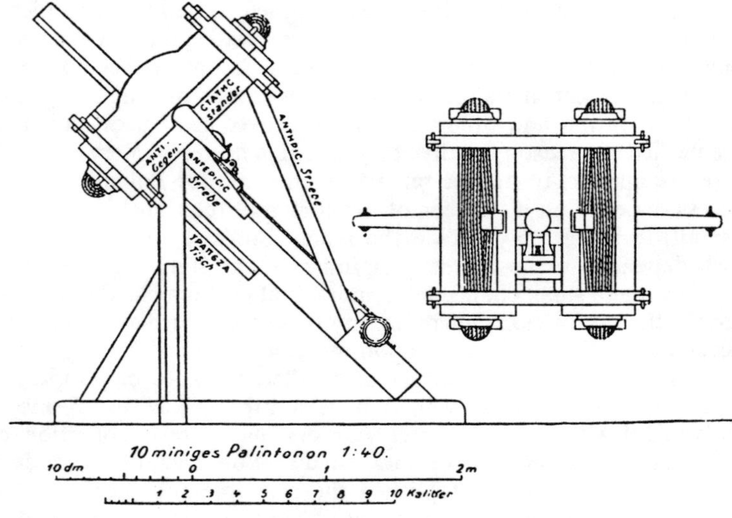

FIG. 5. *Stone-throwing Catapult.*
Left: Elevation of weapon. Right: Tension frame with gut strands. Diameter of the Straining holes = Calibre.—From: Hero, *Belopoiika.* Greek and German by H. Diehls and E. Schramm

peritreta or *perforated beams* and are to be one hole thick, and one and three-quarters wide; at the ends, one and a half.[12]

The design of the ballista varies and its differences are adjusted for the purpose of a single effect. For some are worked by levers and windlasses, some by many pulleys, some by capstans, some by wheels. Yet all ballistae are constructed with a view to the proposed amount of the weight of the stone which such a machine is to let fly. Therefore only those craftsmen can deal with the design who are familiar with the geometrical treatment of numbers and their multiples.

The diameter of the holes which are made in the frames (through the openings of which ropes are stretched, made especially of woman's hair or of the sinews of animals) are taken proportionately to the weight of the stone which the ballista is to shoot, in accordance with gravity, just as in the case of catapults the length of the arrows furnishes the module. Therefore in order that persons who are ignorant of geometry may be equipped and may not be delayed by calculation amid the perils of war, I will expound in

3. Weight-driven clocks. Engraving by Ph. Galle after Stradano, 1570. From Stradano, Nova Reperta

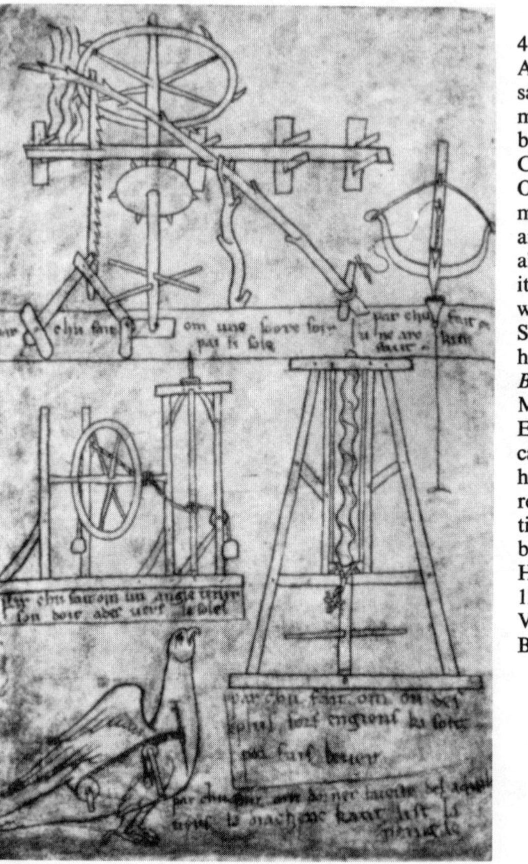

4. *Left*: Automatic sawing machine driven by waterwheel. Crossbow. Operating mechanism for an angel which always points its finger towards the sun. Screw operated hoist. *Below*: Mechanical Eagle which can turn its head in any required direction. Drawing by Villard de Honnecourt, 1235. From Villard's Bauhüttenbuch

Overshot waterwheel, 15th century, for driving a pump. Note the crank, connecting rod and beam. Coloured pen drawing from the medieval Hausbuch, 1480

accordance with my own knowledge gained in practice and also in accordance with the instructions of my teachers. Further I will show the proportions given in the writings of the Greeks between the weights and the measurements of size, I will expound these relationships in terms corresponding to our weights.[13]

The greatest and most indispensable department of worldy wisdom is that which concerns peace of mind; to it have been devoted and are even at the present day devoted most of the philosophers' enquiries; moreover I believe that there will never be an end to theoretical examination of the subject. But Mechanics passed beyond theoretical teaching on peace of mind, and brought knowledge to all men, of how to live in peace of mind by means of one single, minimal part thereof which treats of the so-called manufacture of artillery. For, through the knowledge gained thereby concerning these machines, a man may remain unperturbed either in peace-time by attacks from internal or exterior enemies, or by the outbreak of war. To that end, this part of Mechanics must be studied industriously at all times, and every precaution must be taken. Even in the profoundest peace, a man must realise that he will be more secure if he busy himself with the construction of artillery. Then, in the consciousness of strength, not only will peace of mind be preserved, but those with evil intentions, seeing such preoccupation with this machinery, will not dare to proceed to the attack. But if that is neglected, and if the inhabitants of a town have made no preparation, then every attack, however insignificant, will meet with success.

Now our ancestors composed many works on the construction of propulsive weapons, on their size and form. But since not one of them has systematically described the construction and utilization of these machines, and they have written as though all that were obvious, therefore I think it well to begin with this in my instruction concerning artillery, and to explain the machines used for making propulsive weapons as though they were not at hand to be seen. Thus will all be able easily to follow the instruction.

We will consider the construction of these machines in general and of their individual parts, with their names, how they are assembled and held firmly together and after we have observed the differences between the machines, their construction from the very beginning, we will consider the use and the relative size of each part.

Of these machines, some are called *Euthytona*, or straight, and some *Palintona* or bent backwards. Some call the *Euthytona Scorpions*, since they resemble them in shape. Euthytona can discharge only arrows. The Palintona are called by some stone-throwers since they discharge stones; they discharge either stones or arrows, or a mixture of both.

The art in the construction of artillery is that the missile may be discharged over maximum distance with maximum force against an object. Every effort must be directed toward that purpose.

Missile is the name given to everything discharged by artillery or by any other force such as bows, slings or any other weapons.

The construction of these machines was originated in bows worked by hand. But it became necessary to discharge larger missiles over a greater distance. Larger bows were therefore constructed with increased tension, I mean the resistance to bending by each end of the bow, that is to say, the strength of the horns. As these now became more resistant to bending, a greater force than the hand was needed to bend them.[14]

Most parts of the weapon are removable, so that when necessary it can be taken to pieces and conveniently transported; only the two halves of the frame are not separated, in order that the fibres may be easily threaded through them.[15]

Certain parts which are subject to special strain should be furnished with iron covering fixed with nails; strong woods should be used, and these parts should be secured in every possible manner. But those parts which will not be subject to such strain are made of lighter wood, of less strength, to save both bulk and weight. For since they are not made for immediate use they must be easily dismantled and not costly.[16]

It must be known that the determination of size is based on experience. For since our predecessors paid attention only to the shape and the assembly of the weapon; they achieved no great range for the missile since they did not provide the necessary harmonious proportions. But more recent workers, reducing some parts and enlarging others, have thereby made each weapon accurate and workable. These weapons, that is to say, each individual part, was designed according to the diameter of the orifice for the stretched fibres. The stretched fibre thus determines the size.

The calibre of the Stone-thrower must be estimated as follows:

Multiply by 100 the weight—in *MINAS*—of the stone missiles;

Take the cube root of the product and add $\frac{1}{10}$ of the number thus produced. This gives the calibre in finger-lengths [daktyls].

$$d\,(\text{in finger-lengths}) = 1{\cdot}1 \times \sqrt[3]{100\ \textit{minas}}.$$

For example: If the weight of stones is 80 *minas*

$$80 \times 100 = 8000$$
$$\sqrt[3]{8000} = 20$$
$$\tfrac{1}{10} \text{ of } 20 = 2$$
$$2 + 20 = 22$$

Therefore the calibre will measure 22 finger-lengths. If the cube root is not an exact number it is rounded off by the addition of one tenth. (Cf. Fig. 5).[17]

THE MECHANIC OF ALEXANDRIA

We have already emphasized that, with a few exceptions such as the erection of war machines in Greek and in Roman times and great undertakings for the storage of water which was pumped by the archimedian screw in the mines of Spain under Roman rule, there was in antiquity almost no development of machine craft. In the circumstances, since there was no general need for large machines, the impulse toward technical achievement was partly diverted toward the construction of small apparatus and mechanisms more in the nature of toys. It was in Alexandria in the third century B.C. that the art of building delicate mechanical apparatus first flourished. Here at that period worked Ktesibios, who, in addition to toys, constructed numerous objects of practical utility, a few of which served a scientific purpose. The water organ and the water pump were important discoveries of Alexandrian mechanics. In the succeeding years, this craft of constructing apparatus more or less as playthings but popular for that very reason, was carried further by men who were attracted toward delicate mechanical craftsmanship and were also scientifically gifted.

Remarkable in this respect were Philo of Byzantium, about 200 B.C., and in the first century A.D. Hero of Alexandria who was partly dependent on Philo, but was in many matters an original-minded mechanical craftsman. Hero describes for us in his works numerous devices, such as the so-called pressure engines which utilize the pressure of compressed or heated air or of steam, and work with well-finished levers, valves, stopcocks, gear-wheels, screws, and cylinders fitted with pistons; or again the puppet theatres in which the figures are moved by weights attached to strings carried over pulleys. Much of Hero's apparatus served the purposes of religious ritual, as, for example, the holy-water machine and the temple door-opener. He exercised great influence on the Middle

Ages, the Renaissance and the Baroque period. The elaborate medieval clocks, the fountains with mysteriously moving figures in the gardens of Renaissance princes, the thermometers of Santorre Santorio (Sanctorius), Drebbel and Galileo in the early seventeenth century, all derive ultimately from Hellenistic mechanics.[9] It is not easy to distinguish between the achievements of Ktesibios, Philo and Hero.[10] A few passages from Hero's works on *Pressure Machines* and on *Puppet theatres* (first century A.D.) will illustrate the delicate mechanical technique of Alexandria:

If an incense-offering be kindled on certain altars, then figures standing close by shall bring a drink-offering.

If into certain sacrificial vessels a coin of five drachmas be thrown, water for sprinkling shall flow out from them.

If of two vessels standing on a pedestal, one be full of wine and the other empty, whatever quantity of water be poured into the empty vessel, as much wine shall flow from the other.

Syphons used in outbreaks of fire are arranged as follows. Two bronze hollow chambers (cylinders) are used . . . with a bore suitable for the passage of the piston . . . These are connected to one another by a tube open at both ends. Outside the cylinders, but within this tube . . . clack-valves shall be fixed in such a manner . . . that they can open towards the exterior of the cylinders. (Fig. 6.)

Construction of a temple of which the doors will open spontaneously when the sacrificial fire has been lit and shut when it has been extinguished. (Fig. 7.)

On a pedestal is a small tree, around which is coiled a snake. Nearby stands the Archer Hercules. An apple also lies on the pedestal. If the apple be raised a little by hand Hercules shall then discharge his arrow at the serpent, and the serpent shall hiss.

Construction of an organ such that, when the wind blows, one of its pipes will emit a note.

Many vessels emit a spray of water . . . if one blows into them. If a fire is lit on a certain altar figures will be made to dance. The altar must be transparent, made either of glass or of horn.

Over a heated boiler a ball shall revolve around a pivot. (Fig. 8.)

Construction of a cupping-glass, that draws without being warmed.

To construct a lampstand such that if the oil in a lamp that has been placed on it becomes exhausted, the desired quantity of oil shall pour into the lamp from its own handle, and there shall be no need to place on the lamp a container from which oil can flow into it.[18]

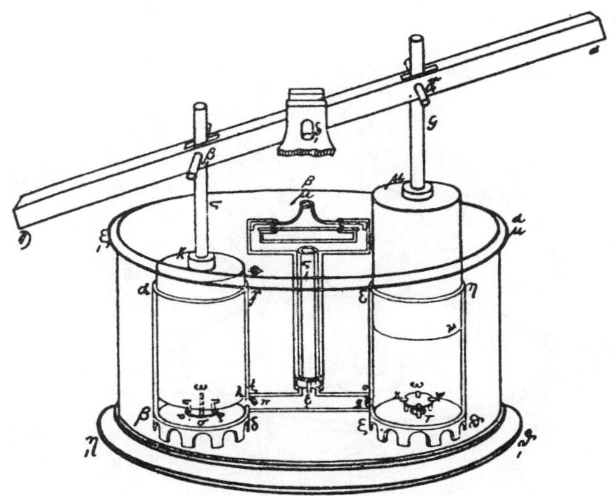

FIG. 6. *Hero's Fire Engine.*
From: Hero, *Opera.* Vol. 1

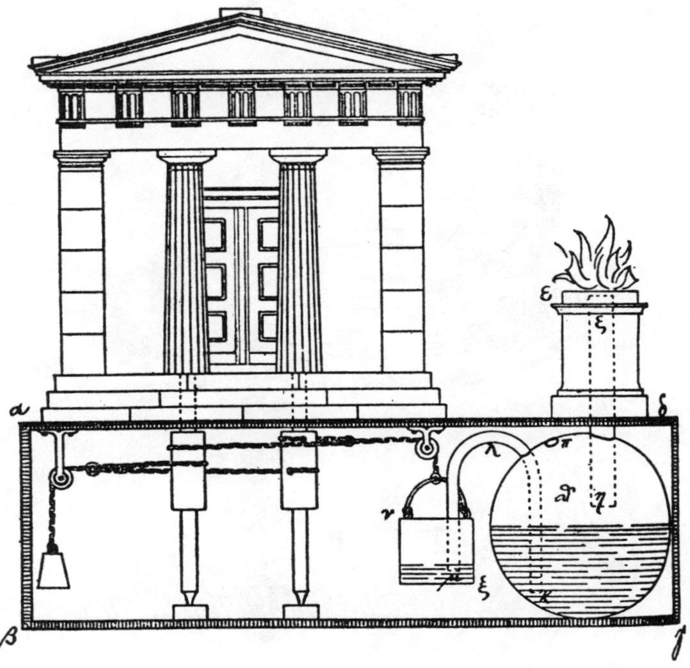

FIG. 7. *Hero's pneumatic Temple Door opener.*
The doors open by themselves when the sacrificial fire is lighted.—From: Hero, *Opera*

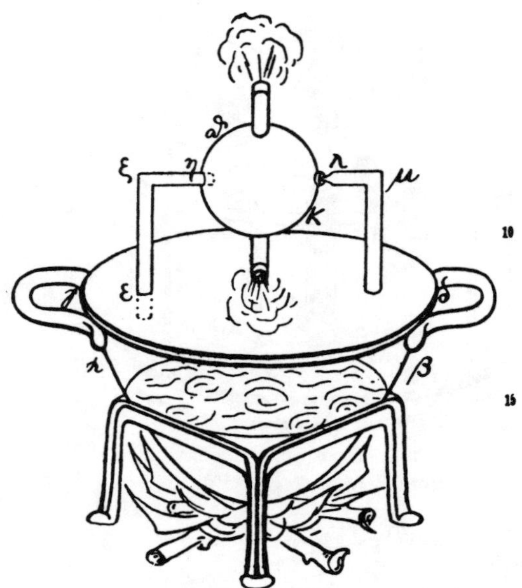

FIG. 8. *Hero's Aeolipile (Reaction Steam Rotor).*
From: Hero's *Opera*

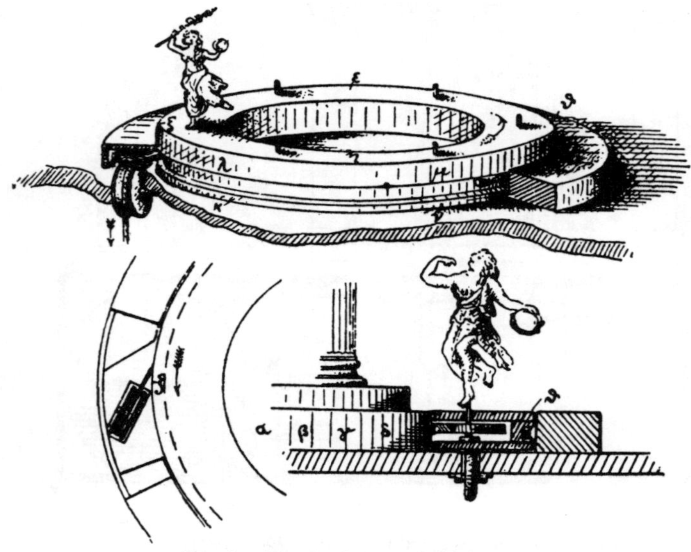

FIG. 9. *Hero's Automatic Theatre.*
The dance of the Bacchantes.—From: Hero's *Opera*

The puppet theatre enjoyed great popularity among our predecessors, partly because it developed such varied artistic skill and partly on account of the astonishment aroused by the spectacles it presented (Fig. 9). For, to put it shortly, every branch of mechanics is employed in the manufacture of the puppets and is indeed necessary for the performance of each single one of them. Here is a sample of the programmes promised from the puppets. A temple or altars of medium size are constructed and they move spontaneously and then halt at a prearranged spot; then each of the figures within moves in accord with the same plan or with its own story; finally temple and altar return to their original position. Puppets constructed for this performance are named travellers. There is, however, another kind named stationary. These undertake the following performance. On a low pedestal is placed a tablet, with doors that open. On the tablet is seen a representation of figures ranged on a definite plan. To begin with the doors of the tablet are closed; they then open spontaneously and the grouping of the figures in the representation is visible. If after a short time the doors again spontaneously close and then reopen, the figures appear to be differently grouped, though in a manner corresponding to that of the earlier representation. If the doors are yet again closed and reopened, the figures are revealed once more in their original grouping and either this closes the performance or a fresh representation is disclosed until the performance is concluded. Each one of the visible figures painted on the tablet appears to be in motion whenever this is needed for the story. Some, for example, can be sawing, others can have a bill hook in their hand, yet others can work with hammer or adze, and at every stroke they produce a noise corresponding to reality. Yet more movements can be accomplished on the stage. For example, a fire can be kindled; or figures formerly invisible may suddenly be seen and then disappear again. In short every desired motion may be produced though no one approaches the figures. The working of the stationary figures is more certain, and less risky than that of the travellers, and allows greater variety in the representation. Our elders, since such things aroused their amazement, named those occupied with them wonder-workers.[19]

Our predecessors showed us only one way to produce motion whether forward or backward, and only in a straight line, and even that was laborious and uncertain. For seldom has success been based on their methods recorded in writing in the case of those which are known to have been tested in practice. But we will show how with less labour and more certain success forward and backward movement may be achieved in a straight line, and we will further demonstrate the possibility of a box (or stage), or a single figure being moved around a given circle, and even along a given right-angled parallelogram.[20]

The construction of the Stationary figures is much surer and more reliable, and their representation much more convincing than is the case with the travelling puppets. For the former there is placed on any little wooden pedestal a box shaped tablet as a stage, having doors that open automatically, and it will then be seen that the figures move automatically according to a definite design. If the box-theatre then spontaneously closes, only a short time will elapse before it reopens, and different scenes are discovered. And all this, or according as is possible, a part thereof, will again move. This can be repeated many times. And outside the stages are visible either suspension mechanisms which are raised on high and can be carried around or some other kind of motion. This is the general idea. The more elegant the performance the craftsman has devised, the more assured is the applause of his fellow-craftsmen. We will show only one of the performances which we consider the most suitable, and we will explain its mechanism. It will suffice to discuss a single one of the better displays. For the same procedure with the same devices may be followed repeatedly, as we have shown for the travelling puppets.[21]

The Alexandrian mechanics of the late Greek period such as Hero whom we have considered above, were at the same time scientists and technicians. Hero wrote not only on delicate mechanical apparatus for utilization of compression and for the puppet theatre, but also on Geometry and on the science of Mechanics. The typical Alexandrian combination of theory and practice is encountered also in the Introduction to the Part treating with Mechanics in the great encyclopedic work produced in the third century of our era by the famous Alexandrian mathematician and mechanical engineer, Pappus.

Mechanical knowledge is highly valued by philosophers because it finds application in matters important to our life; and it is pursued with special zeal by all mathematicians because it provides our first introduction to the nature of matter, and to the elements of which the world is built. For while it describes in general the situation and weight of bodies, and their motion in space, it further not only examines the causes which impel bodies to natural motion, but it also teaches us how bodies may be forced to unnatural motion to change their position. For this purpose use is made of Propositions, which are readily deduced from material at hand.

Those who follow Hero are of the opinion that one part of Mechanics embraces mathematical demonstrations and the other, manual work; moreover the former part, which they term *rational* includes Geometry, Arithmetic, Astronomy and demonstrations of Physics; while the other part which comprises manual skills should teach the art of the metallurgist, of the iron-worker, the builder and the wood-worker as well as the painter, and everything which concerns hand-work. They say that he who has devoted himself to these disciplines from an early age, is versed in these arts and has an active mind, will in the end prove the best inventor (and constructor) of mechanical apparatus. But since it is not possible for one person completely to absorb the whole wide range of mathematical knowledge as well as to learn all these arts, they advise those who wish to occupy themselves with mechanical apparatus to learn that part of such work which is necessary from those who are expert in the relevant special skill.

Of all the skills which are based on Mechanics, the following are the most important for the needs of practical life. One is the art of the constructor of pulleys, who in ancient times were also called mechanics; for these raise on high great loads which would be totally immovable by natural means, and that is done with weights lighter than the loads themselves. Then there is the craft of those who build projecting apparatus; so necessary in war; they also are called mechanics, since missiles of stone, iron or other material are propelled for long distances from the catapulting apparatus constructed by them. Then too there is the skill of those who are actually called builders of machines since by means of the water scoops built by them, water from a great depth is raised on high with great ease.

Mechanics were, however, also called by our elders magicians. Some of them, as Hero in his *Pneumatics*, diligently applied knowledge concerning the air; others, like Hero in his teaching on Puppets and on the principles of equilibrium, sought with catgut and fine threads to imitate the movements of living beings; yet others used water for the purpose, as Archimedes in his work *On floating Bodies*; or water clocks, as Hero in his work *On Waters* in which the teaching seems related to sun-dials and (our) water-clocks. Finally, those also are called Mechanics who understand the manufacture of globes, and produce a representation of the motion of the heavens by a regular circular motion of water.

Some say that since Archimedes of Syracuse the causes and laws of all these things have been known. For among those whose memory has survived to our day, he above all handled every subject with extraordinary penetration as has been testified by Geminus among others, in his book on the quality of mathematicians. But Carpus of Antioch wrote that Archimedes produced only the book on the construction of celestial globes because he considered all the rest of the subject not worth the trouble of writing on. But this divine man, who is so famed among most people for his intellect and for his mechanical knowledge that his memory will live for ever among human beings, established the foundations of Geometry and of Arithmetic in the briefest and most exact fashion; and he so loved these disciplines that he could not make up his mind to introduce aught into them. Carpus himself, however, and several others, rightly introduced Geometry into certain crafts and uses in daily life. For Geometry, which can assist in many crafts and in many demands of daily life, is far from suffering injury thereby; nor, on account of its contribution to these crafts is aught detracted from the honour that is due to it or from its adornment.

Now that I have described the position and classification of the mechanical arts, I think I shall be undertaking a praiseworthy task if I describe, more briefly and clearly than has been done by previous writers on such matters, those foundations of geometry that the ancients proved to be necessary in order to move weights, as well as those propositions which I myself have found useful for such tasks.[22]

Although we must describe Philo and Hero as thoroughly technical specialists for the construction of apparatus, nevertheless as Rehm[11] has justly emphasised, there was in Antiquity little technical specialization. Technical work was the affair of slaves, and the free citizen usually despised manual labour and even activity devoted to the furtherance of discovery. With few exceptions therefore, science could hardly enrich technique.

THE ARCHITECT

Although the Roman world showed more interest than the Greek in technical achievement, yet they attempted no transformation of technical accomplishment or specialization in the technical crafts. The whole realm of engineering belonged in fact to the architect, whose work comprised not only building but also the manufacture of water-clocks and the construction of cranes, military machines and many other technical appliances. The Roman architect Vitruvius in his work *On Architecture* written between 25 and 23 B.C. and dedicated to the Emperor Augustus, clearly reveals the diverse duties of the architect in antiquity. His book is based partly on his own experience, but more on Greek sources. He portrays a theoretically happier relationship between science and practical work than among the Greeks, but there was in fact even among the Romans no real application of science to technical achievement.

We will quote from the *Architecture* of Vitruvius passages concerning the knowledge expected from an architect, concerning architectural classification and concerning the character of the machine, designated to begin with as a *wood joinery*.

Neither talent without instruction, nor instruction without talent can produce the perfect craftsman. He should be a man of letters, a skilful draughtsman, expert in geometry, not ignorant of optics, well grounded in arithmetic; he should know considerable history, have listened to the philosophers diligently, should be acquainted with music; he should not be ignorant of medicine, he should be learned in the findings of the legal experts and should be familiar with astrology and the laws of astronomy.

The reasons why this should be so are these: an architect must be a man of letters that he may keep a better record with notes. Then he must understand the art of drawing that he may be able the more easily to show with painted pictures what will be the intended appearance of the finished building. Geometry again furnishes many resources to architecture. First it teaches the use of ruler and compass, whereby the plan of buildings may be more easily shown on a flat surface with the directions of right angles, plane surfaces and straight lines. By optics, the light to the windows in buildings is duly drawn from certain aspects of the sky. By arithmetic indeed the costs of building are calculated, the divisions of the whole are explained; and difficult problems of symmetry are solved by geometric laws. Moreover the architect must have various historical knowledge, because he often proposes decorations for his work, about which he should be able to render an account to inquirers.[23]

Philosophy, however, makes the architect high-minded, so that he should not be arrogant but rather urbane, fair-minded, loyal, and what is most important, without avarice; for no work can be truly done without good faith and clean hands. Let him not be greedy nor have his mind busied with acquiring gifts; but let him with seriousness guard his dignity by keeping a good name. And such are the injunctions of philosophy. Philosophy, moreover, explains the "nature of things" (and this in Greek is *physiologia*), a subject which it is necessary to have studied carefully because it presents many different natural problems, as, for example, in the case of water-supply. For in the case of water-courses, where there are channels or bends or where water is forced along on a levelled plane, natural air-pockets are produced in different ways, and the difficulties which they cause cannot be remedied by anyone unless he has learnt from philosophy the principles of Nature. So also the man who reads the works of Ctesibius or Archimedes and of others who have written manuals of the same kind will not be able to perceive their meaning, unless he has been instructed herein by philosophers.

A man must know music that he may have acquired *acoustic* and mathematical relations and be able to carry out rightly the adjustments of *ballistae*, *catapultae* and *scorpiones*. For in the cross-beams on right and left are holes 'of half-tones' (*hemitonia*) through which ropes twisted out of thongs are stretched by windlasses and levers. And these ropes are not shut off nor tied up, unless they make clear and equal sounds in the ear of the craftsman. For the arms which are shut up under those strains, when they are stretched out, ought to furnish an impetus evenly, and alike on either side. But if they do not give an equal note, they will hinder the straight direction of the missiles.

In theatres, also, are copper vessels and these are placed in chambers under the rows of seats in accordance with mathematical reckoning. The Greeks call them *echeia*. The differences of the sounds which arise are combined into musical symphonies or concords: the circle of seats being divided into fourths and fifths and the octave. Hence, if the delivery of the actor from the stage is adapted to these contrivances, when it reaches them, it becomes fuller, and reaches the audience with a richer and sweeter note. Or again, no one who lacks a knowledge of music can make water-engines or similar machines.

Again, he must know the art of medicine in its relation to the regions of the earth (which the Greeks call *climata*); and to the characters of the atmosphere, of localities (wholesome or pestilential), of water-supply. For apart from these considerations no dwelling can be regarded as healthy. He must be familiar with the rights or easements which necessarily belong to buildings with party walls, as regards the range of eaves-droppings, drains and lighting. The water-supply, also, and other related matters, ought to be familiar to architects: so that, before building is begun, precautions may be taken, lest on completion of the works the proprietors should be involved in disputes. Again, in writing the specifications, careful regard is to be paid both to the employer and to the contractor. For if the specification is carefully written, either party may be released from his obligations to the other, without the raising of captious objections.

By astronomy we learn the east, the west, the south and the north; also the order of the heavens, the equinox, the solstice, the course of the planets. For if anyone is unfamiliar with these, he will fail to understand the construction of clocks.[24]

The parts of architecture itself are three: Building, Clock making, Mechanics. Building in turn is divided into parts; of which one is the placing of city walls, and of public buildings on public sites; the other is the setting out of private buildings. Now the assignment of public buildings is threefold: one, to defence; the second, to religion; the third, to convenience. The method of defence by walls, towers and gates has been devised with a view to the continuous warding off of hostile attacks; to religion belongs the placing of the shrines and sacred temples of the immortal gods; to convenience, the disposal of public sites for the general use, such as harbours, open spaces, colonnades, baths, theatres, promenades, and other things which are planned, with like purposes, in public situations.

Now these buildings should be so carried out that account is taken of strength, utility, grace. Account will be taken of *strength* when the foundations are carried down to the solid ground, and when from each material there is a choice of supplies without parsimony; of *utility*, when the sites are arranged without mistake or impediment to their use, and a fit and convenient disposition for the aspect of each kind; of *grace*, when the appearance of the work shall be pleasing and elegant, and the scale of the constituent parts is justly calculated for symmetry.[25]

'A machine is a wooden structure having special fitness for the moving of weights. It is moved by artificial means by circular rotation, which by the Greeks is called *kyklice kinesis*. The first kind of machine is the ladder principle (in Greek *acrobaticon*); the second is moved by air pressure (in Greek *pneumatika*); the third is the hoist (in Greek *barulkon* or equilibrium). Now the ladders are so arranged that when the uprights are raised to a fair height and cross-pieces are tied to them, men may safely ascend to inspect military engines.

FIG. 10. *The seven mechanical Arts.*
by Hugo von St. Victor: 1. Weaving. 2. Weapon forging. 3. Navigation. 4. Agriculture. 5. Hunting. 6. Medicine. 7. Acting.—Woodcuts from: Rodericus Zamorensis, Speculum humane vite. Augsburg: Zainer, 1475

But we have wind instruments operated by compressed air from which musical beats and vocal sounds can be produced by instrumentalists.

Machines of draught draw weights mechanically or raise them and place them in position. The design of the ladder prides itself not only on artifice but on military daring. It depends on using tie-pieces and the support of stays. But the design which gains an impulse by the power of compressed air only seeks satisfaction from the scientific refinement of its expedients. The traction machines offer in practice greater advantages in use, and when they are handled carefully, supreme excellence.

Of these machines, some are moved mechanically others are used like tools. There seems to be this difference between machines and instruments, that many people work at machines which need more power, for example, projectile engines or wine presses. But instruments carry out their purpose by the careful handling of a single workman, such as the turning of a hand balista or of screws. Therefore both instruments and machinery are necessary in practice and without them every kind of work is difficult.

Now all mechanical principles are predestined, and the revolution of the Earth teaches both master and mistress. For first indeed, unless we could observe and contemplate the continuous motion of the sun, moon and also the five planets; unless these revolved by the device of Nature we should not have known their light in due season nor the ripening of the harvest. Since then our fathers had observed this to be so, they took precedents from Nature; imitating them, and led on by what is divine they developed the comforts of life by their inventions. And so, they rendered some things more convenient, by machines and their revolutions, and other things by handy implements. Thus what they perceived useful in practice they caused to be advanced by their methods, step by step, through studies, crafts, and customs.

Let us first consider necessary inventions. In the case of clothing, by the aid of the loom, the union of the warp to the web not only covers and protects our bodies, but also adds the beauty of apparel. Again, we should not have plentiful food, unless yokes and ploughs had been invented for oxen and other animals. If windlasses, press-beams and levers had not been supplied to the presses, we should not have had clear oil or the produce of the vine for our enjoyment. And their transport would have been impossible, unless the construction of carts or waggons by land, and of ships by sea had been devised. The equilibrium of balances and scales has been applied to free human life from fraud by the provision of just measures. Besides, there are innumerable mechanical devices about which it does not seem needful to enlarge (because they are to hand in our daily use), such as millstones, blacksmiths' bellows, waggons, two-wheeled chariots, lathes and so forth, which are generally suitable for customary use. Hence we will begin to explain, so that they may be known, machines which are rarely employed.'[26]

Engineering in the Roman Empire was principally State engineering, which accomplished much with the traditional technical means, especially in organizing the making of streets, bridges, aqueducts, war machines and lofty buildings. There were in the Empire more than 186,000 miles of good roads. Ten aqueducts provided Rome daily with some 220 million gallons of water (Plate 1). And the dome of the Pantheon, built in Rome in the second century of our era had a span of over 142 feet 6 inches.

Though the Romans usually constructed gravitational water supplies which necessitated extremely costly aqueducts, yet it must by no means be concluded that they were ignorant of the fundamental principle of communicating pipes for the supply of water by pressure as constructed by the Greeks before them. Perhaps indeed the difficulties of firm control of a water supply system worked by pressure may have partly determined the choice of gravitation. But it was undoubtedly also a question of different styles of building technique. The Roman wished his technical works to express the political power of the Empire. Here was a reason for the gravity supply system, borne on high through the countryside on mighty arches, whereas the pressure system merely adheres closely to the natural contour of the land. Similarly the arrangement of Roman State buildings was on an axial and symmetrical plan which was an expression of the Idea of Power. This was alien to the thought of classical Greece, who erected her buildings in freedom and without constraint.

ROMAN MINING

Roman State engineering enterprises were carried out, like those of the great private factories, by the labour of armies of slaves. For building in peace-time soldiers were also used. Especially hard was the lot of the slaves in mines which were usually State enterprises. We know that in Spain the Romans obtained silver ore from a depth of more than 650 feet. Water removal was effected by means of Archimedian screws, nearly 16 feet 6 inches in length, ranged transversely over one another, each screw raising the water nearly 5 feet. These screws were driven by the feet of slaves (Fig. 11). Pliny (d. A.D. 79) tells us much concerning the mines of antiquity (Fig. 12):

FIG. 11. *Archimedean screws for raising water* from a depth of 210 metres in a Roman lead and silver mine in Spain, A.D. 200. Each of the 5 metre long screws raises the water 1·5 metres. Slaves turned the screws with their feet.—Drawing: German Museum, Munich

FIG. 12. *Greek underground mine workers.* Right, rock cutting at the face; left, loader; centre, water container (Amphora); below, earth collector with leather container. From a Corinthian terracotta vase of the sixth century B.C.—From: Rickard, *Man and Metals*

Gold . . . is found in our whole Earth, and is obtained in three fashions. Firstly it is in grains of gold in rivers, as in the Tagus in Spain, the Padus in Italy, the Hebrus in Thrace, the Pakteolus in Asia, and the Ganges in India; and this is the purest of all gold, as by constant friction it is completely cleansed. The second method is by digging in shafts that resemble wells, and the third is in caved-in mountains. I will now give a more detailed description of these last two methods of extracting gold. . . .

The third method will have outdone the achievements of the Giants. By means of galleries driven for long distances the mountains are mined by the light of lamps—the spells of work are also measured by lamps, and the miners do not see daylight for many months.

The name for this class of mines is *arrugiae*; also cracks give way suddenly and crush the men who have been at work, so that it actually seems less venturesome to try to get pearls and purple-fishes out of the depth of the sea: so much more dangerous have we made the earth! Consequently arches are left at frequent intervals to support the weight of the mountain above. In both kinds of mining masses of flint are encountered, which are burst asunder by means of fire and vinegar, though more often, as this method makes the tunnels suffocating through heat and smoke, they are hewn out in pieces weighing 150 lb. and the men carry the stuff out on their shoulders, working night and day, each man passing them on to the next man in the dark, while only those at the end of the line see daylight. If the bed of flint seems too long, the miner follows along the side of it and goes round it. And yet flint is considered to involve comparatively easy work, as there is a kind of earth consisting of a sort of potter's clay mixed with gravel, called *gangadia*, which it is almost impossible to overcome. They attack it with iron wedges and the iron hammers; and it is thought to be the hardest thing that exists, except greed for gold, which is the most stubborn of all things. When the work is completely finished, beginning with the last, they cut through, at the tops, the supports of the arched roofs. A crack gives warning of a crash, and the only person who notices it is the sentinel on a pinnacle of the mountain. He by shout and gesture gives the order for the workmen to be called out and himself at the same moment flies down from his pinnacle. The fractured mountain falls asunder in a wide gap, with a crash which it is impossible for human imagination to conceive, and likewise with an incredibly violent blast of air. The miners gaze as conquerors upon the collapse of Nature. And nevertheless even now there is no gold so far, nor did they positively know there was any when they began to dig; the mere hope of obtaining their coveted object was a sufficient inducement for encountering such great dangers and expenses.

Another equally laborious task involving even greater expense is the incidental operation of previously bringing streams along mountain-heights frequently a distance of 100 miles for the purpose of washing away the débris of this collapse; the channels made for this purpose are called *corrugi*, a term derived I believe from *conrivatio*, a uniting of streams of water. This also involves thousands of workmen; the dip of the fall must be steep, to cause a rush rather than a flow of water, and consequently it is brought from very high altitudes. Gorges and crevasses are bridged by aqueducts carried on masonry; at other places impassable rocks are hewn away and compelled to provide a position for hollowed troughs of timber. The workman hewing the rock hangs suspended with ropes, so that spectators viewing the operations from a distance seem to see not so much a swarm of strange animals as a flight of birds. In the majority of cases they hang suspended in this way while taking the levels and marking out the lines for the route, and rivers are led by man's agency to run where there is no place for a man to plant his footsteps. It spoils the operation of washing if the current of the stream carries mud along with it: an earthy sediment of this kind is called *urium*. Consequently they guide the flow over flint stones and pebbles, and avoid *urium*. At the head of the waterfall on the brow of the mountains reservoirs are excavated measuring 200 ft. each way and 10 ft. deep. In these there are left five sluices with apertures measuring about a yard each way, in order that when the reservoir is full the stopping-barriers may be struck away and the torrent may burst out with such violence as to sweep forward the broken rock. There is also yet another task to perform on the level ground. Trenches are excavated for the water to flow through—the Greek name for them means 'leads'; and these, which descend by steps, are floored with gorse—this is a plant resembling rosemary, which is rough and holds back the gold. The sides are closed in with planks, and the channels are carried on arches over steep pitches. Thus the earth carried along in the stream slides down into the sea and the shattered mountain is washed away; and by this time the land of Spain owing to these causes has encroached a long way into the sea.[27]

. . . .

We will now consider iron, the most precious and at the same time the worst metal for mankind. By its help we cleave the earth, establish tree-nurseries, fell trees, remove the useless parts from vines and force them to rejuvenate annually, build houses, hew stone and so forth. But this metal serves also for war, murder and robbery; and not only at close quarters, man to man but also by projection and flight; for it can be hurled either by ballistic machines, or by the strength of human arms or even in the form of arrows. And this I hold to be the most blameworthy product of the human mind. In order that death may reach men the more speedily, we attach wings to it; we deck iron with feathers and thus the fault is not Nature's, but ours. A few examples prove that iron could in fact be an innocent metal. Thus in the alliances which Porsenna established with the Roman people after the expulsion of the kings, it was established that iron should be used for no purpose except agriculture.[28]

The Romans in the first century B.C. understood the water-wheel. And Vitruvius gives us a fairly detailed description of a water-mill. But the water-wheel was at first not widely used, since the labour of slaves was available. The water-mill gained ground only slowly, when the abatement of Roman expansion progressively diminished the numbers of prisoners of war who had furnished the supply of slaves. Here and there, from the third and fourth century A.D. we encounter water-mills, especially in lands north of the Alps with their more favourable water supplies. We know more detail of a great Gallo-Roman establishment for grinding corn of about A.D. 200 at Barbegal, near Arles (Fig. 13). Here 16 water-wheels drove 16 mills, capable of producing 28 tons of meal in 24 hours. But this establishment was exceptional. The water-wheel did not come into general use as a source of power until the medieval period.

FIG. 13. *Gallo-Roman mill of the second/third century* A.D. at Barbegal near Arles in France. 16 waterwheels drive 16 mills.—From: *Isis*

PART II

THE MIDDLE AGES

PART II

INTRODUCTION

THE Roman Empire collapsed from its own inner weakness as well as from the attacks of the Germanic peoples. Civilization slowly drifted to the North. Roman centralized State government was replaced by the rule of local lords. Life shifted from the cities to the countryside. Monetary economy yielded to an economy based on the land. The awakened Gallo-germanic peoples were the heirs of antiquity which, however, in technical culture as in other departments was by no means the sole influence. Besides the treasure of technical knowledge inherited from antiquity, the young Romano-germanic peoples had many ancient basic handicrafts from the northern plains which made their mark in due course on medieval material culture. But above all, Christianity which always allowed to Man his dignity and ultimately admitted also the value of the things of this world, also contributed considerably to a firm foundation for technical achievement capable of development. Finally, medieval technical craftmanship was affected by influences, conveyed chiefly by the Islamic world, not only from classical antiquity but also from the Far East. The cleavage of early medieval Europe was confronted by the unity of the universal Church. The little independent economic units that resulted from decentralization, the little towns, domains, abbeys, were fused later into larger bodies. The failure, in the post-Carolingian period, of the centre of both narrow and more extensive political power to combine with the religious centre, the universal church, left ample scope for the activity of productive forces in the sphere of religious and also of material culture. And the many varieties of free economic units and political powers stimulated cultural work through fruitful competition. And if until the tenth century technical achievement was limited mainly to monastic handicrafts which achieved high standards in the production of cultural objects there followed with the growing prosperity of the towns a development of civic manual craftsmanship that produced works no less inspired than the monastic handiwork.

During all the medieval post-Carolingian centuries, including those from the sixth to the twelfth, which is often described as the 'Dark Ages' there were made in western lands important technical discoveries which led slowly to the transformation of social and economic conditions. But medieval technical achievements have hardly been recognized even by historians, and extensive research is needed to substantiate the conclusions already reached. The romanticism of the first half of the last century opened our eyes to medieval art and intellect. Neo-thomism has reincarnated for us the zenith of medieval philosophy. And the fundamental research of Duhem in the early years of the twentieth century and of Anneliese Maier in our own day have familiarized us with medieval science, especially that of the fruitful fourteenth century. Praise is especially due to Lefevbre des Noëttes who since the 1920's has drawn attention to the great technological discoveries of the Middle Ages, although his researches have not been universally recognized. By means of a series of discoveries, and not least by the Christian teaching that men are created free by Nature, and are all equal before Christ, the Christian Middle Ages succeeded in building a civilization which no longer, like that of antiquity, rested on the backs of slaves, but to a greater and greater extent derived power from machinery.

The medieval Christian church is significant for the development of technology from still another point of view. When at the beginning of the thirteenth century, to a small extent by direct transmission, but for the most part indirectly through the Arabs, the whole of Aristotle, and especially the immense quantity of Aristotelian natural knowledge lay at the disposal of medieval Christianity, it was the great achievement of Albertus Magnus and of his pupil Thomas Aquinas to integrate Aristotelian science into the Christian cosmos, and thus to attain a single world-picture, embracing natural knowledge, with God as the summit of the whole. Faith and Knowledge are indeed separate domains; but according to this doctrine, there can be no contradiction between revealed and scientific truth, since that would involve contradiction within God himself. By this Christian Aristotelianism, this attachment to inherited philosophy and knowledge and their harmonious combination with theology, the way was prepared for the development of natural science for the time being within the framework of scholasticism, even though this may not have been the primary intention of the great churchmen of the thirteenth century. And the later scholasticism of the fourteenth century, especially through the strong tendency of Nominalism to turn to the Thing-in-itself developed a critical attitude, based on experience, to the highly scholastic Aristotelian physics and to a fresh view of a modified Aristotelian Nature study which, as has been conclusively shown by Anneliese Maier,[12] planted at least the seed of the new physics which arose in the sixteenth and seventeenth centuries. It is this new physics which has been of special significance for the development of technology since the eighteenth century. The admission of Aristotelianism into scholasticism by Albert and Thomas in the thirteenth century was thus ultimately of extreme importance even for modern technology. That theological development may take very different courses can be well observed in Islam. The great religious teacher Al-Ghāzalī (d. 1111) rejected philosophy and science because he believed that they led to loss of faith concerning the Origin of the World and the Creator. In particular, he could not reconcile faith in the omnipotence of God with the Greek view of a world accessible to human understanding. Muslim science began to decline from about the year 1100, because the way had not been found to a just combination of religion, with philosophy and expanding knowledge.

EARLY CHRISTIANITY AND TECHNOLOGY

We have already observed how important it was for the development of western science and technology that Christianity did not deny Nature but accorded her a certain, albeit a lesser value. This view is clearly expressed in the works of the early Church fathers, though it was threatened in course of time by Oriental influence which preaches of pure spirit. Gregory of Nyssa in the fourth century A.D. proclaimed the association of sensual and spiritual nature established by God, whereby nothing created is worthless, and Man is ordained supreme over Nature which helps him on the road to God. Thus Nature is exalted by and with man.

The Great Catechism

. . . Now these two worlds have been separated from each other by a wide interval, so that the sensible is not included in those qualities which mark the intellectual, nor this last in those qualities which distinguish the sensible, but each receives its formal character from qualities opposite to those of the other. The world of thought is bodiless, impalpable, and figureless; but the sensible is, by its very name, bounded by those perceptions which come through the organs of sense. But as in the sensible world itself, though there is a considerable mutual opposition of its various elements, yet a certain harmony maintained in those opposites has been devised by the wisdom that rules the Universe, and thus there is produced a concord of the whole creation with itself, and the natural contrariety does not break the chain of agreement; in like manner, owing to the Divine wisdom there is an admixture of the sensible with the intellectual department, in order that all things may equally have a share in the beautiful, and no single one of existing things be without its share in that superior world. For this reason the corresponding locality of the intellectual world is a subtile and mobile essence, which in accordance with its supramundane habitation, has in its peculiar nature large affinity with the intellectual part. Now, by a provision of the supreme Mind, there is an admixture of the intellectual with the sensible world, in order that nothing in creation may be thrown aside as worthless, as says the Apostle, or left without its portion of the

Divine fellowship. On this account it is that the commixture of the intellectual and sensible in man is effected by the Divine Being, as the description of the cosmogony instructs us. It tells us that God, taking dust of the ground, formed the man, and by an inspiration from Himself, He planted life in the work of His hand, that thus the earth might be raised up to the Divine, and so one certain grace of equal value might pervade the whole creation, the lower nature being mingled with the supramundane. Since, then, the intellectual nature had a previous existence, and to each of the angelic powers a certain operation was assigned for the organization of the whole, by the authority that presides over all things, there was a certain power ordained, to hold together the earthly region, constituted for this purpose by the power that administers the Universe. Upon that there was fashioned that thing moulded of earth, an 'Image' copied from the superior power. Now this living being was man. In him, by an ineffable influence, the godlike beauty of the intellectual nature was mingled. He to whom the administration of the earth has been consigned takes it ill and thinks it not to be borne, if, to that nature which has been subjected to him, any being shall be exhibited bearing likeness to his transcendent dignity.[29]

On the Making of Man

. . . When, then, the Maker of all had prepared beforehand, as it were, a royal loding for the future king (and this was the land, and islands, and sea, and the heaven arching like a roof over them), and when all kinds of wealth had been stored in this palace (and by wealth I mean the whole creation, all that is in plants and trees, and all that has sense, and breath, and life; and—if we are to account materials also as wealth—all that for their beauty are reckoned precious in the eyes of men, as gold and silver, and the substances of your jewels which men delight in—having concealed, I say, abundance of all these also in the bosom of the earth as in a royal treasure-house), he thus manifests man in the world, to be the beholder of some of the wonders therein, and the lord of others; that by his enjoyment he might have knowledge of the Giver, and by the beauty and majesty of the things he saw might trace out that power of the Maker which is beyond speech and language.

For this reason man was brought into the world last after the creation, not being rejected to the last as worthless, but as one whom it behoved to be king over his subjects at his very birth.[30]

The needful services of our life are divided among the individual animals that are under our sway, for this reason—to make our dominion over them necessary.

It was the slowness and difficult motion of our body that brought the horse to supply our need, and tamed him: it was the nakedness of our body that made necessary our management of sheep, which supplies the deficiency of our nature by its yearly produce of wool: it was the fact that we import from others the supplies for our living which subjected beasts of burden to such service: furthermore, it was the fact that we cannot eat grass like cattle which brought the ox to render service to our life, who makes our living easy for us by his own labour; and because we needed teeth and biting power to subdue some of the other animals by grip of teeth, the dog gave, together with his swiftness, his own jaw to supply our need, becoming like a live sword for man; and there has been discovered by men iron, stronger and more penetrating than prominent horns or sharp claws, not, as those things do with the beasts, always growing naturally with us, but entering into alliance with us for the time, and for the rest abiding by itself: and to compensate for the crocodile's scaly hide, one may make that very hide serve as armour, by putting it on his skin upon occasion: or, failing that, art fashions iron for this purpose too, which, when it has served him for a time for war, leaves the man-at-arms once more free from the burden in time of peace: and the wing of the birds, too, ministers to our life, so that by aid of contrivance we are not left behind even by the speed of wings: for some of them become tame and are of service to those who catch birds, and by their means others are by contrivance subdued to serve our needs: moreover art contrives to make our arrows feathered, and by means of the bow gives us for our needs the speed of wings. . . .[31]

Very different from the Christian Church Father's acceptance of life is the view of the Taoist mystic of the eighth century A.D., author of the *Kwan-Yinn-Tzu.* Following Christopher Dawson, we will give a passage from this work as an example of Oriental rejection of this world and of worldly activity. With such a religious spirit, science and technology received no stimulus from religion.

'Outside the Principle, the Tao, all is nothing. Everything that seems to exist forms part of the unity of the Tao. In this absolute and universal unity, there is no succession, no time, no distance. In the

Tao a day and a hundred years, a furlong and a hundred leagues do not differ. . . . We must not, therefore, speak of laws of nature and of supposed breaches of these laws, such as changes of form or of sex, levitation, fire that does not burn and water that does not drown, monsters, prodigies and so forth. There is no such thing as a prediction, since time does not exist, and consequently there is no future. There is no such thing as levitation, since there is no space. The Tao is Unity which is contained in a single point, and has no past or future. I am one with all beings, and all beings are one with the Tao. Every phenomenon results from the play of the Tao, not from law. For a corpse to rise and walk, for a man to catch fish in a basin, or to come in and go out through a door that is painted on the wall is no anomaly, since there is no rule.' 'To distinguish between cause and effect, agent and product is illusion and fiction. The common herd imagine that noise is produced by a drum, when it is beaten by a man with a drumstick. But in reality, there is neither drum nor drumstick nor drummer. Or rather Drum-drumstick-drummer are the Tao which has produced in itself the phenomenon of drumming. The words signify nothing, seeing that the things signified do not exist.'

'That which is seen in a state of waking is no more real than that which is seen in dreams. And the man who sees is no more real than that which he sees. The man who dreams and the man of whom he dreams are no more real the one than the other.'

'It is because he knows that nobody exists, that the Sage is equally benevolent and indifferent to everybody.'[32]

Even Augustine (A.D. 425) whose whole thought moved almost exclusively around the subject of God and the Soul, yet broke into a paean of praise of the creation which has linked corporeal and incorporeal. And in his praise he included the technical arts as affording testimony of the excellence of the soul.

In praise of Creation

Now I will speak of the good which God has given, and still gives even in the condemned state of man: in which condemnation of him, God took not all from him that He had given him (for so he would have ceased to have had any being), nor did He resign His power over him, when He gave him thrall to the devil; for the devil himself is his thrall, He is cause of his subsistence, He that is only and absolutely essential, and gives all things essence under Him, gave the devil his being also . . .

It is His daily work that the seed unfolds itself out of a secret clue as it were, and brings the potential forms into such actual comeliness. It is He that makes that strange combination of a nature incorporeal (the ruler) and a nature corporeal (the subject) by which the whole becomes a living creature. A work so admirable, that it is able to amaze the mind, and force praise to the Creator from it, being observed not only in men, whose reason gives him excellence above all other creatures; but even in the least fly that is, one may behold this wondrous and stupendous combination! . . .

For besides the disciplines of good behaviour, and the ways to eternal happiness (which are called virtues) and besides the grace of God which is in Jesus Christ, imparted only to the sons of the promise, invention has brought forth so many and such sciences and arts (partly necessary, and partly voluntary) that the excellence of his capacity makes the rare goodness of his creation apparent, even then when he goes about things that are either superfluous or pernicious, and shows from what an excellent gift, he has those his inventions and practices. What varieties has man found out in buildings, attires, husbandry, navigation, sculpture, and imagery! What perfection has he shewn, in the shows of theatres, in taming, killing, and catching wild beasts! What millions of inventions has he against others, and for himself in poisons, arms, engines, stratagems, and the like! What thousands of medicines for the health, of meats for the throat, of means and figures to persuade, of eloquent phrases to delight, of verses for pleasure, of musical inventions and instruments! What excellent inventions are geography, arithmetic, astrology, and the rest! How large is the capacity of man, if we should stand upon particulars! . . .

. . . Come now to the body: . . . mark what great goodness, and providence is shewn herein by God Almighty. . . . You see the other creatures have a grovelling posture, and look towards earth, whereas man's upright form bids him continually look to the things in heaven. The nimbleness of his tongue and hand in speaking and writing, and working in trades, what does it but declare for whose use they were made so?[33]

It has already been shown that the Middle Ages slowly overcame slavery. It was a gradual process, closely linked with technical development. Nor must the contribution of the moral influence of the Church be overlooked. Paul had written in his Epistle to the Galatians (III, 28) 'There is neither Jew nor Greek, there is neither bond nor free, there is neither male nor female; for ye are all one in Christ Jesus.' The dimunition in the supply of slaves after the decline of the Roman Empire caused a higher value to be placed on free manual labour. From the end of the fourth century, there was a transition from slavery to the system of serfdom with bondage to the land. Augustine regarded slavery as the fruit of sin. Slavery was to him a condition that must be endured as unalterable except if the master should grant a liberation.

. . . The providers are the commanders, as the husband over his wife; parents over their children, and masters over their servants; and they that are provided for, obey, as the wives do their husbands, children their parents, and servants their masters. But in the family of the faithful man, the heavenly pilgrim, there the commanders are indeed the servants of those they seem to command; ruling not in ambition, but being bound by careful duty; not in proud sovereignty, but in nourishing pity. . . .

. . . God made man reasonsable, and lord only over the unreasonable, not over man, but over beasts. Whereupon the first holy men were rather shepherds than kings, God shewing herein what both the order of the creation desired, and what the merit of sin exacted. For justly was the burden of servitude laid upon the back of transgression. And therefore in all the Scriptures we never read the word Servant, until such time as that just man Noah laid it as a curse upon his offending son. So that it was guilt, and not Nature that gave origin unto that name. The Latin word *servus* had the first derivation hence: those that were taken in the wars, being in the hands of the conquerors to massacre or to preserve, if they saved them, then were they called *servi*, of *servo* 'to save'. Nor was this effected beyond the desert of sin. For in the justest war, the sin upon one side causes it; and if the victory fall to the wicked (as sometimes it may) it is God's decree to humble the conquered, either reforming their sins herein, or punishing them. . . .

Sin therefore is the mother of servitude . . . therefore many religious Christians are servants unto wicked masters, yet not unto freemen, for that which a man is addicted unto, the same is he slave unto . . .

. . . But take a man as God created him first, and so he is neither slave to man nor to sin. But penal servitude had the institution from that law which commands the conservation, and forbids the disturbance of Nature's order; for if that law had not first been transgressed, penal servitude had never been enjoined. Therefore the

Apostle (*Ephesians*, VI, 5, 7) warns servants to obey their masters and to serve them with cheerfulness, and good will: to the end that if they cannot be made free by their masters, they make their servitude a freedom to themselves, by serving them, not in deceitful fear, but in faithful love, until iniquity be overpassed, and all man's power and principality disannulled, and God only be all in all.[34]

In the subsequent period we encounter more and more frequently the enfranchisement of the bond by ecclasiastical institutions or through ecclasiastical influence. We give below a document of the ninth century.

I HENRY have from fear of God and for the health of my soul set free my bond-woman Reginheid with her children Waldgelt and Folcheid. I have also freed another bond-woman, Zeizbirc, whom the freeman Albrich entrusted to me to set free. They shall be as free as if they had been born and brought up by free parents. They shall have no duty to serve against their will either our heirs or the heirs of our heirs, but only God to whom all are thrall. Such property as they own or may in the future acquire, they shall posses and enjoy. They shall live and work for their own benefit, and that which they produce, they shall possess. This is conceded and granted to them. They shall find refuge and protection in the holy monastery of Weissenburg which has been erected above the *Lauter* to the glory of the holy apostles Peter and Paul and of numerous other saints, and which is now ruled by the worthy Abbot Grimald. In return they shall pay every year on the feast of St Martin as tribute to their chosen priest and confessor two pfennigs or wax to the value of two pfennigs as a pious gift to this holy foundation in the name of their afore-mentioned master. Thereupon they shall remain free like other duty-paying and tributary persons or clerics who have been freed in similar circumstances. Should they, however, be negligent or tardy in payment of their dues, I shall punish them according to the laws. And they shall remain free from generation to generation. . . .

Publicly executed in Biblisheim, on the 16th July, in the year 837 after the Incarnation of our Lord, in the 15th Indiction, when Ludwig governed Gaul, in the 23rd year of his Imperial reign. Witnesses: Henry who has earnestly pleaded for this deed for the tributaries (and eleven men).

I Hugbald have written and signed the deed.[35]

THE MONASTERIES AND TECHNICAL SKILL

The courts of the great feudal lords and the monasteries were the centres of admirable handicrafts prior to the growth of towns. Especially in the service of the Church, medieval technology produced incomparably fine work. Theophilus, probably an eleventh century German Benedictine, described in an extensive and very clear account the wealth of arts and crafts which served churches and religious houses. Handwork held indeed from the time of Benedict of Nursia to that of the early Franciscans an important place in the Rules of the Orders. Thus in contrast to the period of antiquity, free manual labour acquired religious significance and real dignity. 'Happy is he who earns his bread by the work of his hands' declared John Chrysostom in the second half of the fourth century A.D. (Homilies on Genesis, L,2). And Theophilus enjoins silence whilst performing manual labour to the glory of God and for the benefit of sufferers.

I, Theophilus, an humble priest, servant of the servants of God, unworthy of the name and profession of a monk, to all wishing to overcome or avoid sloth of the mind or wandering of the soul, by useful manual occupation and the delightful contemplation of novelties, send a recompense of heavenly price.

We read in the exordium of mundane creation that man, made after the image and likeness of God and animated by the inspiration of the Divine breath, was also, by the excellence of so much dignity, raised above other living creatures; as capable of reason, he merited to participate in the counsel and genius of Divine providence, and, gifted with free-will, he beheld superior to himself but the will of his Maker and the obligation to reverence his decree. Wherefore, miserably deceived by diabolical astuteness, he lost the privilege of immortality through the fault of disobedience, yet so transmitted his power of wisdom and intelligence to his posterity, that whoever would supply care and application might be able to acquire a capability of every art and science, as by an hereditary right.

In this manner, human industry, seizing upon this faculty and applying itself in its divers acts to gain and to pleasure, transmitted it, through the development of time, to the predestined epoch of the Christian religion, and it came to pass that a people devoted to God converted to his worship that which Divine ordinance had, to the praise and glory of His name, created. On this account, the pious devotion of the faithful may not neglect that which the careful prevision of our predecessors transmitted to our age; and may man embrace with all avidity that which God has conferred upon man, as an inheritance, and labour to acquire it.

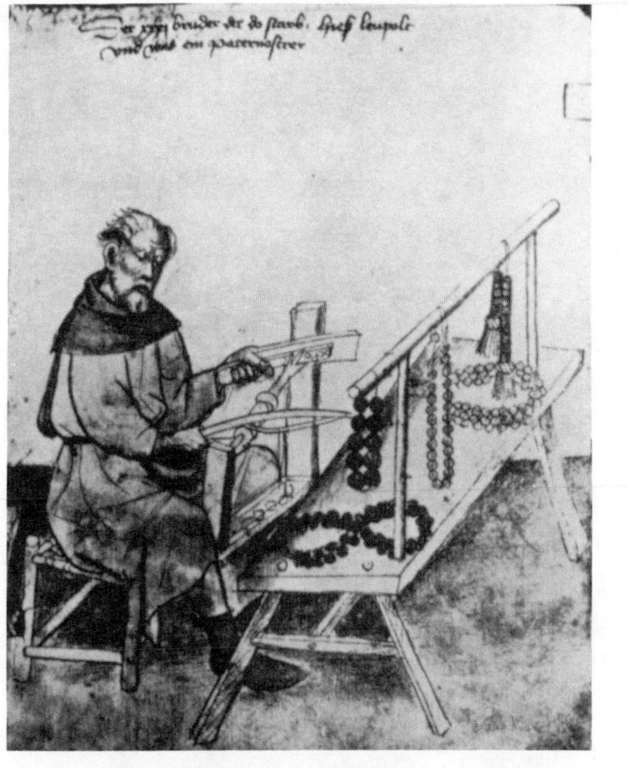

5. Bowstring operated boring machine for preparing pearls for necklaces, 1390. Illustration from the Chronicle of the Mendelschen Zwolf-Bruder-Stiftung

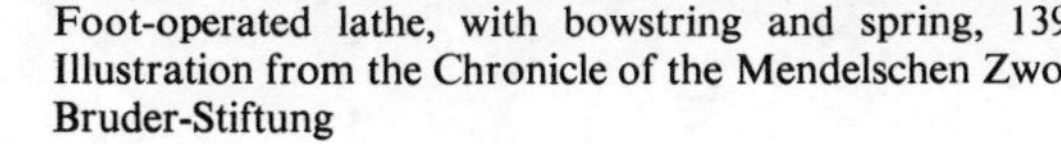

Foot-operated lathe, with bowstring and spring, 1395. Illustration from the Chronicle of the Mendelschen Zwolf-Bruder-Stiftung

6. *Left:* Firearm testing. Firing a cannon for stone shot to test the strength of the barrel. 15th century. Painting from A. Dachsperger's *Kriegstechnische Bilderhandschrift*, 1443

Gunpowder Stamps, 1470. Water colour picture from a collection of original drawings of military arts, 1470

Skill in which let no one glorify himself inwardly, as if received from himself and not from elsewhere, but let him be thankful humbly in the Lord, from whom and through whom all things are received, and without whom, nothing; nor let him wrap his gifts in the folds of envy, nor hide them in the closet of an avaricious heart, but, all jealous feeling repelled, let him with cheerful mind answer with simplicity to those seeking him, and let him fear the judgment of the Gospel upon that merchant, who, failing to return to his lord a talent with accumulated interest, deprived of all reward, merited the censure from the mouth of his judge of 'wicked servant'.

Fearing to incur which sentence, I, frail and unworthy and almost without name, offer gratuitously to all desirous with humility to learn, that which Divine authority, which affluently and not precipitately gives to all, gratuitously conceded to me, and I admonish them that in me they may recognize the goodness and admire the generosity of God, and I advise them likewise that if to this their labours are added, they may believe beyond a doubt that excellence awaits them. . . .

Therefore, dearest son whom God has supremely blest, in that there is freely offered to thee that which many attain only by unbearable toil; when they traverse the seas with great danger to life, suffer hunger and cold, or exhaustion by the daily servitude of study and yet are not destroyed by their avidity for learning; urgently seek thou with eager gaze this sketch of the various arts; study it with faithful memory and master it with burning love.

If thou studiest it with care, thou wilt learn what kinds and mixtures of colours are in Greece; what skilfully executed enamels and many kinds of Niello work are executed in Russia; the manifold decorations performed by Arabs by chased, cast and pierced work; what is accomplished in Italy in fashioning vessels, in stone-cutting and in inlaying gold on bone (i.e. ivory); the many kinds of precious windows treasured in France; skilful Germany's praise of fine work in gold, silver, copper, iron, wood, and stone. When thou hast read this through many times and hast faithfully memorized it, then as often as thou usest my work to thine advantage, pray for me to the mercy of almighty God. He knoweth that I have written what is herein neither to seek praise nor from a desire for worldly reward, nor from jealousy or envy aiming to destroy aught that is valuable and rare, nor have I been silent over aught reserved for myself, but I have afforded help to many and considered their advantage, for the glory of God and for the exaltation of his Name. [36]

. . . through a disposition of sincere affection, I have not hesitated to convey to virtuous disposition how much honour and perfection there is in avoiding indolence, and in contemning ignorance and sloth; and how sweet and agreeable it is to indulge in the exercise of divers usefulness after the word of a certain author, who says:—

> 'To know anything is praiseworthy; it is a fault to be unwilling to learn.'

Nor let any one be slow to understand him, concerning whom Solomon has said, 'He that increaseth knowledge increaseth labour,' because whoever carefully meditates may mark what perfection of mind and body may result from it.

For it is evident; clearer than the light; because whoever gives his mind to sloth and levity, also indulges in vain trifles, and slander, curiosity, drinking, orgies, quarrel, fight, homicide, excess, thefts, sacrileges, perjury, and other things of this kind, which are repugnant in the eyes of God, overlooking the humble and quiet man, working in silence in the name of the Lord, and obedient to the precept of the holy Apostle Paul: '*But rather let him labour, working with his hands the thing which is good, that he may have to give to him that needeth.*' [37]

Through the spirit of wisdom thou knowest that all created things proceed from God, and that without him nothing exists. Through the spirit of intelligence thou hast acquired the faculty of genius, in whatever order, in what variety, in what proportion, thou mayest choose to apply to thy varied work. Through the spirit of counsel thou hidest not the talent conceded to thee by God, but by working and teaching openly, with humility, dost faithfully expound to those desirous to learn. Through the spirit of perseverance thou dost shake off all lethargy of sloth, and whatever with quick diligence thou startest, thou dost complete with full vigour. Through the spirit of science accorded to thee, thou rulest with genius from an abounding heart, and from that with which thou dost entirely overflow thou bestowest with the confidence of a well-stored mind for the common good. Through the spirit of piety thou dost regulate the nature, the destination, the time, the measure and the means of the work; and, through a pious consideration, the price of the fee, that the vice of avarice or covetousness may not steal in. Through the spirit of the fear of God thou of thine own power cannot accomplish aught, since thou recallest that thou neither possessest nor willest aught unconceded by God; but by believing, confiding and giving thanks, thou dost ascribe to divine compassion whatever thou hast learned, or what thou art, or what thou mayest be. . . .

Act therefore now, well-intentioned man, happy before God and men in this life, happier in a future, in whose labour and study so many sacrifices are offered up to God; henceforth warm thyself with a more ample invention, hasten to complete with all the study of thy mind those things which are still wanting among the utensils of the house of the Lord, without which the divine mysteries and the services of ceremonies cannot continue. These are the chalices, candelabra, incense burners, vials, pitchers, caskets of sacred relics, crosses, missals and other things which useful necessity requires for the use of the ecclesiastical order

If thou desirest to fabricate these, in this order thou dost commence.[38]

Of Tempering Files

Burn the horn of an ox in the fire, and scrape it, and mix with it a third part salt, and grind it strongly. Then put the file in the fire, and when it glows sprinkle this preparation over it everywhere, and, some hot coals being applied, thou wilt blow quickly upon the whole, yet so that the tempering may not fall off, and quickly withdrawing it, extinguish it equally in water, and taking it out, dry it slightly over the fire. Thou wilt in this manner temper all things which are made of steel. . . .[39]

Another kind of tempering of iron instruments is also made in this manner, by which glass is cut, and also the softer stones. Take a three year old buck-goat, and tie him up within doors for three days without food; on the fourth day give him fern to eat and nothing else. When he shall have eaten this for two days, on the night following enclose him in a cask perforated at the bottom, under which holes place another sound vessel in which thou wilt collect his urine. Having in this manner for two or three nights sufficiently collected this, turn out the buck, and temper thine instruments in this urine. Iron instruments are also tempered in the urine of a young red-haired boy harder than in simple water. . . .[40]

Of Founding Bells (*Cf Plate* 2)

In making a bell, first cut a dry piece of wood, the length desired for the bell, so that on every side it may protrude beyond the shape to the length of one palm, and let it be square at one larger end, at the other more pointed and round, so that it can be revolved in a hole. And let it be drawn out larger and larger, so that, when the work has been finished, it can easily be taken out. This wood must be cut around in the thicker part, one palm before the end, that a hollow may be made two fingers wide, and let the wood be there round; near this furrow the extremity of the wood is made thin, that it may be joined into another curved wood, by which it is able to be revolved like a lathe. Two planks are also made, equal in length and width, which are joined together and made firm with four pieces of wood, so that they may be held from each other according to the length of the aforesaid wood; a hole should be made in one plank in which the rounded top can be turned, and in the other alike, opposite, an incision must be made two fingers deep, in which the round cutting can be revolved. Which being done take the block itself and apply strongly beaten clay around it, first of all two fingers thick, which being carefully dried, apply another upon it, and do thus until the mould be supplied as thou mayest wish to have it, and beware that thou never at any time superpose clay upon other (clay), unless that below has become perfectly dry. Then set this mould between the planks, and, the boy who can revolve it being seated, thou wilt turn it as thou mayest wish, and holding a cloth moistened in water thou wilt smooth it.

After this, taking tallow, cut it up very finely and macerate it with the hand, and two even pieces of wood being fixed together of the thickness thou mayest wish, thou wilt thin out the tallow placed between them upon an even board, with the wooden roller, as the wax above, water being placed under that it may not adhere, and so thou wilt immediately lift it suddenly, and lay it upon the mould and fasten it round with a hot iron. Again, thinning a piece of grease, in the same manner, thou wilt fasten it next to the first, and do thus until thou hast covered the mould. Make the rim of the bell of the thickness thou pleasest. Turn the grease, when quite cold, with sharp instruments, and shouldest thou wish any ornament about the sides of the bell, of flowers, or letters, hollow them out in the tallow, and fashion four openings near the neck, that it may sound better. Then superpose clay, sifted and carefully mixed; which being dry, add other above it. That again being quite dry, turn the mould upon its side and remove the wood by striking gently, and the mould being again raised, thou wilt fill the opening above with soft clay and impress the curved iron, in which the clapper should hang in the middle, so that its extremities may project outside. And the clay has become dry, make it even with the rest of mould and cover it with tallow, so that the ends of the iron may adhere well in it. After these things form the neck, and the handles, and the air-hole, or funnel, above them, and cover with clay. And when the clay has a third time become dry, place iron hoops around so closely that there may not be more than the breadth of a hand between two hoops, upon which hoops place two layers of clay. These being dry, turn the mould upon its side and cut a large hollow in the inside, in the circumference and in depth, that it may not remain thicker than one foot, because, were the mould whole within, it could not be raised, on account of the exceeding weight, nor be cooked through, for the thickness.

Then make a cave in the place where thou wish this mould to enter for cooking, deep, according to the size thereof.[41]

The great emphasis laid on manual labour by the early monastics was softened in the succeeding period. Thomas Aquinas in the thirteenth century posed the question: 'Must members of the Order perform manual labour?' And his reply limited the obligation.

Manual work is based on four things: Primarily to gain a livelihood . . . Secondly to overcome idleness which is responsible for many evils . . . Thirdly to curb appetite, in so far as the body is mortified thereby . . . Fourthly and finally it facilitated the distribution of alms.

If a man could support life without food, he would not find it necessary to work with his hands; the same applies to those who have other legitimate means of livelihood.

Insofar, however, as handwork is performed to overcome idleness or to mortify the body, it does not of itself come within the scope of a command, since the body can be mortified and idleness can be overcome by many methods other than by handwork.

Finally, insofar as handwork is designed for the giving of alms, it again does not fall under the command, except only in case someone is for some reason bound to give alms and knows of no other way to help the poor.

If, therefore, nothing in their Rules enjoineth any special handiwork, then are members of the Orders no more bound to execute handwork than are those in the outside world.[42]

If, latterly, intellectual pursuits tended to take the place of manual work in the Orders, even among the Benedictines who so greatly stressed hard physical work, nevertheless attention must be drawn to the extensive industrial and economic activity which developed from the twelfth century within the Cistercian Order, which undertook the task of bringing to Eastern Europe the more advanced technical development of the West. Some examples will be given below.

Monkish scholarship in the early Middle Ages expounded traditional ancient and theological learning as well as their own knowledge of the world around them in a number of comprehensive Encyclopaedias. Handicrafts were often treated in these works, since the monasteries were at the same time centres of diverse handicraft production. The framework of the teaching in the monastic schools and later in the Cathedral and endowed schools was that of the Seven Liberal Arts, inherited from late Roman times, comprising Grammar, Rhetoric, Dialectic and Geometry, Arithmetic, Astronomy and Music. In the succeeding period, an effort was made to include in this scheme some natural history, and something of the mechanical arts. In the first half of the twelfth century, an ingeniously speculative system of science and art was developed by the German Hugh of St Victor, who in 1133 became Director of the School of St Victor in Paris.

Man, declared Hugh, is destined to Completion but he is incomplete and needs to develop. Nature, which is at the bidding of Man, must help in this development. She is the means, but the goal and end of this development is God. But the technical Arts are an imitation of Nature. The imitation of everything in Nature which human reason discovers to be useful serves him for this Divine evolution. In this way technical work becomes anchored in religion.

The goal and object of all human activity and effort governed by wisdom should be directed either to recovery of the primal purity of our Nature, or to assuagement of those evils to which our life today is necessarily subject. . . .

If therefore we give thought to the restoration of our Nature, that is a divine activity; if we are concerned for our own weakness, and take pains to produce what is necessary, that is a human activity. Every activity therefore is either divine or human. We may not unsuitably name the first *understanding*, since it is performed by the higher kind of being; the latter, however, we will call *knowledge*, since it is performed by those of a lower kind, and because it to some extent needeth counsel. If then, Wisdom, in conformity with the above statement governeth all reasonable activities, there followeth the further statement that Wisdom containeth within herself these two elements, understanding and knowledge. Understanding again may be divided into two kinds, since she is concerned with the search for Truth by Theory, that is to say speculative understanding, and the consideration of Morals by Practice, that is to say active understanding. This last is also called Ethics, or Morals. Knowledge, because it embraces the works of mankind, may suitably be called Mechanics, that is to say the imitative art. . . .

It would take too long and be too laborious a task to discuss here in detail how far the artist's work is an imitation of Nature. But we may prove this briefly by way of example: he who casts a statue has in his mind's eye the image of a man. He who builds a house bears in mind the conformation of a mountain. . . .

The summit of a mountain offers no resting place for waters; similarly a house must be built to a certain height in order to afford security against the impact of invading rain. The discoverer of the use of clothing had previously observed that every form of life has its own method of protection to ward off damage from Nature. Bark surrounds the tree; feathers cover the bird; fish are enwrapped by scales; sheep by wool; hairs cover draft cattle and also wild animals; the shell-fish is sheltered by the mussel-shell; his tusk preserves the elephant from fear of a projectile. And yet there is good reason why man entereth the world without protection or covering, while each animal bringeth with it into the world the protection and weapon fitting to its nature. For it was necessary that Nature should care for those who know not how to care for themselves. But it is precisely in order that Man may be given full opportunity to make his own discoveries that he is left to discover by his own Reason those things that Nature bestoweth on other creatures. . . . Thus were invented those glorious arts which thou observest Man is zealous to practise today. This is the origin of painting, weaving, sculpture, foundry and countless other arts so that we bestow our admiration both on Nature and on the artists. . . .

There are, we have maintained, only four sciences; and these comprise all the others. They are: Theory, which is concerned with the discovery of truth; Practice, which considereth the culture of morals; Mechanics which ruleth the activities of our life; and Logic which giveth power to speak correctly and to examine with sagacity. . . .

Theory extendeth to Theology, Physics and Mathematics; Mathematics to Arithmetic, Music and Geometry. Practise is divided into Personal, Domestic and Public. Mechanics is divided into the arts of weaving, of weapon foundry, navigation, Agriculture, the Chase, Medicine, Drama. Logic is divided into Philology and Oratory. Oratory is divided into that which convinceth by sound truth, and that which deceiveth through false conclusions. Convincing Oratory is divided into Dialectic and Rhetoric. . . . Mechanics comprises seven branches: Weaving, Weapon foundry, Navigation, Agriculture, the Chase, Medicine and Drama (Fig. 10). Of these, three are related to the external equipment of Nature, by which Nature protecteth herself against injuries; four are related to inner equipment, that by nourishment and nurture she may grow and prosper. Thus there is a definite similarity to the Trivium and the Quadrivium; since the Trivium (Grammar, Dialectic, Rhetoric) is concerned with the designation of words, a somewhat superficial matter, while on the contrary the Quadrivium (Arithmetic, Geometry, Astronomy, Music) deals with ideas which are inwardly grasped. . . .

Certain Arts are called mechanical, that is *adulterine* or imitative, because they are concerned with manual activity, which doth borrow the manner of Nature, just as those other seven 'liberal' arts which are called Free Arts either because they require an unconstrained and well-trained mind to enable them to advance in turn with sagacity arguments and counter-arguments concerning the causes of things; or because in olden days only freemen, that is, persons of noble descent were accustomed to devote themselves to these arts, while the common folk and those of lower birth occupied themselves with the mechanical arts, since these require experience of bodily labour . . . Mechanics is a form of knowledge which must embrace the methods of production of all things. The art of weaving includes all sorts of weaving, sewing and spinning, whether by hand, with needle, spindle, awl, winch, weaver's reed, shuttle, encaustic iron, or any of the other tools . . . The second mechanical art is weapon foundry. The word *arm* or weapon is also used with the meaning instrument, tool, implement. The art of weapon foundry is therefore also denoted as the art of implements, not so much because its practice involves the use of implements as because it produceth as one might say implements from a given material. The material may be stone, wood, metal, sand, clay. Related to weapon foundry are two sub-species of art—that of the building hand-worker and that of the smith. The art of the builder by hand is divided between masonry—which is carried out by the mason proper and by the brick-layer—and the art of the wood-worker—by which is meant the joiner, the carpenter and other such manual workers who with pick and hatchet, with file and axe, with saw and drill, with plane, trowel and ruler, polish, hew, chisel, file, cut, join together, and paint material of every kind, whether clay, brick, stone, wood, bone, sand, chalk, plaster or other such stuffs. The work of the smith is divided into hammering, whereby a definite form is bestowed on the substance with hammer strokes, and foundry which bestoweth a given shape by casting. . . .

Navigation comprehendeth all trade, buying, selling, and exchange of home and foreign wares. It holdeth so to say a rightful place among kindred arts as *rhetorical*, since rhetoric is especially needed in this calling. . . . Navigation searcheth distant parts of the world, and seeketh unknown shores; it traverseth terrible wildernesses and maintaineth with barbaric peoples and tribes speaking an unknown tongue traffic that is governed by laws of noble humanity. . . .

Agriculture hath four divisions: tilling the field which it is believed will yield a sufficient crop; planting shrubs and trees, for example in vineyards and vegetable gardens; cultivation of woodland and meadow as may be seen in fields, alpine meadows, moors and pastures; cultivation of flowers as may be seen in flower gardens and rose hedges. . . .

The chase is divided into hunting wild beasts, obtaining birds and fishery. Hunting wild beasts is carried on by many methods: with nets, snares, traps, pitfalls, bow, javelin, lance, encompassment, smoking out, dogs and falcon. For catching birds there are used the trap, sling, net, bow, lime-twig, and hook. Fish are caught with trawling-nets, lines, weir-baskets, hooks, spears. To this craft appertaineth also the preparation of all meals, delicacies and drinks. The name came from one of the sub-species of the craft, because in antiquity nourishment depended more on the products of the chase, just as even today in some districts where bread is very rare, food usually means meat and drink usually means honey-mead and water . . . To 'the Chase' belong all activities of baker, butcher, cook and innkeeper.

Medicine compriseth two divisions, dealing respectively with Causes and Cures. There are six causes: Air, exercise and rest, elimination and replenishment, food and drink, sleeping and remaining awake, emotion. These are called Causes, because they produce and maintain health. . . . Every medical cure occureth either internally or externally. Internal cures are those which are introduced into the body by mouth, nose, ears or anus. . . . External cures are for example bandages, poultices, plaster. Surgery is of two sorts, either cutting, sewing or cauterising the flesh; or setting and bandaging the bones. . . .

Dramatic Art deriveth its name from the theatre, where people were accustomed to assemble to delight in the plays,—not because plays were performed only in the theatre, but because performances were more frequent in the theatre than elsewhere.

Logic, as the science of Reason, also called the science of distinction and judgment, compriseth Dialectic and Rhetoric. Logic as the science of speech, thus belongeth to Grammar, Dialectic and Rhetoric. It containeth within itself the science of well-ordered speech. And it is this Logic, as the Science of speech, that we enumerate as the fourth division, after Theory, Practice and Mechanics. [43]

ISLAMIC TECHNOLOGY

From the eleventh century, but especially during the twelfth and thirteenth centuries, Islamic science gained great influence over the Christian West. The prime significance of Islamic culture for the western world lay in the transmission and systematic arrangement especially of Greek but also of Iranian and Indian knowledge, and beyond this in the enrichment of science and technology through some independent contributions, whereby accuracy of measurement was improved in Astronomy, and simple quantitative experiment was introduced into Physics.

In technology, masterly activity was manifested in the construction of apparatus, especially among the Benū Mūsā in the ninth century, and by al-Gazarī about the year 1200. Thereby these men joined the Greeks Philo and Hero.

Benū Mūsā Instrument for Extraction of Objects from Water

We will show how an instrument (Fig. 14) can be produced with which a man, if he lets it down, can raise pearls from the sea, and can obtain objects that have fallen into wells or have sunk deep down in rivers and seas. For that purpose we construct the two identical halves (*abgr* and *dhwe*) of a (hollow) copper cylinder; but it is an advantage if one is a trifle larger than the other, so that one can contain the other which can be pushed up a little way inside the first. Each half cylinder will be 1 Ell or a little longer, and the diameter of the cavity will be $\frac{1}{11}$ Ell and a trifle more. If it is constructed weaker than this, the instrument will not be serviceable. One half-cylinder will be adjusted to the other so that there is not the least space between them. Then two hinges, *lf* and *tm* are soldered on to them in such a manner that the line *ar* on the half (*abgr*) is not separated from the contiguous line *wh* on the half (*hwde*). When the two half-cylinders are fitted together, the line *de* falls on and touches the line *bg*; if the two halves are separated, then the lines *bg* and *de* are also separated from one another. Teeth are then soldered on to each half of the cylinder along the lines *bg* and *de* as may be seen in the figure, so that, when the two halves of the cylinder are closed together, tooth will engage with

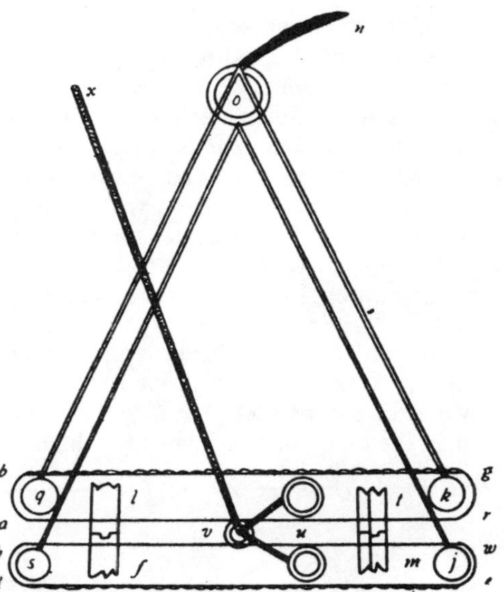

FIG. 14. *Arabian instrument for lifting objects out of the water.* Ninth century.—From: Hauser, *Über das Werk über die sinnreichen Anordnungen der Benū Mūsā*

tooth. The teeth should be fixed to follow the shape of the cylinder, for that is the best plan. Next, on to the backs of the half-cylinders four rings, *k*, *j*, *s*, *q*, should be firmly soldered, and tested to be sure that they hold fast. To each of these rings is attached, as shown in the figure, a separate chain, 2 Ells in length though it matters not if they are a little longer or shorter. The ends of each of these four chains are joined to one another at the point *o*, to which is attached also a long chain (*on*) whose length will depend on the depth to which the instrument will be lowered. Another chain is made fast at the point *u*, midway between *ah* and *rw* and they are fastened on to each of the two half-cylinders near to the lines *ar* and *hw* respectively. This chain is 4 fingers long. Its centre is at the point *v*, where it is fastened and joined to a long chain *vx*.

It follows from the above description that when the chain *vx* is pulled, the two half-cylinders close together; and that when the chain *on* is pulled, this will pull the four chains *ko*, *qo*, *so* and *jo*, whereby the two half-cylinders will separate. If we wish to haul out the pearls or anything else that has sunk down, we pull on the point *o* where the four chains are bound together, and the instrument opens as described above. Then we let it down over the desired spot. When it has reached bottom and remains stationary, we slightly loosen the chain *on* whereby the four chains slacken a little, and then pull on the chain *vx*, so that the instrument is withdrawn and can be seen and everything which it has gripped can be removed.[44]

We will show how a trough can be set up anywhere near a river, and will always be full, so that men can draw water and animals drink from it, and the water in it will remain always of the same depth, neither increasing nor diminishing (Fig. 15).

Let *F* be the river. From it we will lead a pipe *R* to the point where we wish to set up the trough. Into this pipe a stop-cock *H* is ground. To the cock-key is fixed a bar *s* such as is usually employed. The point of penetration into the pipe must be level with the bar. When the bar

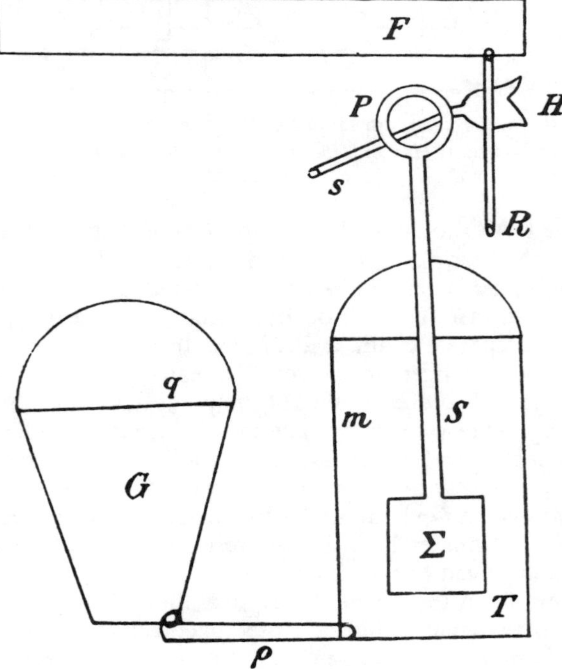

FIG. 15. *Arabian device for keeping water at a constant level in a container.* Ninth century.—From: *Der Islam*, 1918

is turned, this turns the stop-cock until it is opened. Below the pipe *R* we instal a trough *T*, in which we place a float Σ. On the upper end of the float we fix a rod *S* which reaches to the bar *s*. There we affix to it a ring *P* through which the bar *s* is inserted; so that, when the float is raised in the trough by the inflowing water, the stop-cock turns and thus closes itself. Let *m* be the point where the rise of the water in the trough *T* closes the stop-cock. We now place another trough *G* in the position desired. Its upper rim must be on a level with that of trough *T*. At ground level or close to it, we lead pipe *p* from one trough to the other (Fig. 15).

It is clear from the above description that when float Σ is at the bottom of the trough *T*, water flows from the river *F* through *R* into *T*, and thence on through the pipe *p* to the trough *G*. The float Σ will then rise continuously until the water has reached *m* and *q*. The stop-cock then closes itself and no more water flows through it. But if some water is removed from *G*, or if an animal approaches and drinks the water from *q*, the float Σ will sink, the stop-cock will open, and the same quantity of water will flow into *T* as has been removed from *G*. And thus the process continues—and that is what we wished to prove.[45]

The West also owes to the Orient the wind-mill, which the Arabs probably brought to the West from Iran in the twelfth century—a new source of power, which contributed together with the water-power and the use of animal power to replace human labour and thus to influence economic and social conditions. As examples of the use of windmills in Islamic culture between the tenth and fourteenth centuries, we give these three quotations taken respectively from al-Mas'ūdī (about 950), al-Qazwīnī (thirteenth century) and al-Dimaschqī (fourteenth century).

Segistan is the land of winds and sand. There the wind drives mills and raises water from streams, whereby gardens are irrigated. There is in the world (and God alone knows it) nowhere where more frequent use is made of the winds.[46]

The wind there is never still, so in reliance on it, mills are erected. They do all their grinding with these mills. It is a hot land, and has mills which depend on the utilization of the wind.[47]

There is a district in Segistan where winds are very frequent, as are great masses of sand. The inhabitants use the winds to drive their mills and also to convey sand from one place to another, so that the winds are subject to them as they were to Solomon (Peace be with him). For the construction of windmills, they proceed as follows: They erect a building as high as a minaret, or they utilize a mountain-top, or a corresponding hill or a city tower. They then pile one building above another, and in the top one is the mill which turns and grinds. Below is a wheel turned by the wind which has been harnessed (Fig. 16). If the wheel below turns, then the mill above turns with the wheel. Whatever wind may blow, these mills will turn, though there is only a single mill-stone.

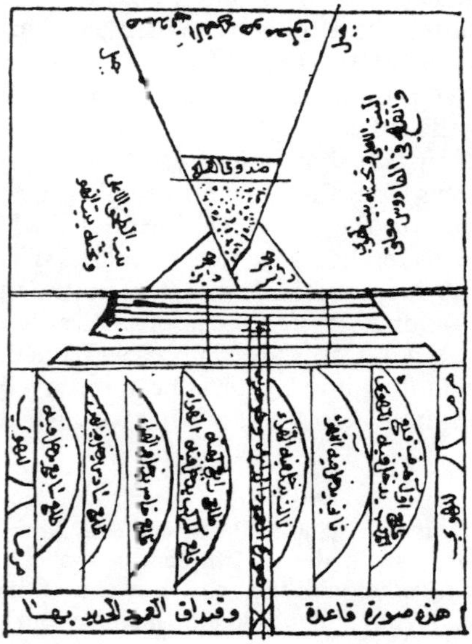

FIG. 16. *Arabian windmill, early fourteenth century.*
Above, the mill; below, the sails.—From: E. Wiedemann, *Beiträge zur Geschichte der Naturwissenschaften*

When the buildings are completed, four loop-holes are made in the lower building, like the loop-holes in walls, but reversed; as their wider part is outside, and their narrower part directed inward, a channel is formed through which the air-current penetrates powerfully, as through the goldsmith's bellows. The wide end is at the opening and the narrow end within, that it may thereby be more adapted to favour the passage of the air-current which must penetrate the mill building, no matter in what direction the wind is blowing. When the air has penetrated the mill-house through the entrance thus prepared for it, it finds a *sarīs* prepared for it, like a weaver's shuttle, which lays the threads over one another for him. This appliance has 12 sides or ribs, or one can reduce them to as few as six. Cloth is firmly nailed on them, like the casing of a lantern, only the cloth is distributed on to each separate rib, so that each is covered. The cloth has a pad which the air fills and pushes forward. Then the air fills the next pad and pushes it forward, and then the third. Thus this *Sarīs* rotates; and as a result of its rotation, the mill-stone turns, and grinds the corn.

Such mills are needed on high mountains and in regions where there is little water but strong wind.[48]

Among the remarkable technical achievements of the world of Islam must certainly be reckoned the erection of great irrigation works in Western Asia, in North Africa and in Spain. Al-Muqaddasī has left us a description of the water-gate of Ahwāz in Persia in the tenth century.

The town of Ahwāz is like Ramla; it lies on both sides of the river, though most of the markets are on the Persian shore.

. . . The river water flows in raised conduits to the city reservoirs. Part of it flows to the gardens, and near the pillar behind the island, within earshot, reaches the dam. This has been wonderfully constructed from blocks of rock in order to hold back the water behind them. Fountains and wondrous works are there. The dam holds back the water and divides it into three canals which extend to their suburbs and water their seeds. They say that if the dam were not there, Ahwāz could not be cultivated and could make no use of its canals.

In the dam are gates which are opened if the water rises too high; if that were not done, Ahwāz would be drowned. The rushing water creates a roar which makes sleep impossible for most of the year. It is most powerful in winter, for it is fed by rain, not snow. The river Maschruqān [modern Nahr] flows through the lower part of the town, but it is usually no more than a dry bed. At a place named al-Dauraq its waters are dammed. As a result of the canals [from the dam], Ahwāz is fertile and ships navigate the river as at Baghdad. The canals separate at the highest part of the town and rejoin lower down at a place called Kārschnān. From there ships can pass to Basra. And they have wonderful mills on the water.[49]

TECHNICAL ACHIEVEMENTS OF THE MIDDLE AGES

We do not propose to give in this work a narrative history of technology, but we should not like to omit at this stage of our survey a review of the series of technological inventions made between the Carolingian period and the thirteenth century. The Middle Ages were indeed richer in technological discovery than is usually supposed. In particular the Middle Ages succeeded in utilizing the elemental forces of beast, water and wind to a far greater degree than was possible for Antiquity, which depended mainly on the muscles of slaves. The use of wind-power by means of a wind-driven wheel was unknown in Antiquity, with the exception perhaps of the design of Hero of Alexandria in the first century A.D. to utilize wind to drive an organ pump. The medieval change-over to the application of natural sources of power betokened technical progress which had important results, comparable in modern times only to those following the introduction of the steam engine in the eighteenth century and the utilization of atomic energy in our own day. We cannot here give contemporary documentary evidence of the medieval technical achievements which laid the foundation for modern technological development. Technology is creation, and this creation has not always been

FIG. 17. *Ancient Egyptian double yoke with neck and body girths.*
From: Ginzrot, *Die Wägen und Fahrwerke der Griechen und Römer*. Vol. 1

recorded in writing. From the early and middle medieval period we lack sufficiently characteristic writings on just those questions which occupy us now. We must, therefore, let some contemporary pictorial representations serve as sources.

From the tenth century, the Christian Middle Ages effected an improvement in horses' harness, which made it possible to increase the draught power of the horse three or fourfold, as compared with horse-power in antiquity.

In olden days a pair of horses was harnessed to a vehicle by a double yoke which lay along the horse's neck. The yoke was held firm by a girth passing round the neck and breast of each horse. The vehicle's pole was made fast to the centre of the yoke. When the horse pulled at the vehicle, the throat girth pressed directly on his wind-pipe, and this hindered the exertion of his full power (Figs. 17, 18, 19). In place of the yoke of antiquity, there was now introduced a kind of horse-collar resting on the animal's shoulders, with long ropes attached. The horse now pulled with his shoulders and could use his full strength (Figs. 20, 21). Transport of heavy weights which in antiquity, owing to the nature of horses' harness was necessarily performed by slaves could now be carried out by horses. By the ninth or tenth century, the West had adopted the modern saddle, stirrup and bridle, and nailed horse-shoes. The stirrup, which was perhaps invented on the eastern steppes, transformed cavalry tactics in war, for it gave the rider a firmer seat and therefore increased the effectiveness of his

7. 16th century Billet Furnace. Part of a painting by Lucas von Valckenborgh, 1575

8. Excavation work has disclosed ore outcrops; mining operations ensue; the discovery is celebrated. Water colour from the *Schwazer Bergbuch*, 1556

Overland connecting rods for transmitting the power of a water wheel to two mine pumps. Engraving by M. Thym from Löhneyss, *Bericht vom Bergkwerck*. Hamburg 1690

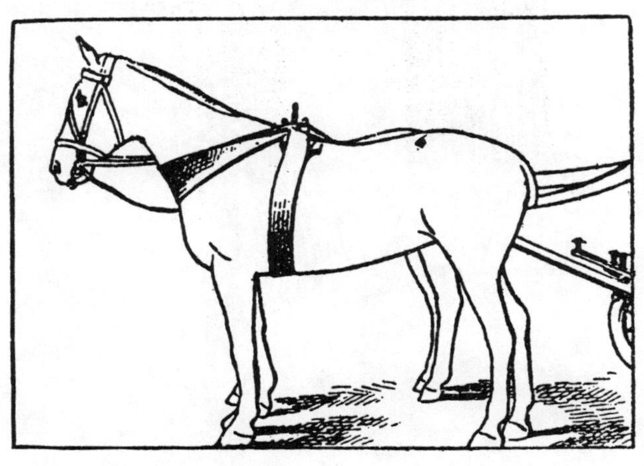

FIG. 18. *Ancient yoke harness with neck and body girths.*
From: Lefebvre des Noëttes, *L'attelage et le cheval de selle à travers les âges* (Paris 1931) p. 163

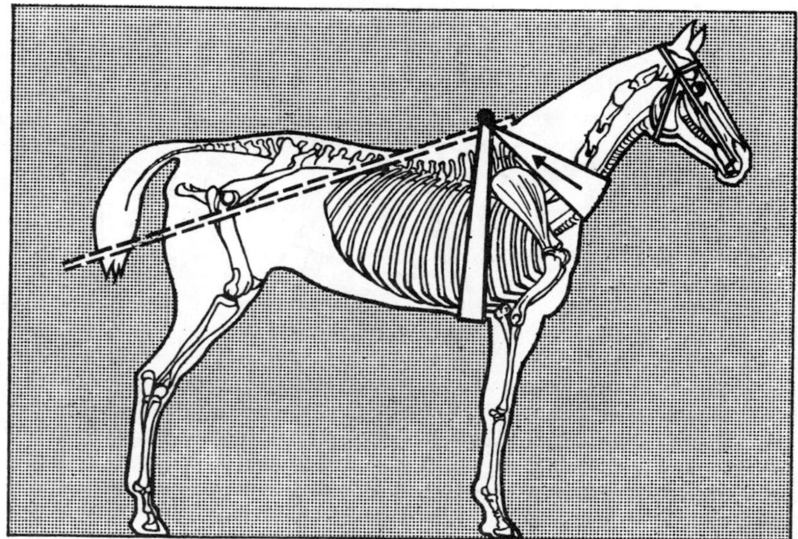

FIG. 19. *Ancient harness with neck and body girths.*
The neck girth presses on the animal's windpipe.—From: *Histoire de la locomotion terrestre*

attack. The iron horse-shoe made possible a much better exploitation of the horse than in antiquity, when at best the hoof was protected by a horse-shoe tied on the hoof, that impeded the animal's progress. These developments, however, took place very slowly. But it gradually became possible to some extent to free men from the heaviest labour. The horse

FIG. 20. *Mule and horse working in the fields.*
The horse (right) in the new collar harness. From the Bayeux Tapestry, eleventh century

also entered into agriculture. The greater draught-power obtained with the new harness made it possible to use the horse for the heavy plough (Fig. 22). The horse could perform agricultural work much faster than the slow ox. The heavy wheeled plough, with vertical blades, horizontal ploughshare, and board to turn the loosened furrows was an invention of

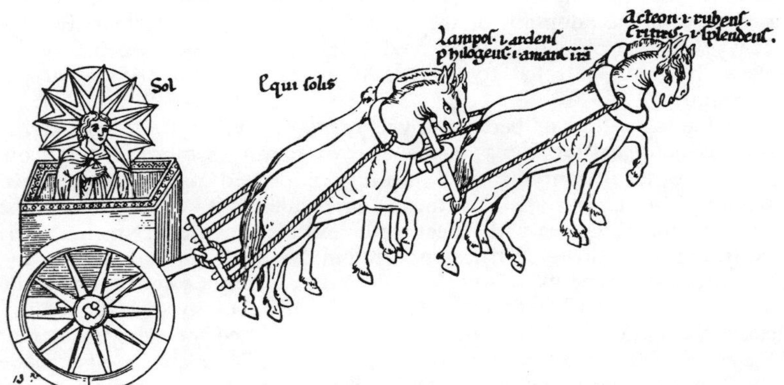

FIG. 21. *New medieval harness.*
Drawing from: Herrad von Landsperg, *Hortus deliciarum.* About 1180

FIG. 22. *Horses with new harness working in the fields.*
Woodcut from: Virgil, *Opera.* Strassburg: Grüninger 1502

the middle medieval period. Perhaps the wheeled plough was sometimes used in antiquity, but the heavier type only came into use in the Carolingian period with the adoption of the three-field agricultural system. But the heavy wheeled plough as illustrated (Fig. 22) was not used before the eleventh century. It led to a marked increase in the productivity of agriculture.

At the same time as better use was made of animal power, better use for technological work was made also of power from wind and from flowing water. From the ninth century improved rigging encouraged sailing ship navigation. But the Square-sailed Viking ships and the Mediterranean galleys with lateen sails had also to use oarsmen. In the twelfth century, however, there appeared in the North the fully seaworthy, broad-beamed type of sailing-ship without oarsmen. Once more a battle against slavery had been won. Then, too, in about the twelfth century came the important invention of the easily worked swinging rudder at the stern of the ship (the stern-rudder) (Figs. 23, 24, Plate 2). This made the ship far easier to manoeuvre than when equipped with a simple rudder at the side; and it became much easier to beat to windward. The compass, which came originally from the Far East, appears equipped with magnetic needle, wind-rose and suspension for the first time in Europe in the thirteenth century as an aid in navigation at sea. Together with the other above-mentioned devices it created the conditions for the conquest of the open sea. Of great importance as well in this respect was the popularization

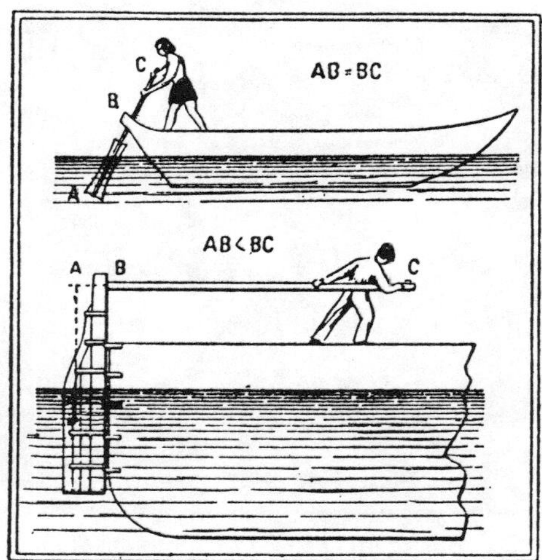

FIG. 23. *Old and new rudder with vertical axis.*
By Lefebvre des Noëttes.—From: U. Forti, *Storia della tecnica Italiana* (Firenze 1940) p. 196

FIG. 24. *Hinged vertical rudder*, 1242.
Miniature in a MS. of 1242 (Alexandri Minoritae Apocalypsis explicata, University Library, Breslau)

in the fifteenth century of the carvel style of sailing-ship. We will in anticipation refer briefly to this point. Whereas the pinnace type of ship, such as the Viking boats and the Hanseatic 'Koggers' were clinker built with wooden planks laid together and overlapping, the carvels were distinguished by their butt jointed planks. The general introduction in the North of this method of timber-planking, which had been used to some extent in the Mediterranean since olden times, permitted an increase in the size of ships.

We have already shown that wind-mills, probably originating in the East, were introduced to the West in the twelfth century, and likewise helped to replace human muscles as sources of power. And with the wind-mills, there were developed larger forms of water-mills. These had indeed been known in antiquity, and were sometimes used in the larger establishments in Roman provinces across the Alps, but they did not gain recognition or importance until the twelfth century. In that century we find the water-wheel used to drive grain-mills, saw-mills, fulling-mills, bellows, and later to spin yarn and to work the forge hammer (Plate 4 and Fig. 34). In East Germany, the Cistercians, who liked to establish themselves in solitary wooded valleys accomplished pioneer technical work of great value, especially in the use of water-power for corn-mills, for foundries, iron-works, saw-mills and for fulling work for felt-manufacture.

The great improvement in transport resulting from the better utilization of horse-power may well have contributed considerably to the gradual replacement of the hand-worked private mill, by water-mills with a far greater capacity and dealing with the grain of a wide region. And the conditions were similar for saw-mills, foundries and a few other technological works.

The improvement in traffic lanes that we observe in the late Middle Ages—we refer to roads, bridges and canals—necessarily proceeded hand in hand with these developments. It was the utilization of water-wheels to drive bellows that first made possible the production of cast iron in the late Middle Ages. And mining was greatly stimulated by pumps and ore transporting devices driven by water power. After the Roman period had reinstituted the ancient technique of building arches, expanded to the scale of church architecture, then in the twelfth and thirteenth centuries the Gothic style brought the technical innovation of buttresses and flying buttresses (Fig. 25), and with the labour of a community of free manual workers, as an expression of the western Christian spirit created those marvels the Gothic cathedrals, which—as it were a *Credo ut intelligam* turned to stone—presented a unique combination of symbolic architecture and structural technique. The medieval building-yards achieved much technical progress beyond the scope of building technique.

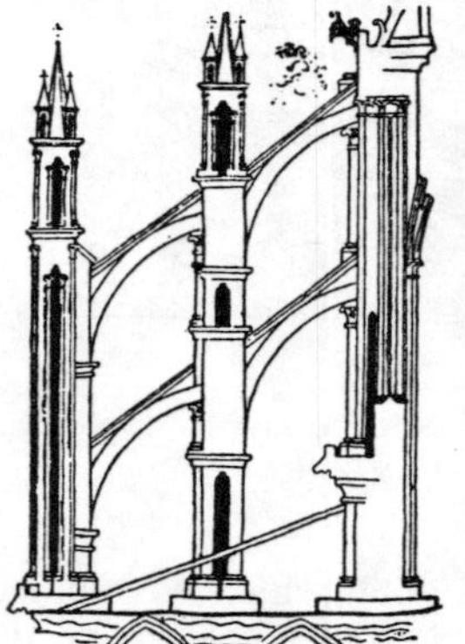

FIG. 25. *Gothic buttresses, thirteenth century.* Drawing by Villard de Honnecourt, 1235.—From Villard: *Bauhüttenbuch*

In textile technique, the late Middle Ages achieved important progress in the direction of mechanization of the rudimentary processes. The twelfth century developed the treadle loom, and probably in the thirteenth century came the so-called hand spinning-wheel (Figs. 26, 27). The late Middle Ages saw in several places development of a capitalist economy for textile work. A remarkable example is the establishment in Florence of the woollen cloth industry. A contribution to its earliest stages was perhaps made by the Humiliaten Monks who lived by the Rule of St. Benedict and brought their familiarity with the superior basic processes of Flemish textile production from the north to Florence in 1239, where it developed in the two following centuries in the hands of an ambitious people. This industry presented a combination of large centralized workshops with wage-earning workers, and of widely scattered home workers. As a result there soon arose a far-reaching division of labour.

In 1338 there were in Florence more than 200 textile workshops, where 30,000 persons lived by the production of cloth. But great industries of this sort, which already indicate the beginning of the breakdown of the medieval economy, were exceptional.

The Middle Ages also achieved important progress in the development of chemical technology. The improvement of distillation apparatus made it possible to distil liquids with low boiling points. Thus in Italy in the eleventh century alcohol was produced from wine. This liquid, contrary to the *watery* element of Aristotelian cosmology, was not distinguished by

FIG. 26. *Pedal loom with two heddles, thirteenth century.*
Drawing from a MS. in Trinity College, Cambridge

FIG. 27. *Wheel for spinning and bobbin winding, fourteenth century.*
Spinning and bobbin winding cannot be done at the same time but must be done alternately. From the Manuscript Royal 10E IV, f. 137, British Museum, London

the pair of qualities *WET* and *COLD*, but presented instead the quality of heat—and indeed of burning. Even more important for chemistry and chemical technology was the discovery in the twelfth century of the powerful acids such as sulphuric acid and nitric acid. These acids, which were first described in the thirteenth century, were produced by dry distillation of alum and vitriol, or of alum, vitriol and saltpetre. Saltpetre provided a means to separate gold and silver from one another. This discovery constituted an important stimulus to metallurgy.

Yet more important for posterity than the above-mentioned achievements, and especially so for town-dwellers, were two inventions of the end of the thirteenth century, the wheeled clock with weights and spectacles (Plate 3). The former introduced hours of equal length, whereas previously day and night had been regarded as independent units, each divided into twelve hours. As a result, hours had been of different length according to the season. In summer there were long day-hours and short night-hours, and in winter the reverse. The length of the hour now became independent of changing seasons. That implied a certain withdrawal from Nature; the mechanism of the wheeled clock had ousted Nature. The commercial life and the extension of trade in the later medieval towns may have favoured this development. Similarly with spectacles inasmuch as here also a technical medium was interposed between subject and object. The spectacle, at first consisting merely of two convex lenses to aid the sight of the elderly, doubtless played its part in the great spiritual upheaval of the succeeding centuries. Men could now continue reading in old age. The external conditions had been created for the enjoyment of an enlarged spiritual realm.

We will consider later the technological gains of the fruitful fourteenth century and of the fifteenth century, in part already touched by the spirit of the Renaissance, together with the problems arising from them.

VILLARD DE HONNECOURT, PIERRE DE MARICOURT AND ROGER BACON

The extensive field of work of a thirteenth century architect and engineer is brought vividly before us by the Office record of Villard de Honnecourt of Picardy. This album of thirty-three parchment sheets dates from about 1235 and probably comprises a collection of designs intended for use in teaching. It contains architectural designs, geometrical constructions for the builder, surveying exercises, drawings for sundry apparatus and machines for use in peace-time and in war, a plan for a *Perpetuum mobile*, studies in artistic anatomy and in the proportion of bodily measurements. Quite in the spirit of Vitruvius, the building of machinery is included in the architect's sphere of work. Next to drawings of large machines built of wood, such as a crane and a sling for missiles, are those of smaller apparatus for use by priests, such as the warming apple with the brazier in gimbals and the angel whose finger always points to the sun, and the mechanical eagle whose head is always turned toward the priest. We recognize here the influence both of the Alexandrian and of the Arabian builder of apparatus. But an original medieval technical achievement is the saw-mill driven by a water wheel with automatic feed of the material worked. Here we encounter for the first time in Western technology the *Perpetuum mobile*, a wheel with an uneven number of moveable hammers which is supposed to turn continuously. Perpetual circular motion is a phenomenon of the heavenly spheres which appertains, according to the medieval Aristotelian view, to the translunar world. Man's attempt, since the thirteenth century, to produce such perpetual motion here on our earth may perhaps be regarded as a desecration of the Aristotelian conception of perpetual circular motion which proceeded only in heaven. And just from the beginning of the thirteenth century, when the West came, through the Arabs, to know the whole of Aristotle, and especially Aristotelian cosmology and physics, just in that period we first encounter this profaning of the idea of the circular motion of the heavens. Western man, with his urge to mould Nature, wished to make himself the initiator of perpetual circular motion on earth, the image of the divine circular motion of the heavenly spheres.

We will show a few of Villard's sketches, with the short texts accompanying them.

Villard de Honnecourt greets you, and begs that all those who work with the help given in this book will pray for his soul and will keep him in memory. For good counsel can be found in this book on the great art of masonry and the construction of hand carpentry; and you will find therein the art of drawing which both requires and teaches the main features and the discipline of geometry.[50]

For many days, Masters have discussed how to make a wheel that will turn by itself (Fig. 28). Here is one that can be made from an odd number of hammers or with quicksilver.[51]

If you wish to make a Hand-warmer, you should make in copper a sort of apple, with two halves which can be locked into another. Within the copper apple must be six copper rings. Each ring has two pivots and at the centre must be a pan with pivots. These pivots must so reciprocate with one another that the fire-pan shall remain horizontal. For one pivot supports the other, and if you have constructed it correctly according to the text and the drawing, then you can turn it to whichever side you wish, and the fire will never fall out. This apparatus is suitable for a bishop; he can cheerfully hold High Mass, for if he bears this device in his hands, he will remain protected

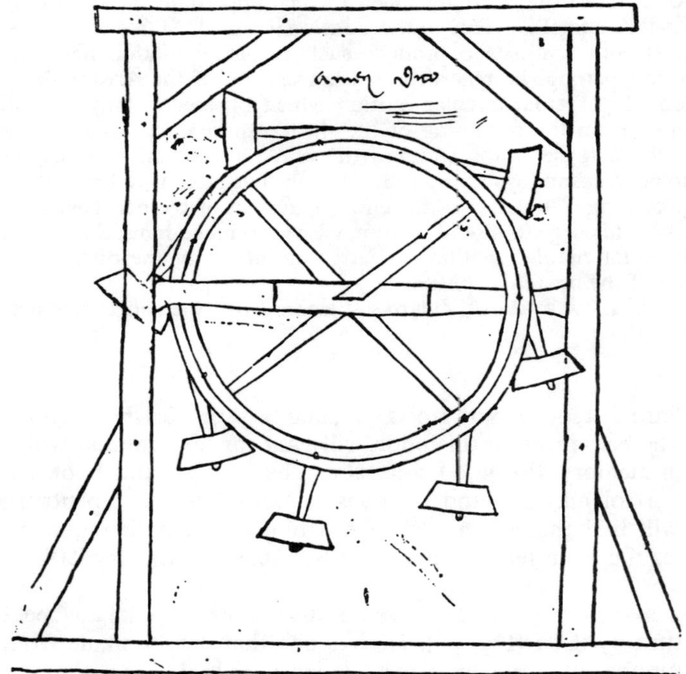

FIG. 28. *Perpetual motion device of* 1235.

Drawing by Villard de Honnecourt.—From Villard: Bauhüttenbuch. Pub. by H. R. Hahnloser, (Vienna 1935) Plate 9. The old French text runs: Maint jor se sunt maistre desputé de faire torner une ruée par li seule; vès ent ci com en puet faire par maillès non pers u par vif argent. Naturally the wheel is at right angles to the axis; but is shown turned into the plane of the drawing from cold so long as the fire lasts. There is nothing better than this device. It is so constructed, that in whichever direction it is turned, the fire-pan remains level.[52]

In this fashion one makes a saw work by itself (Plate 4)
In this fashion one constructs a faultless bow.
In this fashion, one builds an angel whose finger points always toward the sun.
In this fashion one constructs one of the strongest machines in existence for raising weights.
In this fashion, one makes an eagle that always turns his head toward the deacon when he is reading the Gospel.[53]

Pierre de Maricourt, a fellow-countryman of Villard de Honnecourt, wrote in 1269 on a crusade, before Lucera in Italy, to his friend Siger de Foucaucourt an *Epistle concerning the magnet*. This contains a masterly experimental study of the magnet and speculates concerning its possible utilization.

The poles in the heavens are for him the seat of the magnetic directional power, which he regarded as divine. He described a series of excellent experiments which led him to a clear understanding of several properties of magnetism, an understanding to which nothing further was added until the work of William Gilbert about 1600. Pierre emphasized the value of experiment and demanded of the natural scientist manual dexterity as well as familiarity with the general nature of things and with the motion of the heavens. And in the second part of his *Epistle* he sought, just as Villard de Honnecourt had done, to produce a machine with perpetual motion which should be the earthly counterpart of the ceaseless circular motion of the stars. He believed that the lodestone mysteriously followed the heavenly directional powers, and thus revealed the connexion between Macrocosm and microcosm; and he thought that he had found in it the Philosopher's Stone which would enable him to attain the realization on our globe of perpetual circular motion. The Letter of Pierre de Maricourt bears significant witness to the urge to technical production of western man during the Gothic period, albeit the realization of the idea, in spite of his knowledge of certain important experimental facts was bound to miscarry owing to its highly speculative foundation.

How the Master of this work should be equipped

Know then, dearest friend, that an investigator of this subject must have understanding of Nature, and not be ignorant of the celestial motions. He must also be clever in the use of his hands in order that, by means of the lodestone he may produce wonderful effects. For by his carefulness he will be able in a short time to correct an error which in an age he could never do by his knowledge of natural sciences and mathematics, if skill were lacking in the use of his hands. For in occult matters, we investigate many things by manual industry, and in general without it we are unable to bring anything to completion. Many things, however, are subject to the realm of reason, which we cannot fully investigate by the hand. From the above, therefore, it is clear what qualifications are required in an investigator of this subject.[54]

On the construction of the wheel

. . . In this chapter I shall reveal to you the way to construct a continually moving wheel of wonderful ingenuity (Fig. 29), in the

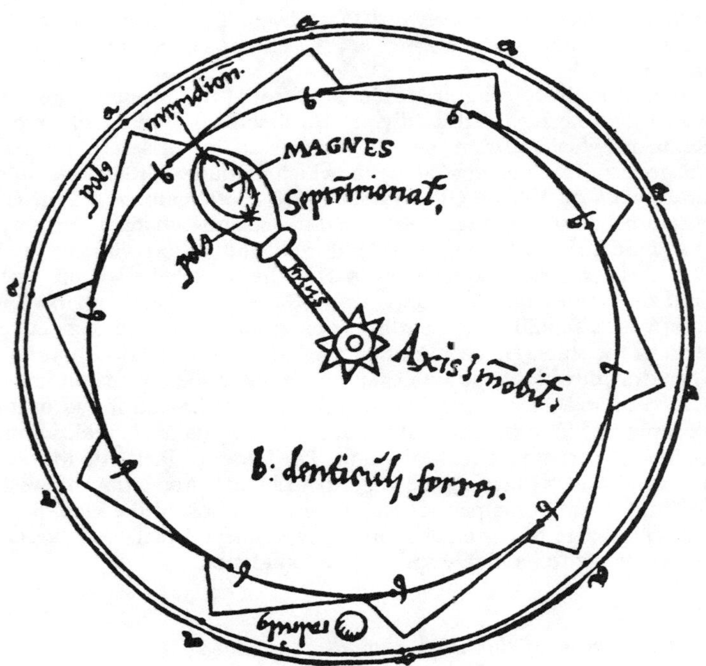

FIG. 29. *Magnetic perpetual motion device*, 1269.
From Pierre de Maricourt, *De magnete*

invention of which I have seen many engaged in vain attempts and wearied with much labour. For they did not perceive that they could affect the accomplishment of this work by the virtue of power of the lodestone.

For the composition or construction of this wheel, construct a silver case, like that of a concave mirror embellished with clever workmanship, with carvings and perforations which you will make both for the sake of beauty and of lightening its weight; because the lighter it is, the more rapidly it will turn. You will then perforate it so that the eye of the ignorant shall not perceive what is cleverly inserted inside the case. Let there be inside, however, small nails or teeth of iron, of one weight, inserted on the edge, bent towards each other so that the distance separating them is not more than the thickness of a bean or a pea. Let the above-mentioned wheel be uniform in the weight of all its parts, then fix an axis through its centre upon which the said wheel may revolve, the axis remaining quite immoveable. Let a silver bar be attached to the axis and placed between the two cases, on the end of which let a magnet be set prepared as follows:

Let it be rounded and the poles established, as has been described. Then let it be shaped like an egg, with poles intact, and let it be filed somewhat in two intermediate and opposite parts, so that it may be compressed and occupy less space, and that it may not touch the walls through the motion of the wheel. When it has been thus fashioned, let it be placed on the silver bar like a stone in a ring, and let the north pole be somewhat inclined toward the small teeth of the wheel so that its virtue may flow, not diametrically, but at a certain angle, in to the iron teeth, so that when any tooth comes near to the north pole, and owing to the impetus of the wheel, passes a little beyond, it may approach the south pole which will flee rather than attract it, as is obvious from the rule above given. Thus every little tooth will be in a perpetual state of attraction and repulsion. And in order that the wheel may perform its duty more rapidly, insert between the cases a small round pebble of brass or silver, of such size that it may be caught between any pair of teeth; so that when the wheel is raised, the pebble may fall on the opposite side. Therefore, when one side of the wheel is always rising, the fall of the pebble on the opposite side, caught between any two of the teeth, will likewise be perpetual because, as it is drawn towards the centre of the Earth by its weight, it will prove to be an assistance and will not let the teeth come to rest in a direct line with the stone. Let there be, moreover, spaces between the teeth conveniently hollowed out, so that they may properly catch the pebble in its fall as the present figure shows.

Farewell! Finished in camp at the siege of Lucera on the 8th August, 1269 A.D.[55]

That elusive thirteenth century figure, Pierre de Maricourt's pupil Roger Bacon (*circ.* 1214–1294) appears, if we consider only certain passages of his works, to be spiritually akin, through his strong advocacy of observation and experiment, to the founders of the new experimental science; but if we regard his work as a whole, he appears to us in a different light. Experiment did not play for Bacon the same role as in the new Natural Science. He considered that experiment should demonstrate that knowledge transmitted by tradition or gained by Reason could also be of practical use. He did not envisage a systematic questioning of Nature by means of experiment, but rather the use of experiment in the attempt to find practical application of knowledge already gained in traditional fashion by Reason, and thus to dominate and surpass Nature. Naturally this often led to new discoveries to which speculation had not led the way. Thus Bacon's experimental science was in fact a contribution to technology. Thus manual workers and alchemists were more involved with it than pure natural scientists. His creative impulse fed by his rich imagination led Bacon to dream of future technological achievements.

For there are two modes of acquiring knowledge, namely, by reasoning and experience. Reasoning draws a conclusion and makes us grant the conclusion, but does not make the conclusion certain, nor does it remove doubt so that the mind may rest on the intuition of truth, unless the mind discovers it by the path of experience; . . .

He therefore who wishes to rejoice without doubt in regard to the truths underlying phenomena must know how to devote himself to experiment. For authors write many statements, and people believe them through reasoning which they formulate without experience. Their reasoning is wholly false. For it is generally believed that the diamond cannot be broken except by goat's blood, and philosophers and theologians misuse this idea. But fracture by means of blood of this kind has never been verified, although the effort has been made; and without that blood it can be broken easily. For I have seen this with my own eyes.[56]

For the things of this world cannot be made known without a knowledge of mathematics. For this is an assured fact in regard to celestial things, since two important sciences of mathematics treat of them, namely theoretical astrology and practical astrology. The first . . . gives us definite information as to the number of the heavens and of the stars, whose size can be comprehended by means of instruments, and the shapes of all and their magnitudes and distances from the earth, and thicknesses and number, and greatness and smallness, . . . It likewise treats of the size and shape of the habitable earth . . . All this information is secured by means of instruments suitable for these purposes, and by tables and by canons . . . For everything works through innate forces shown by lines, angles and figures.[57]

Vessels can be made which row without (the force of) men, so that they can sail onward like the greatest river or sea-going craft, steered by a single man; and their speed is greater than if they were filled with oarsmen. Likewise carriages can be built which are drawn by no animal but travel with incredible power, as we hear of the chariots armed with scythes of the ancients. Flying machines can be constructed, so that a man, sitting in the middle of the machine, guides it by a skilful mechanism and traverses the air like a bird in flight. Moreover, instruments can be made which, though themselves but small, suffice to raise or to press down the heaviest weight. They are but one finger-length in height and of the same width; and by their means, a man can raise himself out of prison. Furthermore, an apparatus can be constructed by means of which a single man can draw a thousand men up to himself despite their resistance. Similar instruments can be constructed, such as Alexander the Great ordered, for walking on the water or for diving, without the least danger.[58]

URBAN INDUSTRY

The blossoming of city life and of free civic institutions since the Middle Ages was closely linked with the development of a comprehensive civic hand-craftsmanship of which the products proclaim the work of free men under the influence of Christian culture. The fourteenth century was the most prosperous period for the guilds and corporations. In some towns we find as many as fifty or sixty different guilds. Such numbers testify to the far-reaching specialization of manual work. Nuremberg which, unlike other medieval towns, had no actual guild organization, sheltered no less than fifty groups of handicraftsmen, comprising 1217 Masters. From a register of the year 1363 we learn that they were distributed as follows:

		Masters
1.	Shoemakers	81
2.	Tailors	76
3.	Bakers	75
4.	Cutlers	73
5.	Butchers	71
6.	Leather-workers	60
7.	Furriers	57
8.	Cobblers	37
9.	Tanners	35
10.	Coopers	34
11.	Dyers	34
12.	Braziers, Belt-makers, tin-founders, tinkers	33
13.	Mantle-makers	30
14.	Clothiers	28
15.	Locksmiths	24
16.	Mirror-makers, glazier beyond the gate	23
17.	Needle-makers and wire-drawers	22
18.	Farriers	22
19.	Trunk-makers	22
20.	Mail glove makers	21
21.	Coach-builders	20
22.	Hatters	20
23.	Fishermen	20
24.	Bridle, spur and stirrup makers	19
25.	Tool-makers	17
26.	Money-changers	17
27.	Saddlers	17
28.	Goldsmiths	16
29.	Carpenters	16
30.	Tin-smiths	15
31.	Pewterers	14
32.	Ironers	12
33.	Ribbon manufacturers? (Pandberaiter)	12
34.	Glovers	12
35.	Bag-makers	12
36.	Glaziers	11
37.	Potters	11
38.	Cabinet-makers	10
39.	Cloth shearers	10
40.	Rope-makers	10
41.	Sheet metal worker, Buttonsmith, Polisher	9

	Masters
42. Stone-masons	9
43. Boiler-smiths	8
44. Blade-makers	8
45. Sword-polishers	7
46. Helmet-smiths	6
47. Nail-makers	6
48. Painters	6
49. Hollow ware makers	5
50. Mail-shirt makers	4

50 groups of artisans comprising 1,217 Masters.[59]

Contemporary with the flourishing handicrafts of the late medieval towns came a series of important inventions, though sometimes limitations imposed by the Guilds hindered technical progress. The lathe which in a simple form goes back to the bronze age received its characteristic development in the fourteenth century (Plate 5, Fig. 30). It is indeed Leonardo da Vinci about 1500 whom we first find working a lathe by a treadle; but already in the fifteenth century foot-power with a crank was used for a grindstone. In the late Middle Ages, when the expanding bourgeosie experienced an increasing need for culture, the highly developed skill in metal working led to the invention of movable type for printing books. Gutenberg's achievement lay in the realm both of technology and of

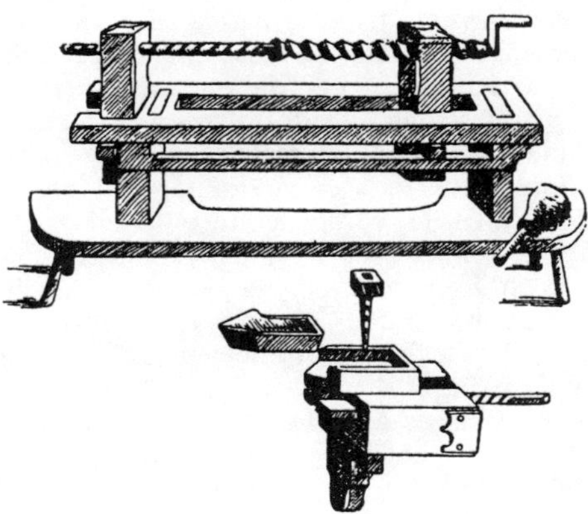

FIG. 30. *Screwcutting lathe.*
Below: Toolrest with screw feed. Drawing from *Das Mittēlalterliche Hausbuch*, 1480

craftsmanship. His great technological feat was the invention of an apparatus for type casting capable of producing any desired quantity of standardized types. For shaping the type he used a copper matrix in which the letter was struck by means of a steel letter-punch cut in relief. Several metal techniques were here combined. As has been so well shown by Franz Schnabel, success was only possible to a man with constructive ability such as the patrician Gutenberg who, familiar with the various metallurgical techniques, was able to press beyond the technological limitations of single groups of craftsmen and was far-sighted enough to recognize the needs both of his own day and of the future. And no less than his technical success was Gutenberg's artistic achievement, whereby in masterly fashion he adapted the forms of the letters to the technological requirements of metal-work. For the letters, instead of being written, were now cut in metal, and the different material necessitated different shapes. A further condition for the wide dissemination of the art of printing books, in addition to the contemporary eagerness for books to serve cultural needs, was a material suitable for printing that was cheaper than parchment. Paper, which had been discovered in the Far East in A.D. 105, became known to the West through the Arabs and was produced in Europe from the thirteenth century. Here again, it was the increase in the material needs of the flourishing city bourgeosie that led to the greater use of linen and created the conditions necessary for increased production of paper.

MILITARY TECHNOLOGY IN THE LATE MIDDLE AGES

The urge of the West toward technical achievement led between 1320 and 1330 to a further invention fraught with extraordinary consequences, artillery that could use the propelling force of gun-powder, which led in the succeeding centuries to a transformation of both attack and defence. With the development of the independence of towns and the establishment of territorial states, military matters in general underwent expansion in many directions, manifested in the fifteenth century in a series of illustrated manuscripts of military technology emanating from Italian and German cultural circles. Foremost is the German military engineer, Konrad Kyeser of Eichstätt in Franconia. He completed in 1405 a large illustrated manuscript of military technology which greatly influenced writings on the subject in the succeeding period. Kyeser discusses vehicles of war, guns and gunpowder, battering rams, lifting-gear, pumps and water works, pontoons, life-belts, hot air balloons and much else involved in war and in the use of gunfire (Figs. 31, 32). In his dedication to Rupert Count Palatine, and to the princes and the Estates of Christendom generally, he declares:

> If India boasts of her precious stones, Arabia of her gold, Hungary of her swift steeds, Italy of her cunning, England of her riches, France of her refinement and friendliness, then indeed Germany is

FIG. 31. *Medieval gun for lead shot*, 1405.
From a picture in: Konrad Kyeser von Eichstätt, *Bellifortis*

FIG. 32. *Medieval cannon for stone shot, under a roof shelter*, 1405.
Fired by a red hot iron hook.—From a picture in: Konrad Kyeser von Eichstätt, *Bellifortis*

justly famed for her determined, strong and courageous soldiery. As heaven is adorned with stars, so does Germany shine by her free crafts and is honoured for mechanical knowledge and distinguished by her many industries of which we are justly proud. Furthermore, our army is famed throughout the world. For when the rise of many nations aroused attention, disturbed law and order, and upset the balance of even-handed justice, we Germans acted differently. We are not out of our minds nor are we so spiritually weak that we do not prefer to be led by Truth rather than suffer ourselves to be betrayed by falsehood.[60]

The epitaph which Kyeser composed for himself testifies to the unstable wandering life of such a military engineer in the late Middle Ages.

Reckon six thousand and six hundred years since Adam was formed,
Since the world was created;
Since Man was born of the spotless Holy of Holies;
It was the fourteenhundredth year;
There perished in a strange land Konrad Kyeser of Eichstätt,
A man friendly, open-handed, gentle, sociable in his dealings;
He was eloquent too, and determined in action,
A follower welcomed to their palace by many princes,
Known as a finely educated and sound man.
Well-disposed to him was Wenzel of Bohemia, the King of Rome,
Sigismund too, who still doth hold the sceptre of Hungary,
The famed Duke of Lausitz whose name is John,
Stephen the Elder of the Banate too, and Austria's Duke,
William Albert, the Elder, as also the younger Albert,
The Lord of Oppeln too, known as Duke John;
And Francis, Lord of Carrar and Padua liked him well.
The gorgeous court of Apulia and Sicily's isle
Raised him to highest honours; Poland's quiet plains knew him well;
So did the Campanua country that lieth so small by Capua,
Also Milan, Tuscany and Lombardy,
Denmark, Norway, Sweden and blossoming Franconia,
France and Burgundy, Spain and Wallachia.
Russia too doth know him, Lithuania, Moravia and Meissen,
United Ukraine, Steiermark and Great Carinthia,
And the men of Svendborg and of Stettin. Weep for him, ye citizens of Eichstätt.

Princes of Saxony, ye loved him well, ye too, Dukes of Silesia;
Him, the man of renown his fame shone forth everywhere
When as Bellifortis he overcame whole armies.
His experience hath gained him a mastery of the art of war equalled by no man.
Now hath horrible death, most horrible,
Cast him down to Wrath. Thus lieth he bound,
Far from his home, confined deep down in his coffin.
Weep for him, ye Princes! Weep, ye nobles, and weep, ye who are poor.
Mourn for him, ye winged creatures, and animals with three feet, and with four!
And all who go on two feet, and even the myriad worms.
All creation, endowed with life or no,
Every soul, living or dead,
Whether visible or invisible, guilty or in salvation,
All God's works, coming into being, growing or complete.
Let all be filled with grief, with sorrowing pity.
Every man who hath Faith, doth pray that God will take him to Heaven.
O thou sacrificial paschal Lamb, have pity on him, O Jesus.
Thou bearest the sins of the world; grant to him too thy Mercy.
O thou who forgivest earthly sins, grant him Peace,
Grant him future life in the realm of Heaven.
O thou most exalted Spirit, grant to this prayer an assenting Amen.[61]

The manufacture and service of gunpowder artillery required many different handicrafts such as casting, forging, carpentry and joinery, and preparation of the gunpowder. There thus arose a new calling, that of the so-called Master-gunsmiths, who practised all these crafts. These men were highly esteemed both by princes and by the townships. From a document introducing Merckiln Gast, a Frankfurt Master-gunsmith of about the last decade of the fourteenth century, we learn something of his ability.

Merckiln Gast, the gunsmith, can perform the following:
First he can restore spoilt gunpowder to its original state, rendering it again efficient.
Item he can separate and refine saltpetre and salt.
Item he can make powder that will last 60 years.

Item he can shoot with large or small guns.
Item he can cast from iron small-arms and other guns.[62]

This is the first document in which we hear of an iron founder. Iron casting was one of the greatest inventions of the Middle Ages. It developed in the Rhineland about the beginning of the fourteenth century. It followed the more efficient use than in the early and high Middle Ages, of water-power to work the great bellows needed to produce sufficient furnace heat to smelt iron. Only in the second half of the fifteenth century was the iron cast direct from the cupola. The blast furnace for iron began, slowly to develop from the fourteenth century, but not until the sixteenth century can we speak of a true blast-furnace. The introduction of the blast-furnace was, to begin with, not directly linked with the invention of cast-iron.

Several of the Master-gunsmiths mentioned above wrote descriptions of their technological experiments. There have survived a number of drawings that illustrated such books on Guns and Fireworks. They are akin to such illustrated manuscripts of military technology as that of Kyeser. The German Fireworkbook of about 1420 especially, exerted great influence and was widely copied; moreover it was printed from the sixteenth century onward. It discusses in detail what a Master-gunsmith should be able to do, and especially what should be his disposition. He who deals with powder and gun, those devilish means of annihilation, says the unknown medieval writer, should ever be conscious of his responsibility and keep God always in mind (Plate 6).

When princes, counts, lords, knights, villeins or towns are troubled, besieged by enemies, assaulted in their castles, fortresses or towns, it is above all needful that they have pious and reliable servants who will freely oppose to the enemy body, life, property and such goods as God hath bestowed on them; who dare to remain steady rather than to flee, and to hold firm rather than to yield, who would be shamed by aught of evil or despair; wise men, who know the damage that can be done to them by the use of shot, projectile and assault and who take precautions against them, who know how to build a defence and make good bulwarks . . . they should know to use their implements with adroit skill against the enemy, and above all that, without overwhelming advantage, they should undertake no mischievous skirmishes without the castles in which they are engaged; they shall maintain good friendship together without strife or dissension and will fulfill their tasks according to the best advice. And those princes, counts, lords, knights, villeins or towns that have such pious, trusty and wise servants, must have good confidence in them.

FIG. 33. *Fifteenth-century muskets.*
Woodcut from: *Rudimentum novitiorum.* (Lubeck 1475)

But they need men who can and will work as smiths, masons, carpenters, cobblers, and also good guards and master gunsmiths. And if great comfort may be derived from good gunsmiths, then it behoveth all princes, lords, knights, velleins and towns, and all who have master-gunsmiths that their gunsmiths shall be good Masters, and shall know how to prepare well and to manufacture all waters, oils and powders that belong to their art, and other things for the guns, for fire-darts, fire-balls, and random fire-works, and other things with which one may hold back the enemy and work harm to them; wherefore hereafter in this book is described how these things

FIG. 34. *Smithy with water driven hammers.* (Camshaft acting on tail of Hammer.) Woodcut (section) from Spechtshart, *Flores musicae* (Strassburg, J. Prüss 1488)

shall from beginning to end be well and truly made. And because there are so many things that every good master-gunsmith should have, but which he cannot hold in his mind without the written word, therefore all that is useful and necessary thereto hath been written below. . . .

Now here are described what sort of life and good habits must be followed by every man who would be a good master gunsmith and also the skill that he must command. These things appertain to every master gunsmith and he must have them. First he must honour God and he must above all keep him in mind more than do other travellers. Moreover when he is dealing with powder or with gun, he has under his hand his great,—indeed his greatest enemy,—(the Devil). Therefore must he exercise perpetual and triple caution.

He must conduct himself with modesty toward those whom he encountereth in the world. He must be honest and unafraid. And in wartime he must offer comfort, since great comfort is gained from such men.

The master must be able to read and write, for it is not possible that he retain in his mind all matters pertaining to his art, which are described hereafter in this book: be it separation or division; sublimation or purifying; reinforcing or strengthening, and many other things that appertain thereto whether of random or of controlled fire and such things. Wherefore a Master must understand writing if he desire to be a good Master. He must know how to set all in order to construct a firm emplacement and walls, for battering rams, for shelters, for culverins and all that pertaineth thereto from start to finish, as thou wilt find it set down earlier in this book. And first of all he must know three things—the weights of solids, the weights of liquids and the method of measuring.

He must show himself honourable, kind in word and deed, and circumspect in all things, he must ever especially guard against drunkenness. . . .

A certain and true instruction, how to make gunpowder properly and well. Thou shalt take a good balance, and place on each of the scales, opposite to one another, an equal weight of good, purified saltpetre. Then take one part of the saltpetre off the scales; lay this aside, and, opposite the remaining saltpetre place an equal quantity by weight of good, fresh sulphur. This done, lay aside the remaining saltpetre; then take the sulphur from the scale, and divide it into two parts of equal weight; and this done, set aside one part of the sulphur, leaving the other on the scales. Next, add to the remaining part of the sulphur an equal quantity by weight of fir or lime charcoal; this must not be very hard, nor have been quenched in water. This done, lay the portion of charcoal and the remaining sulphur opposite one another, each weighing the same as the above-mentioned charcoal. This done, then take all the above-mentioned parts of the saltpetre, the sulphur and the charcoal, and mix them all together, as thoroughly as thou mayest be able. And when it is thus mixed, then pound it very thoroughly; the more thou poundest, and the finer thou makest the powder, the more powerful and quick acting the powder will become, and the more quickly it will be discharged when fired.

And dry the powder well, in a good vessel in a warm room, and guard thyself carefully from fire, and do not place it too near the oven, for the powder is kindled so readily by heat as by fire. So now thou hast prepared a definite weight, and thou knowest how the material should be weighed out; and thou shalt start again to weigh out more material and proceed just as before. Mark this teaching well, for it is the best and most certain that existeth for this art. For if there is more of one material than there should be according to this instruction, then thou hast worked entirely in vain, and hast spoilt the material. And for him who understandeth the above teaching, for him it is sufficient. . . .

It should be known that many a pious prince, lord, knight, villein and also towns frequently and often suffer siege and invasion for their destruction, and have not known how to take precautions beforehand and had not with them such men as by their knowledge and skill could bring counsel and help to resist and hold back the enemy. And it hath often been obvious that many pious princes, dukes, lords, knights and villeins have paid dearly for this and have suffered ruinous damage, and the worthy nobility established to be the defenders and servants of the holy Roman Empire, and favoured by God for that purpose have been thereby abused and crushed.

Therefore doth the faithful counsellor advise all princes, dukes, knights and villeins, as well as townships and many others who own castles, fortresses or towns, that they take care by providing themselves with such men as are described earlier in this book. And they must have a plentiful supply of provisions and materials and especially of the things needed for those implements described above and also hereafter in this book: saltpetre, sulphur, good wood for charcoal and also for bastions and terraces, and quicksilver, spirits, camphor, arsenic, salammoniac, oils and reinforcements, all these being useful and good for powder and for unconstrained fire. And he warneth that they bethink themselves of the saying: He who letteth himself be found unarmed will be conquered by an army of invalids. And let them recall that when the enemy, who has invested and besieged them think that they command the situation, then can they begin various operations to raise the siege. But if a man hath provided with foresight the things described above, then, whatever his enemy may attempt, that man, so long as he hath well-skilled help will be able to hold off the enemy until relief cometh or until the enemy is driven to abandon the siege with a good box on the ear.

Furthermore it is shown in the next chapter how, if the enemy have come close to the walls of a fortress or a town and will attack with good battering rams and shields and ladders, the assault may be averted, the castle or town may hold against the attack and the enemy be overcome.

If thou art in a fortress or a citadel, and the enemy is before thee and approacheth the walls with good protection or with battering rams, as thou seest in the picture in this book, and if he striketh on the wall with ladders or breaketh them, so that thou, from within knowest not from where the attack and the break-through are launched, take then a dice and proceed within the wall from stone to stone; and where the dice bounceth, break through the wall against the enemy, and be sure that thou hast a good gun well loaded and shoot outward through the break in the wall. But first take thirty pounds of resin, ten pounds of sulphur, six pounds of charcoal, melt the resin, and mix the saltpetre, the sulphur and the charcoal, all well pounded together, into the melted resin. Make from the mixture balls the size of an apple and set fire to them. Throw them to follow the charge of shot among the enemy who are without on the walls. The burning balls will cause such thick fog and smoke, and will burn so furiously, that the enemy will be able to do thee no harm until thou hast well re-loaded the rifle.

Thus can everyone defend himself well against battering rams and shields, and can hold his castle or citadel well secure from any assault. Mark this well, and if thou wouldest halt thine enemies, arm thyself with those things aforesaid which this book hath taught thee. For many fortresses are lost and those within are taken prisoner, because they have failed to prepare provisions and implements and therefore they cannot make use of them until they are rescued or the enemy is repulsed with a good cudgelling. Therefore is it necessary to take thought in advance.[63]

PART III

THE RENAISSANCE

PART III

INTRODUCTION

BY successful craftsmanship and commercial activity, bourgeois society in the towns of the late Middle Ages attained power and confidence. The privilege ownership enjoyed by the medieval feudal powers was slowly broken down by the rising bourgeosie, who, conscious of their newly-won independence, sought to break the bonds imposed on them in the Middle Ages. The Middle Ages had bound men's faith, thought and action within the rigid framework of a uniformity determined by church and religion. But now men became more conscious of themselves as individuals. With increasingly open minds, they confronted the gay diversity of things in this world and their interplay. More and more, man sought in the world a field of activity for his independent mind. It was this ardent appreciation of the things of this world that first rendered possible the resurrection of the culture of classical antiquity. The urge was to create something new and not merely to follow tradition and for stimulus to this fresh start, man directed his gaze back to classical antiquity which he regarded as the fount of all civilization. A precondition of this new orientation, this rebirth of antiquity was enthusiastic study of the classical tongues and the discovery of new sources of ancient writings. Humanist effort in this period was also directed at the technical literature of antiquity, and sought to apply it in practice, but always, again, with a view to using classical knowledge as a point of departure for new achievement.

Technical achievement received special stimulus during the Renaissance period from greater general preoccupation with active life and from the increased desire of many a far-seeing craftsman for intellectual enlightenment and a scientific foundation for his customary rule-of-thumb manual work. Witness to this attitude is borne by a number of writings in the vernacular which offer 'to skilled craftsmen, foremen, stonemasons, builders, painters, sculptors and goldsmiths, for their manifold advantage', knowledge both scientific and practical, especially that from Antiquity.[13]

The transition from medieval to modern that is to the Renaissance, is blurred. It is in any event difficult, as Huizinga has amply shown,[14] to reduce the mixed influences of the Renaissance Period to a unified scheme. Thus, while certain modern features are found in fourteenth century medieval natural science, Renaissance physics on the other hand sometimes exhibits a completely medieval character. And within certain limits the same is true of technology. This will not be foregotten when we speak of Renaissance technology.

ENGINEERING ARTISTS AND LEADERS OF RESEARCH

It is in fifteenth century Italy that we first find the association of practical life with a scientific, though still reserved and modest, point of view. Those responsible for technical achievement were at once artists and empirics. For example, Filippo Brunelleschi (1377–1446), one of these great and versatile men of the fifteenth century, was skilled in the art of the goldsmith, in sculpture, in architecture, in perspective and proportion, in the building of fortresses, in hydraulics, mechanics and construction of scientific apparatus. He corresponded with mathematicians. His studies in perspective linked up with Euclid and with Vitellius. Brunelleschi's vaulting of the dome of Santa Maria del Fiore executed in Florence between 1420 and 1436 was primarily a technical achievement.

Among the many-sided men of those days the most eminent was Leon Battista Alberti, the all-sided, as Burckhardt called him. Alberti started out as a scholar, not as a master, as did Brunelleschi, Francesco di Giorgio Martini and Leonardo da Vinci. But his whole effort was directed to the combining of scientific heritage with practical experience. He himself combined scientific achievement with practical activity, especially as an architect. In a learned work on architecture composed between 1451–2 he treated the whole wide province of building practice. Though starting from Vitruvius, he strove to progress far beyond him. Thus Alberti gives us a theory of the construction of domes, an art unmentioned by Vitruvius but which began to assume special importance during the Renaissance. Experience in building bridges yielded certain rules-of-thumb for arches in the form of simple mathematical proportions. Alberti constantly manifests a desire for a thorough scientific understanding of practical technology. We reprint below some significant extracts from his work *On the Process of Building*.

On Architecture

Him I call an Architect, who, by sure and wonderful Art and Method, is able both with Thought and Invention to devise and with Execution to compleat all those Works which, by means of the Movement of great Weights, and the Conjunction and Amassment of Bodies, can with the greatest beauty be adapted to the Uses of Mankind. And to be able to do this, he must have a thorough Insight into the noblest and most curious Sciences, Such must be the Architect. . . . Why should I instance in Vehicles, Mills, Time-measures and other such minute things which nevertheless are of great Use in Life? Why should I insist on the great plenty of Waters brought

9. Pascal's Calculating Machine, 1642. Engraving from *Machines approuvées par l'Académie Royale des Sciences*, Vol. 4. Paris 1735

10. *Above*: Diving apparatus and Submarine boat. Engraving from G. A. Borelli, *De motu animalium*, 1680

Left: Whirling or Worm device with grinding wheels. Design for perpetual motion, 1629. Engraving from G. Strada, *Kunstliche Abriss allerhand* (Muhlen, 1629)

from the most remote and hidden Places, and employed to so many different and useful Purposes? Upon Trophies, Tabernacles, sacred Edifices, Churches and the like, adapted to divine Worship and the Service of Posterity? Or lastly, why should I mention the Rocks cut, Mountains bored through, Vallies filled up, Lakes confined, Marshes discharged into the Sea, Ships built, Rivers turned, their Mouths cleared, Bridges laid over them, Harbours formed, not only serving to Men's immediate Conveniencies, but also opening them a way to all Parts of the World. Whereby Men have been enabled mutually to furnish one another with Provisions, Spices, Gems, and to communicate their Knowledge and whatever else is healthful or pleasurable. Add to these the Engines and Machines of War, Fortresses and the like inventions necessary to Defending the Liberty of our Country, maintaining the Honour and increasing the Greatness of a City, and to the Acquisition and Establishment of an Empire. I am really persuaded that if we were to enquire of all the Cities which within the Memory of Man have fallen by Siege into the Power of new Masters, who it was that subjected and overcame them, they would tell you, the Architect; and that they were strong enough to have despised the armed Enemy, but not to withstand the Shocks of the Engines, the Violence of the Machines and the Force of the other Instruments of War with which the Architect, distressed, demolished and ruinated them. On the contrary, they would inform you that their greatest Defence lay in the Art and Assistance of the Architect.

And if you were to examine into the Expeditions that have been undertaken, you would go near to find that most of the Victories were gained more by the Art and Skill of the Architects, than by the Conduct or Fortune of the Generals; and that the Enemy was oftener overcome and conquered by the Architect's Wit without the Captain's Arms, than by the Captain's Arms without the Architect's Wit: And what is of great Consequence is, that the Architect conquers with a small Number of Men and without the Loss of Troops. Let this suffice as to the Usefulness of this Art.[64]

The types of vault

There are several Sorts of Vaults; so that it is our Business here to enquire wherein they differ, and of what Lines they are composed; in doing of which, I shall be obliged to invent new Names to make myself clear and perspicuous, which is what I have principally studied in these Books. I know *Ennius* the Poet calls the Arch of the Heavens the mighty Vaults, and *Servius* calls all Vaults made like the Keel of a Ship, Caverns. But I claim this Liberty; that whatever in this Work is expressed aptly, clearly and properly shall be allowed to be expressed right. The different sorts of Vaults are these, the plain Vault, the Camerated or mixed Vault, and the hemispherical Vault or Cupola, besides those others which partake of the Kind of some of these. The Cupola in its Nature is never placed but upon Walls that rise from a circular Platform; the Camerated are proper for a square one; the plain Vaults are made over any quadrangular Platform, whether long or short, as we see in all subterraneous Porticoes.

Those Vaults too which are like a Hill bored through, we also call plain Vaults; the plain Vault therefore is like a Number of Arches joined together Sideways, or like a bent Beam extended out in Breadth, so as to make a Kind of a Wall turned with a Sweep over our Heads for a Covering. But if such a Vault as this, running from North to South, happens to be crossed by another which runs from East to West, and intersects it with equal Lines meeting at the Angles like crooked Horns, this will make a Vault of the Camerated Sort.

But if a great Number of equal Arches meet at the Top exactly in the Centre, they constitute a Vault like the Sky, which therefore we call the Hemispherical or Compleat Cupola.

The Vaults made of Part of these are as follows:

If Nature, with an even and perpendicular Section, were to divide the Hemisphere of the Heavens in two Parts, from East to West, it would make two Vaults which would be proper Coverings for any semi-circular Building. But if from the Angle at the East to that at the South, from the South to the West, thence to the North, and so back again to the East, if Nature were to break and interrupt this Hemisphere by so many Arches turned from Angle to Angle, she would then leave a Vault in the Middle which, for its resemblance to a swelling Sail we will venture to call a Velar Cupola (Baldachin).

But that Vault which consists of a Number of plain Vaults meeting in a Point, at Top, we shall call an Angular Cupola.

In the Construction of Vaults, we must observe the same Rules as in that of the Walls, carrying on the Ribs of the Wall clear up to the Summit of the Vault; and according to the Method prescribed for the Forme, observing the same Proportions and Distances: From Rib to Rib, we must draw Ligatures crossways, and the Interspaces we must fill up with Stuffing. But the Difference between the Working of a Vault and a Wall lies in this: that in the Wall the Courses of Stone are laid even and perpendicular by the Square and Plumbline; whereas in the Vault the Courses are laid by a curve Line, and the Joints all point to the Centre of their Arch.

The Ancients hardly ever made their Ribs of any but burnt Bricks, and those generally about two Foot long, and advise to fill up the Interspaces of our Vaults with the lightest Stone, that they might not oppress the Wall with too great a Weight. But I have observed that some have not always thought themselves obliged to make solid Ribs, but in their Stead have at certain Distances set Bricks lying Sideways with their Heads jointing into each other like the Teeth of a Comb; as a Man locks his right Hand Fingers into his left; and the Interspaces they filled up with any common Stone, and especially with Pumice Stone which is universally agreed to be the properest of all for the stuffing Work of Vaults.

In building either Arches or Vaults, we must make use of Centres. These are a kind of Frames made with the Sweep of an Arch of any rough Boards just clapped together for a short Service, and covered either with Hurdles, Rushes or any such common Stuff, in order to support the Work till it is settled and hardened.

Yet there is one Sort of Vault which stands in no Need of these Machines, and that is the *perfct Cupola*; because it is composed not only of Arches, but also in a Manner, of Cornices. And who can conceive the innumerable Ligatures that there are in these, which all wedge together, and intersect one another both with equal and unequal angles? So that in whatsoever part of the whole Cupola you lay a Stone, or a Brick, you may be said at the same time to have laid a Key-stone to an infinite Number both of Arches and Cornices. And when these Cornices or Arches are thus built one upon the other, if the Work were inclined to ruinate, where should it begin, when the Joints of every Stone are directed to one Centre with equal Force and Pressure? Some of the Ancients trusted so much to the Firmness of this Sort of Structure, that they only made plain Cornices of Brick at stated Distances, and filled up the Interspaces with Rubble. But I think those acted much more prudently who, in raising this Sort of Cupola, used the same Methods as in Walling, to cramp and fasten under Cornices to the next above, and the Arches too in several places, especially if they had not plenty of Pit Sand to make very good Cement, or if the Building was exposed to South Winds or Blasts from the Sea.

You may likewise turn the Angular Cupolas without a Centre (scaffolding), if you make a perfect one in the Middle of the Thickness of the Work. But here you will have particular Occasion for Ligatures to fasten the weaker Parts of the outer one tightly to the stronger Part of that within. Yet it will be necessary when you have laid one or two Rows of Stone to make a little light Stays or Catchers jutting out, on which, when those Rows are settled, you may set just Framework enough to support the next Courses above, to the Height of a few Feet, till they are sufficiently hardened; and then you may remove these Frames or Supports, higher and higher to the other Courses till you have finished the whole Work.

The other Vaults, both plain and mixed, or camerated, must needs be turned upon Centres: But I would have the first Courses, and the Heads of their Arches placed upon very strong Seats; nor can I approve the Method of those who carry the Wall clear up first, only leaving some Mouldings, or Corbels, upon which, after a Time, they turn their Arches; which must be a very infirm and perishable sort of Work. The true Way is to turn the Arch immediately, and equally with the Courses of the Wall which is to support it, that the Work may have the strongest Ligatures that is possible, in a Manner all of one Piece. The Vacuities that are left between the Back of the Sweep of the Arch and the Upright of the Wall it is turned from, called by Workmen the *Hips* of the Arch, should be filled up not with Dirt or old Rubbish, but rather with strong ordinary Work, frequently knit, and jointed into the Wall.

I am pleased with those who, to avoid overburdening the Arch, have stuffed up these Vacuities with earthen Pots, turned with their Mouths downward that they might not contain any wet, if it should gather there, and over these thrown in Fragments of Stone, not heavy, but perfectly sound.

Lastly, in all Manner of Vaults, let them be of what Kind they will, we ought to imitate Nature, who, when she has knit the Bones, fastens the Flesh with Nerves, interweaving it every where with Ligatures, running in Breadth, Length, Height and circularly. This artful Contexture is what we ought to imitate in the joining of Stones in Vaults.

These Things being compleated, the next and last Business is to cover them over; a Work of the greatest Consequence in Building, and no less difficult than necessary; in effecting and compleating, of which the utmost Care and Study has been over and over employed. Of this we are to treat; but first it will be proper to mention something necessary to be observed in working of Vaults.

For different Methods are to be taken in the Execution of different Sorts. Those which are turned upon Centres must be finished out of hand, without Intermission; but those which are wrought without Centres must be discontinued, and left to settle Course by Course, lest new Work being added to the first before it is dry, should ruin the Whole. As to those which are turned upon Centres, when they are closed with their Key-stones, it will be proper immediately to ease the Props a little, that those Centres rest upon; not only to prevent the Stones fresh laid from floating in the Beds of Mortar they are set in, but that the whole Vault may sink and close by its own Weight equally, into its right Seat. Otherwise in drying, the Work would not compact itself as it ought, but would be apt to leave Cracks when it came afterwards to settle. And therefore you must not quite take away the Centre immediately, but let it down easily Day after Day, by Little and Little, for Fear if you should take it away too soon, the Building should never duly cement. But after a certain Number of Days, according to the Greatness of the Work, ease it a little, and so go on gradually till the Wedges all compact themselves in their Places, and are perfectly settled.

The best Way of letting down the Frame is this: When you place your Centre upon the Pilasters, or whatever else it is to rest upon, put under each of its Feet two Wedges of Wood; and when afterwards you want to let it down, you may with a Hammer safely drive out these Wedges by little and little, as you shall judge proper.

Lastly, it is my Opinion that the Centres ought not to be taken away till after Winter, as well for other Reasons, as because the Wasting of the Rains may weaken and demolish the whole Structure; though else we cannot do greater service to a Vault than to give it Water enough, and to let it be throughly soaked, that it may never feel Thirst. But of this subject we have said enough.[65]

Of Stone Bridges

It remains now that we treat of the Stone-Bridge, the Parts whereof are these: the Banks of the Shore, the Piers, the Arches and the Pavement. Between the Banks of the Shore and the Piers is this Difference, that the Banks ought to be by much the strongest, inasmuch as they are not only to support the Weight of the Arches like the Piers, but are also to bear the Foot of the Bridge, and to bear against the Weight of the Arches, to keep them from opening in any Part. We ought therefore to be very careful in the Choice of our Shore, and to find out if possible a Rock of solid stone, since nothing can be too strong that we are to intrust with the Feet of the Bridge; and as to the Piers, they must be more or less numerous in proportion to the Breadth of the River. An odd Number of Arches is both most pleasant to the Sight, and conduces also to Strength, for the further the Current of the River lies from the Shore, the freer it is from Impediments, and the freer it is, the swifter and easier it flows away; for this, therefore, we ought to leave a Passage perfectly free and open, that it may not shake and prejudice the Piers by struggling with the Resistance which it meets with from them. The Piers ought to be placed in those Parts of the River where the Water flows the most slowly, and (to use such an Expression) the most lazily; and those Parts you may easily find out by means of the Tides: Otherwise you may discover them in the following Manner: Imitate those who threw Nuts into a River, whereby the Inhabitants of a Town besieged, gathering them up, were preserved from starving; strew the whole Breadth of the River, about fifteen hundred paces above the Place which you intend for your Bridge, and especially when the River is fullest, with some such light Stuff that will easily float. And in those places where the Things you have thrown in Clusters (are) thickest together, you may be sure the Current is strongest. In the Situation of your Piers, therefore, avoid those Places, and chuse those others to which the Things you throw in come the slowest and thinnest.

King *Minos*, when he intended to build the Bridge of *Memphis*, turned the Nile out of its Channel, and carried it another Way among some Hills, and when he had finished his Building brought it back again into its old Bed. *Nicore*, Queen of the *Assyrians*, having prepared all the Materials for building a Bridge, dug a great Lake, and into that turned the River; and as the Channel grew dry as the Lake filled, she took that Time to build her Piers. These mighty Things were done by those great Princes.

As for us, we are to proceed in the following Manner. Make the Foundations of your Piers in Autumn, when the Water is lowest, having first raised an Inclosure to keep off the Water, which you may do in this manner: Drive in a double Row of Stakes, very close and thick set, with their Heads above the Top of the Water, like a Trench; then put Hurdles within this double Row of Stakes, close to the side which is next to the intended Pier; and fill up the Hollow between the two Rows with Rushes and Mud, ramming them together so hard that no Water can possibly get through. Then whatever you find within this Inclosure, Water, Mud, Sand, and whatever else is a Hindrance to you, throw out. For the rest of your Work, you must observe the Rules we have laid down in the preceding Book. Dig till you come to a solid Foundation, or rather make one of Piles burnt at the End and driven in as close together as ever they can stick. And here I have observed that the best Architects used to make a continued Foundation of the whole Length of the Bridge, and not only under each Pier; and this they did, not by shutting out the whole River at once by one single Inclosure, but by first making one Part, then another, and so joyning the whole together by degrees; for it would be impossible to withstand and repulse the whole Force of the Water at once. We must therefore, while we are at work with one Part, leave another Part open, for a Passage for the Stream. You may leave these Passages either in the Channel itself, or if you think it more convenient, you may frame wooden Dams, or hanging Channels, by which the superfluous Water may run off.

But if you find the Expence of a continued Foundation for the whole Bridge too great, you may only make a separate Foundation for every particular Pier, in the Form of a Ship, with one Angle in the Stern, and another in the Head, lying directly even with the Current of the Water, that the Force of the Water may be broken by the Angle. We are to remember that the water is much more dangerous to the Stern than to the Head of the Piers, which appears from this, that at the Stern the Water is in a more violent Motion than at the Head, and forms Eddies which turn up the Ground at the Bottom; while the Head stands firm and safe, being guarded and defended by the Banks of Sand thrown up before it by the Channel. Now this being so, this Part ought of the whole structure to be best fortified against the Violence of the Waters; and nothing will conduce more to this, than to make the Pile-work deep and broad every Way, and especially at the Stern, that if any Accident should carry any of the Piles, there may be enow [sic] left to sustain the Weight of the Pier. It will be also extremely proper to begin your Foundation at the upper Part of the Channel, and to make it with an easy Descent, that the water which runs over it may not fall upon it violently as into a Precipice, but glide over gently, with an easy Slope; because the Water that rushes down precipitately routs out the Bottom, and so being made still rougher, carries away every Thing that it can loosen, and is every Moment under-mining the Work.

Build the Piers of the biggest and longest Stones, and of such as in their Nature are best adapted for supporting of Frosts, and as do not decay in Water, nor are easily softened by any Accident, and will not crack and split under a great Weight. And build them exactly according to the Square, Level and Plumb-line, omitting no Sort of Ligature Length-ways, and placing the Stones Breadth-ways in alternate Order so as to be a Binding one to another; absolutely rejecting any Stuffing with small Pieces of Stone. You must also fasten your Work with a good number of copper pins and clamps so well fitted in that the Joynts of the Structure may not separate, but be kept tight and firm. Raise both the Fronts of the Building angular, both Head and Stern, and let the Top of the Pier be sure to be higher than the fullest Tide; and let the Thickness of the Pier be one fourth of the Height of the Bridge (Fig. 35). There have been some that have not terminated the Head and Stern of their Piers with an Angle, but with an Half-circle; induced thereto, I suppose, by the Beautifulness of that Figure. But though I have said elsewhere that the Circle has the same Strength as an Angle, yet here I approve better of an Angle, provided it be not so sharp as to be broken and defaced by every little Accident. Nor am I altogether displeased with those which end in a Curve, provided it be very much lengthened out, and not left so obtuse as to resist the Force and Weight of the Water. The Angle of the Pier is of a good Sharpness if it is three Quarters of a Right Angle, or if you like it better, you may make it two thirds and thus much may suffice as to the Piers.

If the Nature of your Situation is such that the Sides or Banks of the Shore are not as you could wish; make them good in the same Manner as you build your Piers, and indeed make other Piers upon the Shore, and turn some Arches even upon the dry Ground; to the Intent that if in Process of Time, with the continual washing of the Water and Force of the Tides, any Part of the Bank should be carried away, your Passage may still be preserved safe, by the Production of the Bridge into the Land.

The Arches ought upon all Accounts, and particularly because of the continual violent shaking and Concussion of Carts and other Carriages, to be extreamly stout and strong. Besides, as sometimes you may be obliged to draw immense Weights over them, such as a Colossus, an Obelisk or the like; You should provide against the

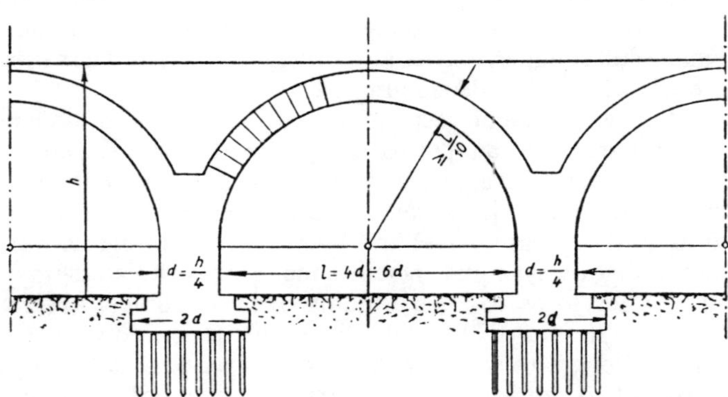

FIG. 35. *Arched bridge in stone according to Alberti, fifteenth century.* Drawing from H. Straub: *Die Geschichte der Bauingenieurkunst* (Basle, 1949)

inconveniences which happened to *Scaurus* who, when he was removing the great Boundary Stone alarmed all the publick Officers on account of the Mischief that might ensue. For these Reasons, a Bridge, both in its Design and in its whole Execution should be well fitted to bear the continual and violent Jars which it is to receive from Carriages. That Bridges ought to be built of very large and stout Stones, is very manifest by the Example of an Anvil which, if it is large and heavy, stands the Blows of the Hammer unmoved; but if it is light, rebounds and trembles at every stroke. We have already said that all vaulted Work consist of Arches and Stuffing, and that the strongest of all Arches is the Semi-circle. But if by the Disposition of the Piers, the Semi-circle should rise so high as to be inconvenient, we may make use of the Scheme Arch, only taking Care to make the last Pier on the Shore the stronger and thicker. But whatever Sort of Arch you vault your Bridge with, it must be built of the hardest and largest Stones, such as you use in your Piers. And there should not be a single Stone in the Arch but what is in Thickness at least one tenth Part of the Chord of that Arch (Fig. 35); nor should the Chord itself be longer than six times the Thickness of the Pier, nor shorter than four times. The Stones also should be strongly fastened together with Pins and Cramps or Copper. And the last Wedge, which is called the Key-stone, should be cut according to the Lines of the other Wedges, but left a small Matter bigger at the Top, so that it may not be got into its Place without some Strokes of a light Beetle; which will drive the lower Wedges closer together and so keep them tight to their Duty. The filling up or Stuffing between the Arches should be wrought with the strongest Stone, and with the closest Joynts that can possibly be made. But if you have not a sufficient Plenty of strong Stone to make your Stuffing of it, you may in case of Necessity make use of a weaker Sort; still provided that the whole Turn of the Arch, and the Course of Work behind both the Sides of it, be built entirely of strong Stone.[66]

In what Measure Weights and large Stones are moved from one Place to another, or raised to a great Height. . . .

. . . I shall not waste Time in explaining any such curious Principles as that it is the Nature of all heavy Bodies to press continually downwards and obstinately to seek the lowest Place; that they make the greatest Resistance they are able against being raised aloft, and never change their Place but after the stoutest Conflict, being either overcome by some greater Weight or some more powerful contrary Force. Nor shall I stand to observe that Motions are various, from high to low or from low to high, directly, or about a Curve; and that some things are carried, some drawn, some pushed on, and the like; of which Enquiries we shall treat more copiously in another Place. This we may lay down for certain, that a Weight is never moved with so much Ease as it is downward; because it then moves itself; nor ever with more Difficulty than upwards, because it naturally resists that Direction; and that there is a Kind of middle Motion between these two which perhaps partakes somewhat of the Nature of both the others, inasmuch as it neither moves of itself, nor of itself resists; as when a Weight is drawn upon an even Plain, free from all Rubs. All other Motions are easy or difficult in proportion as they approach to either of the preceding. And indeed Nature herself seems in a good Measure to have shewn us in what Manner great Weights are to be moved: for we may observe that if any considerable Weight is laid upon a Column standing upright, the least Shove will push it off, and when once it begins to fall, hardly any Force is sufficient to stop it. We may also observe that any round Column, or Wheel, or any other Body that turns about, is very easily moved and very hard to stop when once it is set on going; and if it is dragged along without rolling, it does not move with half the Ease. We further see that the vast Weight of a Ship may be moved upon a standing Water with a very small Force, if you keep pulling continually; but if you strike it with ever so great a Blow suddenly, it will not stir an inch. On the Contrary, some things will move with a sudden Blow or a furious Push, which could not otherwise be stirred without a mighty Force or huge Engines. Upon Ice too, the greatest Weights make but a small Resistance against one that tries to draw them. We likewise see that any Weight which hangs upon a long Rope is very easily moved as far as a certain Point; but not so easily, further. The Consideration of the Reasons of these Things and the Imitation of them may be very useful to our purpose; and therefore we shall briefly treat of them here. The Keel or Bottom of any Weight that is to be drawn along should be even and solid; and the Broader it is, the less it will plough up the Ground all the Way under it, but then, the Thinner it is, it will slip along the Quicker; only it will make the deeper furrows and be apt to stick. If there are any Angles or Inequalities in the Bottom of the Weight, it will use them as Claws to fasten itself in the Plain, and to resist its own motion. If the Plain be smooth, sound, even, hard, not rising or sinking on any Side, the Weight will have nothing to hinder its Motion, or to make it refuse to obey, but its own natural Love of Rest, which makes it lazy and unwilling to be moved. Perhaps it was from a Consideration of these

Things and from a deeper Examination of the Particulars we have here mentioned, that *Archimedes* was induced to say that if he had only a Basis for so immense a Weight, he would not doubt to turn the World itself about.

The Preparation of the Bottom of the Weight and the Plain upon which it is to be drawn, which is what we are here to consider, may be effected in the following Manner. Let such a Number of Poles be laid along, and of such a Strength and Thickness as may be sufficient for the Weight. Let them be sound, even, smooth, and close joined to one another. Between the Bottom of the Weight and this Plain which it is to slide upon, there should be something to make the Way more slippery; and this may be either Soap, or Tallow, or Lees of Oil, or perhaps Slime. There is another Way of making the Weight slip along, which is by underlaying it cross-ways with Rollers. But these, though you have a sufficient number of them, are very hard to be kept even to their proper Lines and exact Direction; which it is absolutely necessary they should be, and that they should all do Duty equally at once, or else they will run together in Confusion, and carry the Weight to one Side. And if you have but few of them, being continually loaded, they will either be split or flatted, and so be rendered useless; or else that single Line with which they touch the Plane underneath, or that other with which they touch the Weight that is laid upon them, will stick fast with their sharp Points and be immoveable. . . .[67]

L. Olschki who has analysed the work of these artist-engineers of the fifteenth century in great detail speaks of them as the 'experimenting Masters'.[15] They did in fact endeavour to solve many technical problems by experiment. Thus Francesco di Giorgio Martini, who worked as military engineer at the court of Urbino in 1475, laboured to discover all sorts of relationships important for fire ordnance such as the relation between the amount of powder needed and the weight of the projectile, or between the length, bore and thickness of the gun barrel.

On Guns. Our modern age has lately invented an instrument of war of such power that any arms, any forethought, any bravery are of little or no use against it. . . . Therefore is this instrument not without justice named by some as inhuman, yea a devilish invention. And though the object of such a weapon is to furnish material, working power and means to fare forth against evil, nevertheless various figures can be found, as will be clearly seen, of the length and bore of these various weapons, and it doth not now appear to me incongruous . . . to designate their form. As the medical faculty contemplates the causes of disease and thus also surveys the causes of good health; so also it is enjoined on our craft to consider not only how one can inflict harm on such a weapon, but also how one can inflict injury by means of such an instrument. The following are the chief kinds of such machines. Firstly that called the *Bombard* whose length is usually between 15 and 20 feet. The stone shot weighs about 300 lbs. One can also calculate the relation between length and bore of the gun as follows: The bombard's powder chamber that is to say the lower chamber of the barrel which holds the powder—should be twice as long as the diameter of the ball-shot; the screw which holds the magazine to the barrel should have a length equal to half the diameter of the shot; and the barrel itself should be of a length five or six times the diameter of the shot . . . the longer the barrel the greater will be its power. . . . The sixth weapon is called the *basilisk*. It is from 22 to 25 feet long. Its shot, which may consist of any kind of metal weighs about 20 lbs. . . . The tenth and last weapon is called *arquebuse* (*Scoppietto*), and is 2 or 3 feet long. Its small shot is made of lead and weighs from 4 to 6 drachms. Every day a variety of similar instruments is invented for the same purpose. . . . For the bombard, 16 lb. of powder must be used for every 100 lb. weight of the shot . . .; for the arquebuse 8 lb. of powder to every 10 lb. weight of the shot. . . .

Concerning gunpowder, and the method of preserving it. For the above-mentioned reasons it is necessary to use different sorts of powder according to the different shapes of the guns. It must be known that powder for a bombard . . . which discharges up into the air a stone shot weighing 200 lb., must consist of 6 parts of saltpetre, 4 parts of sulphur, and 3 parts of charcoal. . . . Powder for the basilisk is . . . composed of 8 parts of saltpetre, 3 parts of sulphur and 2 parts of charcoal. That for the arquebuse consists of 14 parts saltpetre, 3 parts sulphur and 2 parts charcoal.[68]

The experiments of these men exhibited strongly practical traits. They were naturally not yet able to advance to the formulation of universal laws for these relationships. For that there was needed well over a century of work and the genius of a Galileo.

The impulse of the Renaissance period to discuss and to influence the visible world is manifested most forcefully in Leonardo da Vinci. His extraordinarily developed power of illustration not only in the realm of the fine arts but also especially as regards technical structures, his familiarity, thanks to the Florentine workshop tradition, with the properties of different substances and the varied possibilities of their utilization in the workshop, and his labours to discover by experiment simple mathematical laws of nature, all these made him an engineer in the modern sense. Leonardo's greatness was in the realm of art and of technological construction, not in science, where we recognize his far-reaching dependence on antiquity and on the Middle Ages. But none before him succeeded in penetrating so far into the world of machinery. He recognized as essential parts of the machine the individual actuating mechanisms and their elements which he examined separately from the machines.

Many of Leonardo's technological drawings bear the character of practical workshop sketches from which the individual parts of the machines could really be manufactured. Among the large number of Leonardo's sketches, far in advance of their period, most of which he probably did not put into practice (some of which, indeed, could not have been carried out with the means then available) and which later remained buried in the vast treasury of his manuscripts examined by only a few people in the sixteenth century, we call attention merely to the following: the lathe worked by treadle crank (Fig. 36), the automatic file-cutting machine (Fig. 37), the rolling mill with conical rollers, the spinning machine with pinion and yard divider, the windmill with revolving roof, the boring machines to bore well tubes, and the parachute (Fig. 38). Many of the sketches are undeniably Utopian such for example as the bridge over the Golden Horn which was to traverse the inlet between Galata and Istanbul with a single span of 250 yards.[16] Leonardo conducted experiments on friction and strove in this way to go beyond the purely geometric mechanics of Archimedes. He also tried to establish by calculation and experiment the bearing strength of stays and joists. In spite of many definitely modern and practical features, Leonardo's mechanics also reveal much that is fantastic, allegorical and anthropomorphic. Olschki says truly that Leonardo was influenced not only by the Italian workshop tradition, but also by the romantic speculative spirit of the Platonic Academy of Florence. We may take his definition of the concept of power as evidence of this discordance in Leonardo.

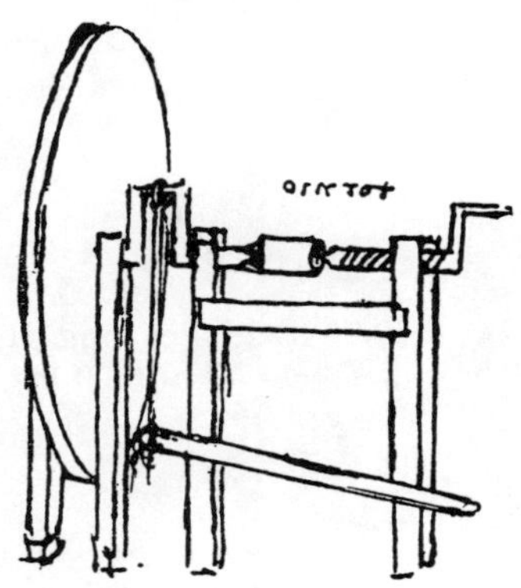

FIG. 36. *Lathe with treadle and crank drive and flywheel for continuous movement in* 1480–82. Sketch from the *Codice Atlantico* of Leonardo da Vinci

Hymn to Force

Force is the same throughout and the whole is in every part of it.

Force is a spiritual power, an invisible energy which is imparted by violence from without to all bodies out of their natural balance.

Force is nothing but a spiritual power, an invisible energy which is created and communicated, through violence from without, by animated bodies to inanimated bodies, giving to these the similarity of life, and this life works in a marvellous way, compelling all created things from their places, and changing their shapes.

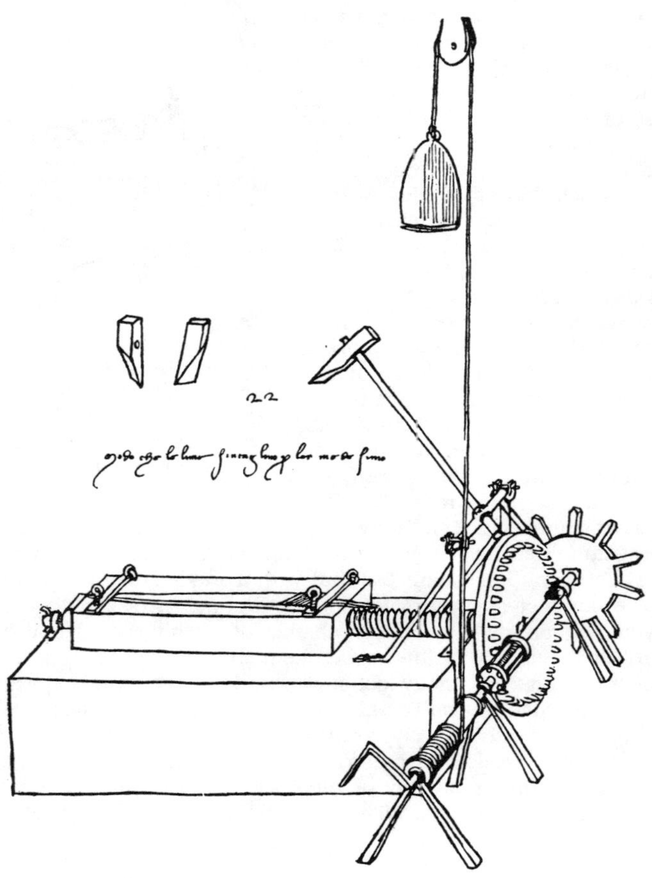

FIG. 37. *File cutting machine driven by weight, in* 1480. Sketch from *Codice Atlantico* of Leonardo da Vinci

It speeds in fury to its undoing, and continues to modify according to the occasion.

Retardation strengthens, and speed weakens it.

It lives by violence and dies through liberty.

It transmutes and compels all bodies to a change of form and place.

Great power gives it great desire for death.

It drives away in its fury whatever stands in the way to its ruin

. . . .

Wherever it is held, it is always ill at ease.

It is always opposing forces of Nature.

It grows slowly from small beginnings to terrible and marvellous energy, and by compression of itself, compels all things.

. . . it lives in bodies which are out of their natural course and state.

. . . it likes to consume itself.

Force is the same throughout, and the whole of it is in the body where it is generated.

Force is but a desire to flight.

It always wants to weaken and extinguish itself.

. . . When compressed it compels all bodies.

Without it nothing moves.

No sound or voice is heard without it.

Its true source is in living bodies.[69]

The following applications for a commission on Milan Cathedral, and for a post at the court of Ludovico Sforza reveal to us the wide artistic and technical field of Leonardo's activities.

You know that when medicines are rightly used they restore health to the invalid, and that he who knows them well makes the right use of them if he also understands what man is, and what life is, and the constitution of the body, and what health is. He who knows these things well will also know their opposite, and he will then be nearer a cure than any one else. The case of the invalid cathedral is similar. It also requires a doctor architect who understands the edifice well, and knows the rules of good building from their origin, and knows into how many parts they are divided, what are the causes that keep together an edifice and make it last, what is the nature of weight and of energy in force, and in what manner they should be combined and related to one another and what effect they will produce when combined. He who has true knowledge of these things, will plan the work to your satisfaction.

Therefore I shall try, without detracting and without . . . abusing any one . . . to convince you, partly by reasoning and partly by my works, sometimes showing the effects by their causes, sometimes sustaining my argument by experiment . . . and bringing in the authority of ancient architects and the proofs afforded by edifices that have been constructed and show the causes of their ruin or their survival, etc.

And thereby to show what is the weight and how many and what are the causes that ruin buildings, and what causes their stability and permanence. But so as not to prolong unduly this discourse to your Excellencies, I shall begin by explaining the plan of the first architect of the cathedral, and show clearly what was his intention, as revealed by the edifice begun by him, and having understood this, you will see clearly that the model which I have made embodies the symmetry, the correspondence, the conformity which appertained to the edifice from the beginning. . . .

Either I, or others who can do better, choose him, and put aside all passions.[70]

Letter to Ludovico Sforza (1482)

Most Illustrious Lord, Having now sufficiently considered the specimens of all those who proclaim themselves skilled contrivers of instruments of war, and that the invention and operation of the said instruments are nothing different from those in common use: I shall endeavour, without prejudice to any one else, to explain myself to your Excellency, showing your Lordship my secrets, and then offering them to your best pleasure and approbation to work with effect at opportune moments on all those things which, in part, shall be briefly noted below.

(1) I have a sort of extremely light and strong bridges, adapted to be most easily carried, and with them you may pursue, and at any time flee from the enemy; and others, secure and indestructible by fire and battle, easy and convenient to lift and place. Also methods of burning and destroying those of the enemy.

(2) I know how, when a place is besieged, to take the water out of the trenches, and make endless variety of bridges, and covered ways and ladders, and other machines pertaining to such expeditions.

(3) Item. If, by reason of the height of the banks, or the strength of the place and its position, it is impossible, when besieging a place, to avail onself of the plan of bombardment, I have methods of

11. Erection of an Obelisk at the Vatican in Rome, 1586. Employing 40 capstans, 140 horses, and 800 workmen. Engraving from N. Zabaglia, *Castelli e ponti*. Rome, 1743

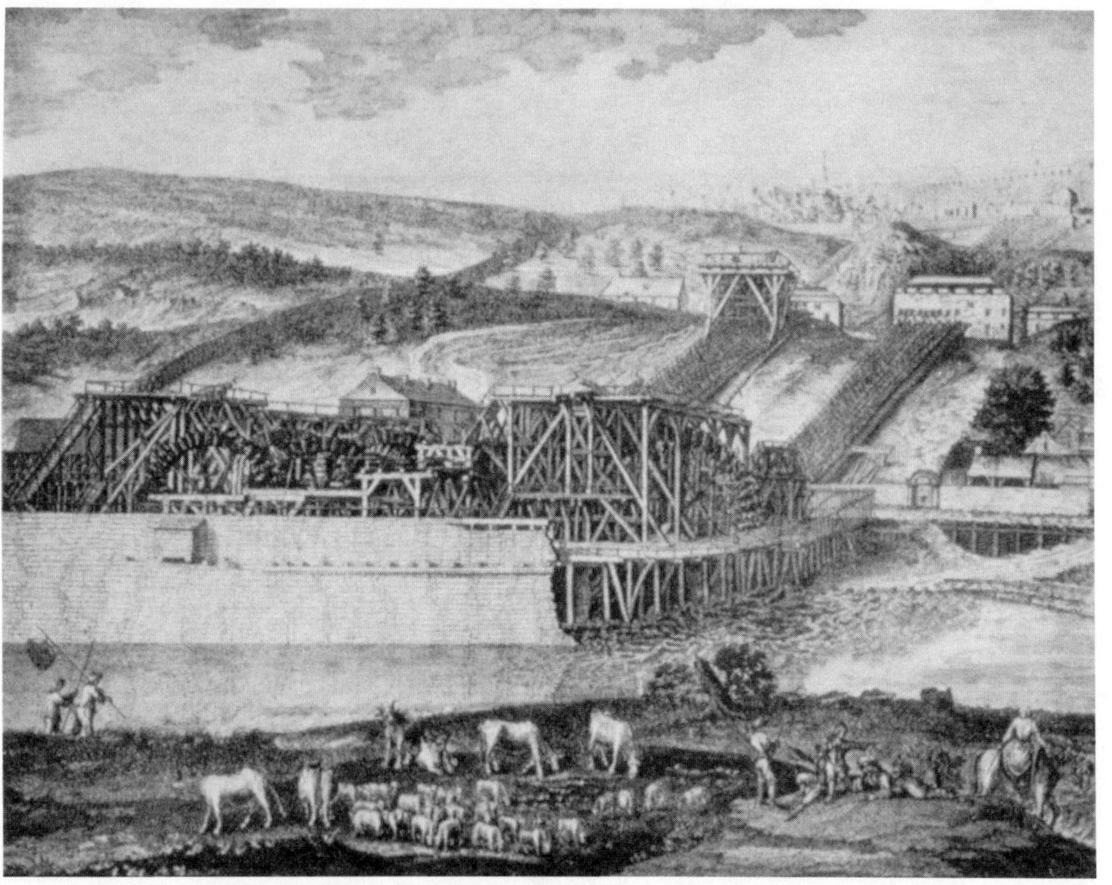

12. Installation for raising water at Marly on the Seine to supply the fountains in the park at Versailles for Louis XIV. Constructed 1681/85. Engraving dated 1715. 14 waterwheels, each 12 m. in diameter, driving 221 pumps, raise 3200 cubic m. of water per day to a height of 162 m. in three stages. Output about 80 H.P.

destroying every rock or other fortress, even if it were founded on a rock, etc.

(4) Again, I have kinds of mortars; most convenient and easy to carry; and with these I can fling small stones almost resembling a storm; and with the smoke of these cause great terror to the enemy, to his great detriment and confusion.

(5) Item. I have means by secret and tortuous mines and ways, made without noise, to reach a designated [spot], even if it were needed to pass under a trench or a river.

(6) Item. I will make covered chariots, safe and unattackable, which, entering among the enemy with their artillery, there is no body of men so great but they would break them. And behind these, infantry could follow quite unhurt and without any hindrance.

(7) Item. In case of need I will make big guns, mortars, and light ordnance of fine and useful forms, out of the common type.

(8) Where the operation of bombardment might fail, I would contrive catapults, mangonels, *trabocchi*, and other machines of marvellous efficacy and not in common use. And in short, according to the variety of cases, I can contrive various and endless means of offence and defence.

(9) And if the fight should be at sea, I have kinds of many machines most efficient for offence and defence; and vessels which will resist the attack of the largest guns and powder and fumes.

(10) In time of peace I believe I can give perfect satisfaction and to the equal of any other in architecture and the composition of buildings public and private; and in guiding water from one place to another.

Item. I can carry out sculpture in marble, bronze, or clay, and also I can do in painting whatever may be done, as well as any other, be he who he may.

Again, the bronze horse may be taken in hand, which is to be to the immortal glory and eternal honour of the prince your father of happy memory, and of the illustrious house of Sforza.

And if any of the above-named things seem to any one to be impossible or not feasible, I am most ready to make the experiment in your park, or in whatever place may please your Excellence—to whom I commend myself with the utmost humility, etc.[71]

In Leonardo's extensive technical work a large part was played by his eager but vain effort to solve the problem of human flight. He studied the flight of birds, industriously attempted to find out about the operation of air-resistance, made projects for many sorts of airplanes utilizing muscular power, invented a parachute and also an air-screw. From the following phrases, we learn of his technical researches concerning flight. We discover especially how, in quite modern fashion, he started his technical work by experiment and measurement and endeavoured—albeit still with inadequate means—to estimate by scales the resistance of a wing against the air.

A substance offers as much resistance to the air as the air does to the substance. See how the beating of its wings against the air supports a heavy eagle in the highly rarefied air, close to the sphere of elemental fire. Observe also how the air in motion over the sea fills the swelling sails and drives heavily laden ships. From these instances, and the reasons given, a man with wings large enough and duly attached might learn to overcome the resistance of the air, and conquering it, succeed in subjugating it and raise himself upon it.

If a man has a tent 12 ells wide and 12 high covered with cloth he can throw himself down from any great height without hurting himself (Fig. 38).

And if you wish to ascertain what weight will support this wing, place yourself on one side of a pair of balances and on the other place a corresponding weight, so that the two scales are level in the air; then if you fasten yourself to the lever where the wing is and cut the rope which keeps it up, you will see it suddenly fall; and if it required two units of time to fall of itself, you will cause it to fall in one by taking hold of the lever with your hands; and you lend so much weight to the opposite arm of the balance that the two become equal in respect of that force; and whatever is the weight of the other

FIG. 38. *Sketches by Leonardo da Vinci*, 1485.
Left: Parachute. Right: Measuring the lifting capacity of a wing. From the *Codice Atlantico* of Leonardo da Vinci

balance, so much the wing will support as it flies; and so much the more as it presses the air more vigorously.[72]

Leonardo, who has left us some thousands of manuscript pages, may have intended at one time to compile a great encyclopedic work in the Italian vernacular taken from authors of antiquity and of the Middle Ages and based also on his own researches. This work would have contained much of great use and interest, especially to the unlearned craftsmen and technologists of his own day. But he never got so far.

The effort to make the treasures of science available to wider circles was, as we have already remarked, an important feature of the Renaissance. Mathematical and technical works of antiquity which the Renaissance presses first issued in Latin and Greek editions, appeared, especially after the turn of the sixteenth century, in the vernacular tongues. Moreover many new discoveries were inserted in the Commentaries and Appendices. Side by side with the Latin versions, there appeared in the sixteenth century many vernacular and especially Italian editions of Euclid, Archimedes, Vitruvius and Hero. Works of antiquity and especially Euclid spread in both learned and popular form to wide circles, and exercised wide and deep influence. Dürer, like Leonardo at once an artist, a handworker and an engineer in 1525 [with his *Instruction concerning measurement with compasses and straight-edge*] presented to his compatriots (and particulary to the artists and craftsmen who in the pursuit of their calling could not dispense with some knowledge of mathematics) a German Introduction to applied geometry as an aid to their daily labours, in which he skilfully popularized the scientific knowledge that he had gathered during his sojourn in Italy and his intercourse with the Nuremberg humanists. Such efforts, which were especially directed to youth, for whom there had hitherto been little concern, contained a social strain, in spite of the fact that social conscience had in general little place in the Renaissance.

And just as Dürer in South Germany, so the mathematician Niccolo Tartaglia in northern Italy worked for the benefit of an aspiring body of hand workers who demanded some scientific foundation for their worshop practice. Master gunsmiths, military engineers, assayers of ore, metal founders, land surveyors, and merchants must often have turned with every variety of request to the experienced Tartaglia. Most pressingly the military man was concerned by the question of the course of the flying missile but the scholastics of Aristotelian mintage could give no satisfactory answer here. It is true that Tartaglia was never a practising gun-smith but he utilized the experiences of practical men and caused several experiments to be carried out. Thus he was able to deduce certain theoretical conclusions and though some error was mixed with truth, new knowledge was acquired. That the course of the missile is curved throughout, and that by raising the barrel 45° it shoots furthest, were substantial results of his work which were of some little use to the practical man. They were a first modest advance into the field of a 'New Science' in which one proceeded from a few experiments to general conclusions which were then confirmed by trials. But as we have said, this was but a beginning. The enquiries of the gunsmiths concerning the true course of their missiles as a practical need were of no less significance for the development of a new physics. Galileo, more than half a century later, also started from practical enquiries. We give below from the year 1538 a Dialogue between Tartaglia (here called Niccolo) and the Duke of Urbino, Francesco Maria della Rovere. We notice that the quadrant (Figs. 39, 40) was already in use before Tartaglia.

First Question, put by the most Illustrious Lord, Francesco Maria, Duke of Urbino. in the year 1538 *in Venice:*

Duke. What reasons are they which you say in your booke dedicated unto me (*O Nuova Scienza*) you have found out concerning the knowledge of shooting in Gunnes?

Niccolo. The proportion and order of shootes not only at marks far of, but also at marks hard by, either what Peece you will, and with what sort of pellet you will.

Duke. Speake more plainely, and give me an example thereof, for I doe not understand what you say.

Niccolo. I am content to shewe yur Excellencie an example of my said invention, but first I must speake of that materiall instrument which I have devised . . . which instrument is made of a square peece of wood or of metall. . . . and containeth a quadrant. . . .

FIG. 39. *Quadrant for measuring the elevation of a gun barrel*, 1538. Woodcut from: Tartaglia, *Quesiti* (Venice 1546)

FIG. 40. *Laying a gun*, 1547.
Woodcut from: W. Ryff, *Der. . mathematischen und mechanischen Kunst eygentlicher bericht* (Nuremberg 1547)

There follows a description of a quadrant furnished with a plummet to measure the elevation of a gun barrel. The quarter-circle is divided into 12 parts, each sub-divided into 12 parts, making 144 divisions in all (cf. Fig. 39). The Dialogue continues:

Duke. What will you infer upon this?

Nicholas. Hereupon I will first infer that a peece of Artillery elevated by one point shootes more farther than it will doe when it lyeth levell, and that a piece eleveted by two pointes will shoote more farther than at one point, and a piece elevated by three points will shoote more farther than at two points. Also a piece elevated by 4 points will shoote much more farther than it will do at 3. points. Likewise a piece elevated by 5. points, will shoote somewhat farther than at 4. points. And a piece elevated by 6. points, shootes a pellet of leade a little farther than it doth at 5. points. For reason teacheth us that the range of a pellet shot out of a piece elevated by 5. points, and the range of a like pellet shot out of the same piece elevated by 6. points doe so little differ, that upon any small advantage happening either by force of powder, or by any other means, the piece being elevated by 5. points will shoote so far as it can doe when it is elevated by 6. points . . . But when one doth mount such a piece, so as they doe mount morter peeces, that is to say at 7. points, with out doubt by mounting the peece at 7. points, he shall not shoot so far as he did when that peece was mounted at 6. points, also at the 8. point, he shall not shoote so far as he did at the 7. point. . . . Finally at the 12. and last point he shall not shoote so far by a great deale as he did at the 11. point. But in this last elevation it may be thought, that by naturall reason the pellet should returne backe into the mouth of the peece, yet by many accidents which do commonly happen in that instant when the peece is discharged, the Pellet will not precisely returne into the very mouth of the peece, but fall downe neare unto the same.

Duke. This is to be graunted as all that which you have besides spoken, but what will you infer upon this?

Nicholas. I will secondarily infer that I have found out in what kind of proportion or order the said shootes do increase by every elevation, and that not onely from point to point of our said instrument, but also from minute to minute, even to the end of the elevation of the 6. point or of 72 minutes, yea and that with every sort of pellet, whether the same be of leade, iron, or stone. Likewise I have found out in what proportion the shootes do decrease when peeces are elevated beyond the said 6. point as mortar peeces are elevated, and that not onely from point to point, but also from minute to minute even to the end of all the 12. points, or of the 144. minutes.

Duke. What profite will come by this your Invention?

Nicholas. The profite of this invention is such, as that by the knowledge of any one shoot out of any peece of ordinance whatsoever, each man may make a table of all the shootes that such a peece will shoote at any elevation, that is to say from point to point, and from minute to minute in our said instrument: the which table shall be of such vertue and propertie, as that any person having the same with him shal not onely know how to shoot, but also be able to teache every unskilfull Gunnar to shoote in such sort of gunnes at any marke so many paces and so far off him as he will.

Duke. If he that hath such a table wil not shoote himselfe, but cause one other person to shoote, shall not that other person learne this secrete?

Nicholas. No, most excellent Lord; but that other person may be likened unto the servants of Appotecaries which continually compound medicines according as they are appointed by Phisitions to doe, and learne not thereby to be Phisitions.

Duke. This seemeth to me a thing incredible, because you say in your said booke that you did never shoote in any gunne, and for that he which will judge of thinges in which he hath had no proofe or experience, is often times deceived. For the eie is that which giveth us a true testimonie of thinges imagined.

Nicholas. It is true that the outer sense doth tel us the truth in particular things, but not in universal things: for universal things are subject only to understanding, and not to any sense.

Duke. You have said enough, and if you can make me to see this which I do not beleeve, you shal work a wonder.

Nicholas. Al things, happening by nature, or art, are thought to be wonders, when no reason is given for the same, but your Excellencie shall find my sayings herein to be true if you will cause them to bee tried with a Peece of Artillerie.

Duke. I must goe now unto Pesaro, but at my returne from thence I will cause all this which you have told mee to be proved.[73]

IRON FOUNDRIES MACHINERY OF THE EARLY RENAISSANCE

Tartaglia wrote, as we have said, in Italian, that is to say, in the vernacular, since he wished to be understood by the practical workers who had no Latin. His versatile Sienese contemporary Vanoccio Biringuccio left a composition on technical Chemistry and metal-work, the *Pirotechnia* or *Work in Fire* of 1540, addressed similarly in the vernacular to technical workers. It is outstanding for the clear and critical fashion in which it is set forth, a sign of the enfranchised minds of the practical workers of Tuscany in those days (Fig. 41). Iron and foundry work may be observed in Biringuccio to have made considerable progress since the Middle Ages. The further development of the use of water-power in mine and foundry is characteristic of the Renaissance. In the *Architecture* of 1464 by Antonio Averlino Filarete, an Italian work of the early Renaissance on the art of building written in fictional form, we have clear evidence of a pig iron furnace to produce molten iron. Filarete describes a journey to the ironworks, and gives an admirable description of the great water-powered bellows.

FIG. 41. *Wiredrawing with water power*, 1540.
Woodcut from: V. Biringuccio, *De la pirotechnia* (Venice 1540)

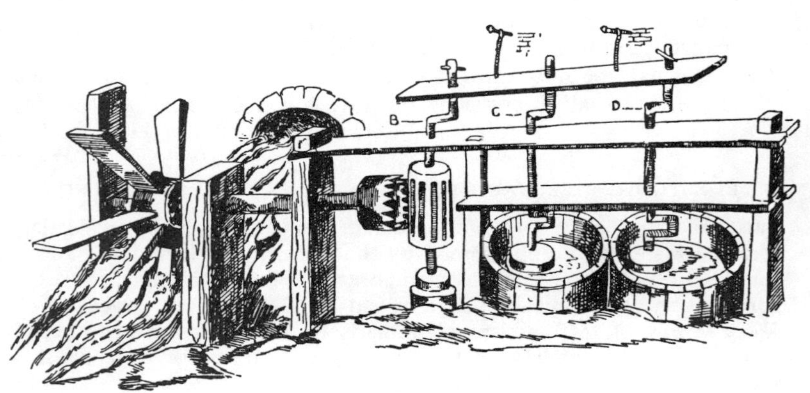

FIG. 42. *Water-driven glazing mill in a Majolica factory*, 1548.
Drawing from the MS.: C. Piccolpasso, *I tre libri dell'arte del vasajo* (London, Science Museum)

The Journey to the Foundry

The prince decided to advance the harbour-city further, and to cause to be erected there all the buildings described in the Golden Book. I was sent out to procure the considerable amount of iron needed for the purpose.

. . . We rode ever upward . . . and reached the summit of a pass whence . . . we perceived a valley dotted with foundries. . . . The land appeared to be very sterile. It belonged to the estate of our leader, and he decided to restore a dilapidated tower which was there. Now we followed down the course of another river, and so reached our goal, the newly erected ironworks. . . On the next day we measured out the site of the prospective buildings, marking it off with ropes. A priest was required to consecrate it, and each one of us dug his three spade-depths. Thereafter I was concerned with finding the necessary iron, and I fared to the foundry which is very hard to describe.

The building in which the iron is prepared lies near to the river, and is square, divided into two unequal compartments by a wall over 15 feet high. The smaller of these compartments is occupied by the smelting furnace, of which, since the floor is raised only the upper surface and opening are visible. The furnace is built of fire-proof stone. In the adjoining compartment are the two bellows which stand on the floor on edge; therefore not flat as is usual elsewhere. They are worked by water power; and each opens into a tube that passes through the partition, penetrates the smelting-furnace and blows on both fuel and ore. The bellows are 12 feet in length and 8 feet wide; the opening through which they draw the air measures 59.5 centimetres square. The bellows are made out of the strongest cowhide bound with good iron. Each blast produces a really thunderous noise. Close by is a basin with running water in which the molten iron is cooled, and from it a strong smell of sulphur arises. The workers are powerful men, blackened, in a shirt or otherwise scantily clad, provided with wooden shoes, and standing close to the furnace which they stoke and release the molten metal; they recall those fellows in the abode of Pluto that torment poor souls. The molten iron is about as liquid as bell-metal; and, like that, it could be poured into buried moulds. And that is what is done. In Milan castle is a cast-iron bombard cannon in the shape of a couchant lion. The molten metal is now conveyed to another factory where it is smelted for the second time; after which it is worked with a hammer until it has assumed the desired shape.

But these works were not yet prepared for the latter process.

Therefore I will describe the hammer that I saw 12 miles from Rome at Grotta Ferrata, a beautiful abbey in a wild mountain district, walled round like a castle, where monks held divine service according to the Greek rite. There a mountain stream, by means of a wheel, drives two cylinders, one of which works the bellows, and the other the hammer. The bellows are the same as those of the foundry. In the factory the metal is smelted for the second time, cast into the desired shape and brought under the hammer. . . . The veins of ore are mined in certain places on the mountain, and subsequently they are vigourously heated in a lime kiln. They are then left to cool, beaten into fragments the size of beans, put through a sieve and then arranged in layers in the smelting furnace in alternation with charcoal. After about twelve hours, they extract the cast iron and thus usually obtain, so they aver, more than 25 loads or 50 tons daily.[74]

In 1517, just over fifty years after the publication of Filarete's volume, the French humanist Nicolas Bourbon completed a Latin poem on the forge hammer which exhibits to us no less vividly the rapid rise of the iron foundries of the Renaissance.

. . . There ariseth a mighty building, near to the Exchange. Square, and of unhewn stone.
It is called the great foundry, is built of ordinary stones,
And the interior is completely clad in pebbles,
Hard stones of amazing resistance to blazing fire. Two powerful bellows blow from the rear wall.
Both are constructed of ox-hide, and driven by a wheel
That ever turneth, being worked by flowing water.
Ever changing in blowing, and ever blowing in changing.
Both are of equal force. Here stands the so-called founder.
Skilfully he seizes the molten mass,
That is named cast iron, and guides the stream of air,
With iron mattock he takes off the slag, controls the heat,
Separates the purified iron from that which is not purified.
Watches throughout days and nights, enduring all.
Sleep for but six hours at a stretch was accounted a sin,
And the whole process lasted for eight weeks
While throughout, the furnace was filled with the same iron.
Powerful bellows hastened to the aid of the exhausted brethren
And renewed the blaze. The iron flowed in streams;
With a loud roar the molten mass rushed forth,

Flames blazed up in the whirling black smoke,
Enormous fiery balls reached almost to the stars,
In their flight; Thus breathes the sulphurous vapour of Etna;
The giant, under the weight of the mountain, seeketh in vain
To free the exhausted member; he howls in fury.
The flames flash thundering forth, the waves pile up on the sea of fire
To maintain the supply of fuel and ore to the ravening furnace;
If the wide open gullet thereof is not completely filled,
Help comes to the founder from the Watchman who dwells above,
Faithful comrade, though black and gloomy as Satan.
At the same time, skilful hands diligently shape vessels,
First forming the moulds from clay, turning them gracefully.
Then the iron is poured out into the moulds.
Here too are cast mortars, those terrible marvels,
An invention of Hell; wroth were the gods with men
When Vulcan bestowed these deadly weapons on the Germans.
Cannon balls too they make, that thunderingly shatter walls;
Towns and citadels are blown down, and giant shapes
Are swept away; like flashes of lightning
These tortures fly forth; their crash is like thunder.
When at first the molten mass is removed from the furnace,
It still merits not the title of true iron.
Yet again a founder smelts it, after he has first destroyed
What he previously made. The second furnace
Improves the iron, makes it supple, shapes it to a cannon-ball.
Powerful smiths must then smooth and polish it.
For this they use the giant iron hammer,
Which is worked by the powerful force of the water.
Patiently they heat the iron again, turn it in the fire,
Grasping it in strong pincers, and quench it still glowing
In the prepared water.[75]

Vanoccio Biringuccio sketches for us a vivid account of a brass foundry in Milan in the first half of the sixteenth century (cf. p. 331). He is discussing a large undertaking for the mass production of objects in daily use.

Now, in short, this work requires infinite labour, and is carried out in various places, for instance, in Flanders, Cologne, Paris, and in many other countries, and also in the city of Milan in Italy, where I have seen a great quantity of it worked and coloured. It is coloured in this way: The masters whom I saw had made a large room a furnace, much longer than it was wide, and built with a certain kind of stone that by its nature resisted continued firing without melting and burning up. Where the fire entered the furnace it was almost entirely open. The body of the furnace was half or more underground; the vault was low; at the top and bottom there were everywhere little airholes; and above in the vault there were two square openings through which the crucibles containing the copper to be coloured could be put in and removed. These were then closed with little fitted clay shutters. The crucibles were made of Valencia clay, or they were brought ready-made from Vienna; they were very large and I believe that those which I saw were about two-thirds of an ell, and I understand they had a capacity of fifty or sixty pounds of metal.

For this process they placed in each one of the vessels twenty-five pounds of German rough copper broken in pieces, and they filled up the rest to within two inches of the rim with a powder of a mineral earth, yellowish in colour and very heavy, that they called calamine. The rest of the empty space in the crucible they filled with powdered glass. Then they put the crucibles in the vault through the above-mentioned opening and arranged them in pairs on the bed at the bottom. They then applied a melting fire for twenty-four hours and after this time they found the material entirely fused, and the copper, which was red before, had become a smooth and lovely yellow, almost like 24-carat gold in colour.

Later I saw in the same shop several apprentices and masters working at this, some of whom were hammering the said brass to make tinsel, some were laminating it to make tips for laces, some were filing it into tailor's thimbles, some were making buckles and other similar objects by casting, and there were some who were working it with the hammer into little bells, spoons, and basins, and others were turning it into candle-sticks or other vessels. In brief, some made one thing and some another, so that whoever entered that shop and saw the activity of so many persons would, I think, believe as I did that he had entered an Inferno, nay, on the contrary, a Paradise, where there was a mirror in which sparkled all the beauty of genius and the power of art. Taking thought of these things with greatest pleasure, there was not a day while I was in Milan that I did not go there and stay an hour or more. In that place I turned my eyes nowhere but I saw some ingenious novelty and beauty of craftsmanship, and thinking of the arrangement and the greatness of the new things that were shown to me, I was sometimes stupefied.

Among other things I saw a group of workmen whose method was entirely new to me. These were eight masters with many others in a room who did nothing but mould in loam and form an infinite number of moulds of all those little objects of everyday use which can be made from brass by casting. They made these in such a fine way that I cannot fail to tell you about it. . . . These masters took the number of patterns of all the things that they had decided to mould, that is, harness buckles, cups, belt buckles, all kinds of chain links, bells, thimbles, window fastenings, and other similar things. On one day they made all of one kind and the next day all of another, and thus they proceeded, changing the pattern every day; and having finished one that they had to mould, they began again at the beginning with another, continuing always with this easy method and process of moulding to accomplish much work.

They took a large quantity of loam mixed with cloth clippings or hemp and, when they had beaten the quantity they wished to use until it was somewhat hard, they spread it out about half an inch thick or less on a little board which was a span in length and somewhat wider than the patterns. After having spread it out well, they dusted it with fine charcoal, and in it they moulded their patterns with the gate, with vents, openings, and all the parts which are needed to make a complete mould. Some of these patterns were of tin and some of brass, accurately made, filed, and well finished, for if once the mould was well made the objects had to come out the same way. Each of these masters had also near him a small square oven of sheet iron. This was lined with bricks or covered with clay and had a little grate underneath and an open mouth the length of the oven. When a little charcoal and fire was put inside on top of the grate it heated the oven and kept it hot. Then they put the fresh half mould that they had just made to dry above the opening where there was a little grid. While it was drying they made another, and in the same manner when it was made they put it near to the first one, and so they continued up to six or eight pieces. Then they again took up the first one (which had had sufficient heat and time to become dry or nearly so) and on top of this they made the other half mould, and on top of this they moulded other patterns. They did the same with all the others and then beginning again from the first one they continued on to all the others in succession, so that when the moulds were finished and were drying one on top of the other they were altogether half an ell or more in height and about half a span wide, or as wide as the board or the particular pattern, for no useless space should be left around them. After these were finished and well dried in an oven like those where bread is baked, they opened them again, layer by layer, and took out the patterns. These gave twenty or more pieces to each mould, resulting in a great number, for some patterns contained as many as fifty or sixty pieces. Finally, after having closed the moulds again and sealed them up, and after having repaired or retouched the gates or any other points where it was necessary, and having given them a wash of fine ashes and water, they put the moulds together again and returned them exactly to their original position, then they bound them very well with iron wires and sealed them with the same loam. Then they took sixteen or twenty of these moulds and, standing them up on the ground all together, they made a circle of rocks around them and covering all the moulds with charcoal they baked them again. After having baked and arranged them well and having made in each part of the mould a gate which would convey the metal to the gates of all the other moulds, they took them to the furnace where the copper was coloured. Then they took one or two of those crucibles from the furnace and filled the mass of the moulds with that well-melted, yellow-coloured copper. They did this singly, or in pairs, or several at a time, according to the number of moulds, filling all the moulds made by the masters whom I have described above. They did this day and night as occasioned by the coloured well-disposed material, or depending on the number of moulds ready.

Pondering on this I thought to myself that this shop alone was sufficient to furnish not only Milan but also the whole of Italy, and surely it seemed to me a splendid and fine undertaking for a single merchant. I thought that he should take great pains to keep alive all the fine undertakings that I saw in that place and continue them and surely it pleased me to see so many things being moulded continuously, and continuously being cast. I believe that similar work is done in Flanders or in other places in Germany where they made candlesticks, basins and many other things of which mass export takes place from there to ourselves.[76]

The deepest insight of all into the technology of mine and foundry in the sixteenth century, its general situation and its special problems is afforded by the work, published in 1556, of the Saxon physician, humanist and mining expert Georg Agricola. But before we turn to him, our attention shall be given to another great German of the first half of the sixteenth century, Paracelsus, who was a year older than Agricola.

THE IDEA OF TECHNOLOGY

Paracelsus, the great physician and student of natural science had, from his wide-flung wanderings among miners, assayers of ores, foundrymen, master gunsmiths, alchemists, master minters and many other hand-workers obtained manifold impressions of a great variety of technical achievement. Against the traditional opinion of authority, he, the great Innovator, sought to attract attention to his own experiences. To chemistry, which he placed in the hands of physicians, thereby incidentally raising its status, he ascribed the task of evolving, from the qualities of various substances, workable principles for the healing of the sick. He raised alchemy from the single-minded attempt merely to create gold to the wide field of all of Nature's creation. So for him foundrymen, smiths, carpenters, master-builders, apothecaries, and physicians as well as hand-workers, technicians, artists and scientists, all alike were in a wide and deep sense alchemists. All these creative men worked by God-given power to the fulfilment of Nature. So for Paracelsus, active life, technical achievement, to which Man the Microcosm has been called by God, was interpreted in the sublime sense of co-operation in the completion of the Universe, the Macrocosm. In numerous passages in his works we encounter this interpretation of natural science and of technical work.

. . . Nature is so subtil and so penetrating in her ways, that she cannot be used except by great Craft; for she does not openly reveal that which may be completed within her, this completion must be accomplished by man. This completion is called alchemy. For an alchemist is the baker while he bakes bread; the vintner while he makes wine, a weaver while he makes cloth. Therefore he who brings that which grows by Nature for the use of man into the state ordained by Nature, is an alchemist. So know then a certain difference in the craft of such a one; that in the same way as a sheepskin may be taken raw for a belt or for a coat, however coarse and clumsy it be as compared to the work of the furrier or the cloth-worker, and coarse and clumsy as it may indeed be, yet one has something direct from Nature not prepared and more coarse and clumsy.[77]

God said: Let it be. And there was everything, but not art, nor the light of Nature. But when Adam was driven out of Paradise, God created the light of Nature for him by commanding him to gain sustenance by the work of his hands; and he also created the light for Eve when he said: In sorrow shalt thou bring forth children; Thus the creatures that had hitherto been like angels became human and earthly. So also Eve was taught to bring up her children, and thus were invented cradles and suckling. Now even as the Word was able to produce creatures, so also it was able to produce the Light that man needed, and his reason and understanding which are necessary to him as his body which he needed not in Paradise.

Now is no more to be known than that we all received limbs at the first creation, after all things had been created. But Understanding was not yet in Adam. It became necessary to man, and was created within him at the expulsion from Paradise. So he received from the Angels their knowledge, but not all knowledge. For he and his children after him had to learn by the light of Nature that which lieth in all things, that that may come forth which previously lay hidden. Although man was made whole at a single creation, it was his bodily half, not his artistic half; for the arts were not recognizable in him, but had to be fathomed by learning.[78]

God . . . wills that we do not simply accept an object as an object but investigate and learn why it has been created. Then can we explore and fathom the use of wool on the sheep and of the bristles on the sow's back; so we can place each thing where it belongeth, and can cook raw food so that it tasteth good in the mouth, and can build for ourselves winter apartments and roofs against the rain. . . . [79]

And since all things from naught were created to their completion, so too nothing is entirely completed since Vulcan must complete it. All things have been completed that they may be to our hand, but not conformable to our hand. Wood groweth toward its consummation, but not into charcoal or cinder. Glue increaseth, but not the harbours. So it is with all growth, that Vulcan may be recognized.

Thus for example, God hath created iron, but not that which can be made from it, not rust or iron bars or sickles; only iron ore, and as ore he giveth it to us. The rest he commandeth to Fire, and to Vulcan, the master of Fire. It followeth that iron itself is subject to Vulcan, and so is the craft thereof. For where that is not correct, there Vulcan performeth not what is needed. Thus it followeth that first the iron must be separated from the slag, then the needed object must be cast. That is alchemy, that is the smith named Vulcan. That which the fire doth accomplish, either in the kitchen or in the furnace, is alchemy. That which regulateth the fire is Vulcan, also the cook and the stoker. . . . It is the same with medicine which is created by God but not completed, for it is hidden in dross. So Vulcan is commanded to separate the medicine from the dross. And this he understandeth to do for medicine, as was the case with iron. Neither in herbs, nor in medicine, nor in stones nor in trees is it shown to the eye. Only the dross can be seen, but within is the medicine. So first the dross must be removed from the medicine, that the medicine be revealed. That is Alchemy, and that is the office of Vulcan. . . .

The arts are all within man. . . . Wherefore they are the alchemists of metals, and being alchemists that work in minerals they made antimony into antimony, sulphur into sulphur; from vitriol they make vitriol and salt from salt. Therefore learn to recognize alchemy, that she alone doth by means of fire change the impure into the pure. Also that not all fires will burn, but every true fire that will remain fire. Thus there are alchemists of wood, such as carpenters who prepare the wood that it may become a house; also the wood-carvers who make of the wood something quite alien to it, and thus is a picture formed from it. Thus too are physicians alchemists, who make from medicine that which is not medicine. The following showeth what sort of an art is alchemy. Similarly there is an art which constructeth the useless from the useful. . . . God hath commanded . . . since God hath created nought to perfection, that Vulcan be commanded to bring it to completion, and not to forge iron and slag together. Observe again an example: bread is created and bestowed on us by God; but not as it cometh from the baker; but the three vulcans, the cultivator, the miller and the baker make of it bread.[80]

We were not originally constituted for labour, but since the Curse, outside of Paradise, it was ordained that we should work. And God laid it on us through the Angel who spoke these words: 'In the sweat of thy brow shalt thou eat thy bread.' That is so; that is settled; that is to say, by labour thou wilt support thyself, with daily lamentation and misery.[81]

GERMAN MINING AND HANDICRAFT

Although Paracelsus regarded manual work—in which he included mining and foundry work—as of such deep import, this was by no means the view generally held in learned circles at the time. Georg Agricola was another of those truly versatile men of the Renaissance who combined with humanist learning a mind directed to the contemplation of Nature and also to practical technological activity. When in 1556 in a comprehensive work *De re Metallica* he described the mining and foundry work of his day, which was particularly well developed in the ore-bearing mountains of his own home, he felt obliged at the opening of his book to apologize for a work that treated merely of mining.

Many are of the opinion that Mining is a casual and unclean activity and above all that it requires less skill and knowledge than bodily labour. But it seems to me when I consider its various parts that the matter is quite otherwise. For a miner must have the greatest experience in his craft that he may truly know what mountain or hill, what point in the valley or plain may be most profitably worked or where he must refrain from working. Moreover he must understand the veins of ore and also the splits and faults in the rock. Then he must know the many and diverse sorts of earth, the solvents, the gems, the ordinary stones, the marbles, the rocks, the metals and their mixtures and he must understand the method by which every process can be completed below the earth. Finally he must understand the art of assaying all sorts of matter and preparing it for smelting. . . . Furthermore the miner must be skilful in many arts and crafts.

There has always been a great difference of opinion concerning mining. While some have praised it highly, others have condemned it bitterly. So it seemed well to me . . . to consider the matter carefully myself, in order to seek the truth. . . . Those who consider that knowledge of mining is useless to those who apply their industry to it assert that hardly one in a hundred of those who prospect for ores or undertake similar work have any reward, but that miners who entrust their whole assured and well laid out fortune to a doubtful and wavering chance are usually confounded in their hopes and, exhausted by their expenses and losses, ultimately lead a most bitter and miserable life. But those who argue thus do not realize how different is a learned and experienced mine expert from an ignorant and inexperienced man. The latter works seams without choice or distinction; the former on the contrary first tests and examines, and if he finds thereupon that they are too narrow or too impervious or on the other hand that they are too spongy and are barren, he thereupon decides that they cannot usefully be worked; hence he chooses only selected seams. . . .

Moreover these same persons who condemn mine-work assert that its profit is by no means permanent, and award their highest praise to agriculture. I cannot see with what justice they make this assertion. The silver mines in Freiberg have already lasted unexhausted for 400 years, and the lead works in Goslar already for 600 years. Both facts can be assured by the monuments of their history. In Chemnitz and Kremnitz gold and silver mining have been carried out together for almost 800 years; this is shown by the oldest Privileges of their inhabitants. . . .

Furthermore, the opponents of mining assert that it is a dangerous occupation because miners are killed by the destructive mine fumes which they breathe, that they waste away from the dust that they absorb which reduces their lungs to pus; then there are accidents; they are crushed by falls in the rock or hurled down in their courses down the shafts and thereby break legs, arms and neck. . . . But no economic usefulness ought to be so highly esteemed that for its sake the lives and limbs of men should be most hazardously risked. These matters are, as I freely recognize, very weighty and beset with alarms and danger. Wherefore I should judge that in order to avoid them, there should be no mining, were it assumed that miners either encounter such dangers more frequently or can by no means protect themselves from them. But then should not the urge to live be mightier than the desire to possess all things of this world, quite apart from metals? Truly one can no longer speak of the 'possessions' of one who relinquishes life under such circumstances, but only of 'what he has left to his heirs.' But since such cases occur but rarely, and only when miners are careless, therefore miners do not cease from mining, just as carpenters are not affrighted from their handwork if one of them who has not exercised care, has fallen from a high building and has lost his life. . . .

Now I come to those who declare that mining is useless to others, because metals, gems and stones which are dug from the earth are useless to themselves. This assertion they struggle to support by evidence and examples and partly to extort from us by abuse. But first they use the following argument. That the earth hides nothing and keeps from sight none of those things which are useful or necessary to the human race, but like a benevolent and kindly mother she expends her goods with the greatest generosity, and brings herbs, legumes, field vegetables and fruits into the daylight before our eyes. On the other hand, those things that have to be dug she has thrust deep down, wherefore they must not be extracted. . . . Their second argument is as follows: Metals afford no fruitful use to man; therefore we ought not to search for them. . . . Since every human consists of soul and of body, therefore neither of these two is in need of those things which are obtained by digging. . . .

Furthermore, the following arguments are stressed. Shafts in search of ore render fields sterile; therefore there was once in Italy a law that none should dig down in the earth for ore, and thus destroy fruitful fields and plantations of wine and fruit. Woods and groves were felled, since unlimited amounts of wood are needed for buildings and for tools as well as for smelting. By the felling of woods and groves, many birds and other animals are exterminated, many of which serve as fine and pleasant food. The ores, they allege, have to be washed, and this washing poisons brooks and rivers and thereby either drives the fish away or kills them. Since therefore the inhabitants of the surrounding countryside, owing to the desolation of the fields and of the woods, the groves, the brooks and the rivers suffer great embarrassment as to how they can obtain the necessities for their livelihood, and since on account of the lack of wood, the cost of building their houses is raised it is clear to all eyes that driving the shafts causes more harm than any use obtained from the ores obtained by mining. . . . Then the metals themselves are abused. First the opponents wilfully revile gold and silver and name them unholy and profligate corrupters of the human race; for those who possess them are in the greatest danger, and those who are without them lay snares for those who own them so that both gold and silver are often the cause of their downfall and destruction. Moreover they revile also the other metals, especially iron. For it has brought great destruction to human life for from it are made swords, javelins, lances, pikes and arrows, with which men are wounded, and murder, robberies and wars ensue. . . . But a projectile or an arrow can be shot into one man's body, whether from a bow, a scorpion or a catapult; on the other hand an iron ball from a bombard can be shot forth through the bodies of many men, and no marble or rock that may intervene is sufficiently hard not to be pierced by the impulse and force. Thus it brings the highest towers to the ground, cleaves the firmest walls, breaking through and overthrowing them . . . but because muskets which can be held in the hand are rarely, and the large ones never, made of iron, but of a mixture of copper and tin, therefore copper and tin are abused even more than iron.

. . . But mankind cannot make without metals the things which serve for livelihood and for clothing. For in the country side which provides the greater part of our livelihood, first of all no work can be accomplished without tools. . . . But what need of further words? If the use of metals were to disappear among mankind, then every possibility would cease either to preserve health or to lead a life compatible with our culture.

. . . Now I must answer the reproaches made against the products of mining. Thus gold and silver are described as destroyers of mankind because they cause the downfall and destruction of those who own them. But what possession can we name that does not cause disaster to man? . . . Therefore the abuses levelled against iron, copper and lead make no impression on wise and worthy persons. For if those metals were to disappear, men would burn with yet fiercer rage, and stirred to unrestrained fury, would fight like wild beasts with fists, nails and teeth. . . .

. . . and thus we see that it is not metals that should be abused but our sins, wrath, cruelty, discord, the passion for power, avarice lust. . . .

But now comes the question whether that which we dig out of the earth should be reckoned good or evil . . . worthy men use them well and find them useful; but evil men use them badly and to them they are useless. . . . For this reason it is not fair or just to rob them of the position and esteem in which they are held among our possessions. If someone uses them for evil ends, that does not make it just to call them evil. For what good things can we not use alike for evil or for good ends?

. . . Having rebutted the arguments and abuses of opponents, we will treat of the uses of mining. Firstly it is of use to physicians; for it provides a number of medicines with which wounds and ulcers and even the plague are wont to be healed. Therefore we should be obliged for the sake of medicine alone to dig into the earth, even had we no other reason for our search. Then it is useful to painters for it produces different colours. If walls are painted with these, they are less damaged than other walls by external damp. Moreover mining is useful to the builder; for it discloses marble, suitable for long-lasting and strong buildings of large size, and serves for decoration and ornament. It is also useful to those whose souls yearn for immortal fame; for it produces metals from which coins, statues and other objects can be constructed, which, next to literary monuments, bestow on man a certain permanence and immortality. It is also useful to merchants; for as I have already remarked, metal coinage is for many reasons more convenient than barter. To whom in fact is it not useful? I will pass over the list of the pleasing, tasteful, skilfully worked objects, useful to the craftsman which goldsmiths, silversmiths, copper-smiths, tin-smiths and iron casters construct in many shapes from metals. What handicraftsman can construct a complete and beautiful work without metals? If he has not used iron or copper tools, he will certainly not have been able to create a work of art either in stone or in wood. From all this the use and convenience of metals is apparent. These things we should not have if the art of mining had not been discovered and used for them. Who does not recognize their great usefulness and indeed necessity for mankind? In short, man could not have dispensed with mining, nor would the goodness of God desire that he should be without it. The question is further asked whether mining is an honourable calling for worthy persons, or whether it is dishonourable and contemptible. We however estimate it among the honourable crafts. For we regard as honourable every craft that does not involve a godless or offensive or unclean method of earning a livelihood. But honourable is the livelihood from mine and foundry; for it increases possession and property in a good and honourable manner. . . .

Moreover, the business of a miner is not dirty How could it be when it is so great, so rich and so honest? . . .

Finally, those who defame mine and foundry point out, in order to belittle them, that once convicts were condemned to work in mines and hewed in gangs of slaves. Now, however, miners work for a wage, and busy themselves like others whose hand-work brings uncleanness. Forsooth, if mining were to be considered unworthy of a free man because slaves once dug through shafts of ore, then must agriculture not be regarded as an honourable calling because slaves once worked on the soil, and do so still among the Turks. The same is true of building, for many builders were slaves; the same is true of medicine, since not a few physicians were slaves; and of many other crafts which were practised by prisoners of war. But agriculture, building and medicine are nevertheless reckoned among honourable crafts. Thus should mining also not be excluded from among them.[82]

In spite of the many opponents of mining work of whose objections Agricola gives us a picture, mining attained vigorous development during the fifteenth and sixteenth centuries. Germany especially was the centre of active mining and foundry technology (Plates 7, 8). From the fourteenth century and especially in the fifteenth and sixteenth many German miners were teachers of their craft in other lands. To be sure the deeper the shafts were sunk in the mines, the harder became the struggle against flood water. Inventors were busily occupied seeking to devise efficient machines for removal of water. Especially the attempt was made to utilize water power more extensively to drive the mine pumps; just as the foundry bellows worked by water wheels were developed to larger and larger size as we heard from Filarete and Nicolas Bourbon. The gigantic whirling wheel that Agricola describes for us by word and by drawing is clear evidence of the desire for motor power on a large scale (Fig. 44). But the necessary water power was by no means always available at those places where it was needed to work pumps or other methods of holding back the flood-water, or to drive the machinery for raising the ore. The technology of the sixteenth century found a solution of this difficulty by rod devices or overland connecting rod systems by means of which power was conveyed from a water wheel in a valley to a height some distance away (Plate 8). These rod systems which were invented in the Erzgebirge in the middle of the sixteenth century and sometimes extended well over four miles fulfilled

FIG. 43. *Hand-operated chain of buckets for raising water.*
Note the cast iron wheel with inserted steel teeth. Woodcut from: Agricola, *De re metallica* (Basle 1556)

FIG. 44. *Wheel for raising water.*
Diameter of wheel 10·7 m. Woodcut from: Agricola, *De re metallica* (Basle 1556)

to a modest extent the function of our modern electric overland transmission. The efficiency of this apparatus for conveying mechanical power was naturally very low.

In addition to the water wheel, and the hand-worked windlass, the man powered treadmill or the horse-capstan which had been used in mining since the beginning of the sixteenth century were still utilized as sources of power. The main problem of mining technology during the Renaissance period, as also during the seventeenth and to some extent during the eighteenth century was precisely to find a practical means of controlling water, as is testified by the *Privileges* granted to the inventors of those days. And in foundries, as we may similarly learn from the inventors' *Privileges*, effort was already directed to the greatest possible economy in the use of the charcoal that served as fuel.

German mining and foundry work which, as described by Agricola, had flourished since the late Middle Ages, suffered, however, a slow decline from the middle of the sixteenth century, resulting from a regrouping of trade routes in the age of discovery, the introduction of metals from newly discovered America and finally from the establishment in England of her own iron industry. Likewise, commercial life in the German and Italian cities was affected by the economic changes which followed the age of discovery. As a result, the guilds and corporations, great as was their significance not only for economic solidarity but also for education and community fellowship, nevertheless here and there hampered technical development. It must, however, be emphasized that we are indebted to the body of highly specialized town handworkers for a whole series of important individual discoveries—to mention only the vice (about 1500), the pocket watch (1510) and the improved slide rest lathe (1561). But limits were set by the corporations to the spread of many an invention. Hans Spaichl of Nuremberg, in 1561 invented an improved lathe slide rest. In 1578 the Council of Nuremberg broke one of these machines to pieces because Spaichl wished to sell it to a goldsmith. Spaichl belonged to the Red-metal turners comprising those handworkers who produced turned objects made of copper and brass. Similarly in 1590, because Wolf Dibler, also a *Red-Metal turner*, had built for the famous goldsmith Hans Petzold a lathe of new design, he was confined for some days in the Tower. No new invention was allowed to be transferred from one specialized form of handwork to another. The following extracts from the Nuremberg Council Decrees throw light on these circumstances. (The lathe here discussed is called indiscriminately *lathe*, *turning mill*, *turning wheel*, *treadle-mill*, *mill work* or simply *work* or a similar phrase.)

Nuremberg Council Decree
2 May 1559:

Since Hans Spaichl, red-metal turner, is a good Master-worker more enquiry shall be made as to whither he proposes to go and what he proposes to do and how he could support himself: etc.

8 May 1559:

Since Hans Spaichl, red-metal turner, wishes to go to Brunswick, and is such a good craftsman, it is not good that he should be allowed to leave here; he should be offered a master gunsmith's wage to remain here and set the Herlin lathe to work and we must enquire whether he makes large guns and what wages his father received as a master gunsmith and what is the condition of the Herlin lathe and report back.

10 May 1559:

Michael Kurzen, the red-metal-smith, shall be commanded by the Council to allow Hans Spaichl to have his extra lathe; and if he will do so, to bring it here and advise upon what pay should be allowed to Spaichl as a master gunsmith.

1 June 1559:

That the request of Hans Spaichl, red-metal-smith and turner, for higher pay be refused.

12 July 1560:

That the loan requested by the applicant Hans Spaichl, red-metal-smith and turner, be refused.

31 May 1561:

That the wheel of the lathe of Hans Spaichl, red-metal-smith and turner, shall, at his request, be inspected by the works staff, in order to see how it can be made to work again. At the same time the ingenious wheel he has offered shall be examined, its use determined, and its value; and report made to us.

4 June 1561:

The ingenious lathe of Hans Spaichl, red-metal-smith and turner, shall be purchased from him for 60 f., and given into the custody of the Master Builder. And in future he shall make no such wheel without my Lord's permission; otherwise a model shall be made for him of a lathe such as he has offered; and report made to us.

1 Oct. 1561:

That the loan of 10 f. shall be granted to Hans Spaichl, the red-metal-smith, for which he has applied to enable him to complete the hand-lathe which he has begun, and the pay office shall be advised.

20 May 1562:

That Hans Spaichl, red-metal-smith, shall be presented with the 10 f. that were lent to him by my Lords for the lathe that he has constructed and he shall be told that he may sell the said lathe to whom he will for as high a price as he can obtain for it.

30 May 1562:

Hans Spaichl, red-metal-smith and turner, shall be paid a further 15 f. for the lathe that he has constructed, and shall be told to sell it as well as he can.

7 July 1562:

The designated five workmen, obtained through the Swedish Master Builder, shall be allowed to go; but they shall be told and required to renounce their citizenship and to take wife and child with them. But an intermediary shall instruct or negotiate with Hans Spaichl, the red-metal-smith and turner, on what terms he would stay here; and shall report to us.

7 July 1562:

Hans Spaichl, red-metal-smith and turner, at his request and on his written undertaking to serve my Lords for the rest of his life, shall now and henceforward be granted a pension of 24 florins as gunmaster, and he shall be instructed that he must again see how he can best sell his lathe; and that if he should suffer a loss he shall be assured that it will be made good to him.

9 July 1562:

Whereas Hans Spaichl, red-metal-smith and turner, cannot use his lathe, it may be bought from him through an agent for 200 f. and inquiry made how it can be put to use again.

8 June 1569:

That the treadle-lathe design of Hans Spaichl, the red-metal-smith and turner, be placed for inspection in the government chamber; at the same time, information shall be sought as to why recently he has left his work, and why he is given a pension.

9 June 1569:

Hans Spaichl's lathe shall be inspected thoroughly by a knowledgeable workman.

15 June 1569:

Their Lordships' severe displeasure shall be indicated to Hans Spaichl the red-metal-smith and turner in that in spite of prohibition he has designed and constructed yet more lathes. My Lords have decided that these models shall not be disposed of but shall be kept under their control. And his answer shall be received and reported to us.

16 June 1569:

Concerning the Report on Hans Spaichl, the contents of which have been read, it shall be shown to him and he shall be told: My Lords are displeased that he thus occupies himself with contrivances, and doth not attend to his trade even though it be of great service to him. But since my Lords do not require his newly constructed lathe, and will not suffer it to be sent forth, and since it is not proper that he shall thus use his skill to the detriment of the whole town, they will therefore keep it under their control. They will subsequently decide what steps shall be taken.

21 June 1569:

Hans Spaichl's lathe design shall at his request be inspected, and he is commanded to answer before the Council also for the horse lathe and the hand-lathe that he hath constructed.

4 July 1569:

Hans Spaichl, the red-metal-smith and turner shall be notified concerning his lathe and reminded what lately happened to him on account of the former model; and that on his undertaking that he would desist from making such models, a pension was granted to him; and that my Lord therefore expected that he would not make more of the same design to the detriment of the whole town since there is no need for them here. But that in order that he should nevertheless suffer no injury, they were willing to give him a reward for this and retain it; but that he must contract to make no more such models which are not in the common interest of the town.

4 July 1569:

Concerning the demand of Hans Spaichl, to grant him 300 f. for the design of his lathe, since it cost him that amount; he shall be told that my Lords consider that this model must be destroyed. And if he will contract to attend to his trade and to conduct it and work like other Masters and throughout his life to make no more designs or models for lathes or for any other implement, then he shall be rewarded for this one with 100 f. And he must sell the horse-lathe that he hath made; but permission shall be conceded to him to make a horse-lathe for the arsenal of Harssdorft, but the other lathe shall not be attached to it.

6 July 1569:

Hans Spaichl red-metal smith and turner, shall at his further request be granted 100 f., as a reward for the design of his treadle-lathe; and to enable him to pay off his debts he shall be lent 150 f. in the following manner: seeing that he now hath annually 20 f. from the Office of Master gunsmith as pension; and he shall now have 20 f. more, so that he will gain annually 40 f.; but 20 f. shall be deducted from him every year toward the refund of the aforesaid 150 f., and the money shall be given not into his own hand, but to a third person such as the secretary for war who will deal most carefully with the creditors to see if the debt can be reduced. But before the money is issued, the necessary bond must be drawn up with him and he must swear to hold firmly to it.

14 March 1571:

Appeal from Erhart Scherl, goldsmith, that Hans Spaichl be permitted to make him a lathe. The master metal-smiths shall be heard thereon, whether this lathe shall be made or whether it would harm the whole town.

24 March 1571:

On the advice of the master metal-smiths we will oppose the appeal of Erhart Scherl the goldsmith, and refuse his request for the lathe.

5 Feb. 1578:

Concerning the report of Sebald Kurtz, red-metal smith, that Hans Spaichel, metal-smith, in defiance of the prohibition hath made a new lathe and hath sold it to Straube the goldsmith. And it shall be required of Spaichl and Straube that they testify on oath concerning this lathe and what they have had to do with it.

7 Feb. 1578:

In regard to the testimony of Christofer Straube and Hans Spaichl, the master red-metal smiths and turners shall be shown the before mentioned lathe, to determine whether injury can be done thereby to trade. And report shall be made to us.

10 Feb. 1578:

On the Report of the red metal smiths and turners that hath been read out to us, the lathe constructed by Hans Spaichl shall be acquired and he shall be paid five gulden for it and he shall swear that henceforward he will make no more of them without the knowledge and permission of my Lords, and the lathe shall be destroyed.

13 Dec. 1578:

On the sworn testimony of the red-metal workers, which has been read, Hans Spaichl the younger, red-metal turner, who for some time has been wandering out of the district, and only returned three weeks ago, undertook and himself assisted in the work of making usable the unusable treadle lathe, which he made formerly in Styria, through the addition of necessary parts, such as bushes, shafts, treadle and other essentials and as he has made it usable, Spaichl shall be summoned, placed under oath and be formally called to account as to how he came to work on the said machine, who instructed him, to whom does it belong, what he has done to it, who helped him and how much he has received for his trouble and work, also whether he has undertaken to fit further parts and, if so, what parts; further he must be prevented from removing either himself or his belongings from here until further orders.

15 Dec, 1578:

The reply of Hans Spaichl the younger, red-metal smith and turner, having been read shall be again shown to the master red-metal smiths who have been sworn. And as they wish to bring something further against Spaichel, they shall state it for the consideration of my Lords. The said Spaichel will be kept under surveillance as previously ordered; and since Hans Winckler, red metal founder, of Styria will shortly be coming here and will be making all sorts of purchases, the masters shall be told to be on their guard against him, that he do naught to prejudice or injure the industry of this town.

16 Dec. 1578:

On the further report of the red-metal smiths concerning Hans Spaichl the Younger, red-metal smith and turner, it shall be required of Spaichl that he pledge himself and swear yet again that he shall never henceforth for the rest of his days leave this town without the permission and prior knowledge of my Lords.

4 July 1581:

Whereas Hans Spaichl, member of the Red-Metal Smith's Guild, and son of a master here, a fitter in Styria, himself assisted in the building of the new lathes and is alleged to have given all kinds of information as to how and with what tools one thing and the other must be done, he shall be placed under arrest and thoroughly questioned; at the same time his father shall be required to state under oath what he knew of his son's action and how he had helped him in it.

24 Dec. 1590:

According to the opinion of the duly appointed Lords in council which have been read aloud, concerning what should or should not be admitted in the dispute between the Red-metal-Smiths and Hans Betzold or Petzold goldsmith, regarding his new lathe, what should be done in future concerning similar lathes and such devices: an honourable committee was appointed to ensure that in order that such and similar advantage and knowledge derived from the apparatus should not be so commonly known, still less—as has in part happened—carried outside the town to other districts, Betzolt shall be required under oath in the proper fashion to enclose his lathe in a case, or otherwise to guard it in such a way that it shall neither be observed nor carried off nor used by his journeymen, or by any others; nor will he place any other man in charge of it, nor allow any other person to work at it, except for his own personal use.

The same applies to all the red-metal smiths and turners here; they must similarly be required to record on oath that henceforth no one outside their craft shall be allowed to use such tools or wheels either hand or water driven, that none shall be made for them, nor shall any person contribute either counsel or action thereto without the special prior knowledge and permission of the committee, and this subject to the severest penalty for perjury. Moreover, all goldsmiths, and other citizens and inhabitants of this town who require or desire special lathes for their handwork or craft are forbidden, under penalty of 50 f. to prepare such apparatus, or to have it made by anyone for them; unless they previously notify the committee or the duly appointed Imperial Lords, so that it can be learned from them how and with what advantage and profit it can be made, also whom they desire to employ to make it and whether it can bring no drawback or injury to other craftsmen or to the community here in this town; And in case anyone should show himself dangerous and disobedient in this matter, then the committee would be obliged to inflict on him the penalty announced above and imprisonment besides.

And because Wolf Dibler, red-metal smith and turner made for Betzolt the screw or mandrel as the most important part of his work, and taught him thereon how to set up the lathe and how to handle the turner's chisel, thereby acting in opposition to his craft, therefore he shall be imprisoned in a barred dungeon for eight days to teach him not to do it again.

15 Jan. 1591:

Wolf Dibler, red-metal smith and turner, was given a penalty of eight days imprisonment in a dungeon; and since he hath served four days thereof already, and hath necessary work to do for the Elector of Saxony; the further 4 days of his punishment shall be commuted with a fine.[83]

SCIENTIFIC MECHANICS AND TECHNOLOGY

Science during the Renaissance as we have already emphasized, breathed life into the writings on mechanics of Archimedes, Hero and Pappus, as well as the pseudo-Aristotelian *Mechanical Problems.* Linking with antiquity and to some extent with the late Middle Ages and especially with Jordanus Nemorarius (thirteenth century), there was in the second half of the sixteenth century an effort toward the further development of Statics by such direct forerunners of Galileo as, especially, the Italians Girolamo Cardano, Federigo Commandino, Guidobaldo del Monte and Giovanni Battista Benedetti, a pupil of Tartaglia, and the Dutchman Simon Stevin, equally important as mathematician, physicist and engineer.

Like Commandino, Guidobaldo del Monte came from Urbino, that town distinguished for its combination of mathematical and humanist studies and for its readiness to tackle technical problems. He was already famed as a patron of Galileo when he wrote in 1577 a work on *Mechanics* in which he improved the theory of simple forces, that is to say of lever, pulley, wheel and axle, the wedge and the screw. The general principle of *relative speeds*, already enunciated in the pseudo-Aristotelian *Mechanical Problems*, Guidobaldo applied successfully to the lever, wheel and axle, and to the block and tackle. Undismayed by the fundamental difference in the approach to Mechanics of Archimedes and of the *Mechanical Problems* (cf. p. 25) Guidobaldo was dependent on both for his statics. The general treatment of statics quite in the Euclidean sense is the bequest of Archimedes. Guidobaldo's conception that technical application of simple machinery is really a violation of Nature, a trick played on Nature, links him with the pseudo-Aristotelian *Problems*. This conception was soon rejected, as will be evident to us by quotations from Francis Bacon (p. 173) and Galileo.[17] We will give now as examples of the technical application of mechanics some significant passages from a later German translation of a work that originally appeared in 1577, the *Six Books of Mechanics* of Guidobaldo del Monte.

There are two things which are wont to be of value to each and every man for the acquisition of riches and great estates, namely UTILITAS and NOBILITAS, great usefulness and noble excellence; these two have apparently collaborated in attending the study of mechanics and in causing it to be widely beloved. And since today we derive and calculate nobility for the most part from first origin and birth, doing so in this instance we meet on the one hand Geometry and on the other hand Physics, and by their willing union and kindly embrace, there emergeth ultimately to the light of day that most noble of all arts, Mechanics. The most noble, I say, in view of the matter she hath conquered and the compulsion of her arguments, which seemeth indeed in one place to be the opinion also of Aristotle. Seeing that she not only, as Pappus testifieth, assisted Geometry to attain her full perfection and completion, but also ruleth with complete power over physical matters and natural objects; therefore all that bringeth help and prosperity to craftsmen, builders, porters, peasants, seamen and many others (sometimes even in opposition to the law of Nature) is subject to the realm of Mechanics and must obey the behest of this noble art. But although some of her numerous achievements are carried out by force in opposition to Nature (as was first believed), often on the other hand they imitate her. She meriteth then the greatest admiration, yet is true and easily credited by him who hath previously learnt from Aristotle that each and

13. Guericke measures the pressure of the atmosphere, 1661. Air is extracted from a cylindrical container; the external atmospheric pressure forces the piston down into the cylinder. Engraving from O. von Guericke *Experimenta nova*, 1672

Atmospheric Steam Engine for pumping water on a Dutch estate, 1781. From a water colour

14. Fulling Mill in the textile factory at Oberleutensdorf in Bohemia, 1728. Engraving by Turner from W. L. Reiner. The Fulling Mill is the only part of the factory which is operated by water power

every mechanical problem and theorem may be conveniently reduced to circular motion; and based on the principle, as clear to the inner understanding as to the external senses that runs. A round appliance moves most easily and the larger in size, the greater its mobility. . . .

. . . Thus in time Mechanics began . . . to drive the plough over the field. Then she progressed further, and with carts and carriages conveyed provisions, wares and all kinds of heavy loads between neighbouring peoples and then brought necessities to us from their place of origin. Moreover, when men came to give thought not only to stark necessity but also to decoration and to convenience, our debt is again to mechanical subtility that hath shown us how to propel a ship with oars, and how, by means of an apparently small and insignificant steering rudder attached behind, a great and powerful . . . freight ship can be controlled and guided in masterly fashion; again, how great heavy stones, beams and timbers can be raised on high by a single man's hand instead of by many others, and can be delivered to the builders; also how to raise water from deep wells to irrigate meadows, gardens and the like. Then again how to press forth wine and oil and all kinds of useful liquors from various sources and render them useful to the master of the house. Also how by two levers drawn to and fro, great trees and considerable blocks of marble may be skilfully divided in two. Then in military matters, how to throw up walls and barricades, how to combat in knightly fashion, how to besiege various kinds of positions, how to storm and in case of need how to defend and to protect them. In addition to all this, I must repeat that infinite and innumerable uses have arisen from the noble art and science of mechanics; By the instruction and constant assistance gained from them by carpenters, stone-masons, marble workers, tenders of vineyards, oil-pressers, apothecaries, locksmiths, goldsmiths and all metal-workers, barbers, surgeons, bakers, tailors, indeed the sum total of all useful hand-workers; they have fulfilled and enriched human life, as indeed we may see daily. So let the latter-day Dedalus-mouthed fellows, wilful scorners of the mechanical arts withdraw and blush with shame (if so be that they are still capable of feeling shame) and henceforth cease in their malignant and false fashion to slander the aforesaid praiseworthy arts for their alleged uselessness. Let them withdraw, I say, and think better of it. But if they will not cease or become wiser, why then let them remain in their raw ignorance where we will leave them. And we will choose rather to follow and imitate herein that Coryphaeus of all philosophers and predecessors of Aristotle, whose ingenious

Mechanical Questions, introduced to worthy posterity, sufficiently attest his love for the art. . . .

But above all mathematicians is to be exalted, with more and yet juster praise, Archimedes, the Unique, who wished to manifest God to us in Mechanics as a unique Idea in conformity with which all lovers of this art should act. . . . For he in such masterly fashion constrained the whole heavenly Globe within a fragile glass that the stars with their infallible constant motion could not appear otherwise than in accordance with Nature. . . .

This very Archimedes, with his Polispasta, and indeed with only his left hand hauled a weight of 500,000 bushels. . . . Finally he relied to such an extent on our Mechanical Art that he became somewhat arrogant, and permitted himself to utter the following phrase opposed to Nature: Allow me to stand still, and I will move the whole Earth.[84]

Among those who associated themselves with the theoretical findings of Guidobaldo on the general principle of relative speed was the practical man Buonaiuto Lorini, military engineer in the service of Cosimo dei Medici and the Venetian Republic, who had a masterly facility in combining practical experiments with theoretical knowledge. Many engineers of the late Renaissance, as for example Agostino Ramelli (1588) indulged, in their published works, in designs on paper which failed to pay regard to the weight of the different parts of the apparatus themselves, to the stiffness of the rope and to friction. So pleasure was taken in the kinetic working of little mechanical models and bold designs were produced which disregarded such problems as the practical effects and the limited capabilities of the material and of the means of manufacture. Lorini however was a true engineer, who was not content with plans of apparatus on paper or with semi-operative models, but always gave consideration to the practical full-sized product. We will reproduce a few passages from Lorini's Italian work *On Fortifications* of 1597, which was much in advance of contemporary technical books (Fig. 45).

I will try to show, condensing the matter as simply and briefly as I can, the action of the lever in pulley blocks, screws, in the axle-tree and in the wheel. Knowledge of these will contribute chiefly to an understanding of what we have to say concerning the investigation and construction of machines, and how these must be assembled and arranged not only with the right proportions (of parts), but also how, with the help of compasses, the power, that is, the increase by the force of the lever, can be clearly ascertained; so that when it comes

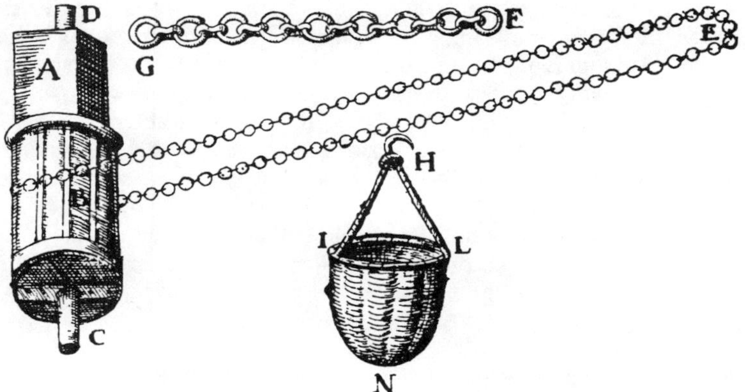

FIG. 45. *Cable railway for fortification construction*, 1597.
Woodcut from: Lorini, *Delle fortificationi* (1597)

to the actual execution of such works, one will not be mistaken in their efficiency such as is the case with those who, not being acquainted with the fundamental principles, rely on the ease with which little models work. But before we proceed further, I must point out the difference between a purely speculative mathematician and a practical mechanic. This difference is based on the fact that proofs and relationships that are deduced from lines, surfaces and from merely imaginary and non-existent bodies are not completely valid when applied to material bodies, since the conceptions of the mathematician are not subject to those obstacles which are inherent in the nature of the material with which the mechanician works. For example, it follows necessarily from mathematical proof that a load can be raised by a force amounting to only a fourth of its own weight if the distance between the pivot and force is four times that between the load and the pivot; but when operating with material bodies, perhaps with a beam as the lever, account must be taken of the weight of this beam. It must be taken into consideration that the longer part of the beam will be on the side of the force, while the shorter part will lie on the side of the load. So the weight of the greater part of the beam will increase the force that is raising the load or holding it balanced. In other cases on the contrary, the material can offer considerable resistance as when it is desired to rotate on their axles heavy wheels which offer resistance owing to the uneven distribution of their weight. And this is particularly the case with wheels whose axles are not truly adjusted and centred, since this can disturb the whole motion. . . . Therefore the judgement exercised by the mechanician who organizes and commands those who will carry out a work consists to a great extent in anticipation of the difficulties resulting from the entirely different characteristics of the materials which must be used. And the less it is possible to formulate definite rules with regard to such difficulties, the greater is the need for care. . . . Therefore I desire by the information given above . . . to show clearly to him who . . . would embark on such undertakings that a knowledge of mathematics will not suffice him, but that he must also be a careful and experienced mechanic.[85]

. . . As regards the effect produced if the forces given are set to work, I believe as has been said above that it may prove to differ in many respects, owing to the weight of the materials with which the mechanician has to deal. This difference may arise especially from the weight of the lower block, and also of the rope, particularly when this is thick and new, that is to say unused, so that it may hinder the running of the pulleys. And this is especially the case if the little axles which bear the pulleys and around which they rotate, are not accurately centred, and are not turned with proper care. Most important, ropes must not rub against one another when the load is raised. But all these things can be remedied. As regards the pulleys, their weight must be brought into correct relationship with their power and with the load that has to be raised. And as for the rope, the thinner it is, although it thereby becomes less durable, so much the greater is the power transferred to the lifting force. But the rope must be strong enough to bear the loads safely, having regard to the number of pulleys which are on each side in the two blocks; for the more there are of them, the less will be the weight which each single rope has to bear.

And in order that the ropes cannot rub against one another, the lower pulley D (in the upper block) must be smaller by double the thickness of the rope than C which lies over it; and the third one E must similarly be smaller than D. It is also to be noted that these pulleys must be perfectly round, and made of bronze, . . . and must have in their exact centre a hole in which the carefully worked axles fit.

As regards the speed of the work, it will be seen that with greater power there is reduced speed, and on the other hand with speed there is less power, in proportion to the added length of the lever-arm, or the increased number of ropes of which each, in the manner described above, bears its share of the weight of the load. And this is true of every kind of instrument and machine that can be made—of which more will be said later.[86]

The more practical attitude toward technology which characterized the Renaissance—of which the works of Biringucci, Agricola and Lorini are examples—declined more and more in the succeeding period. Already in the late Renaissance there was a more markea trend toward the building of fanciful projects which sometimes lost touch with reality, though a few positive results cannot be denied. These considerations lead us to the Baroque period.

PART IV

THE BAROQUE PERIOD

PART IV

INTRODUCTION

THE Baroque period was a time of preparation for important new developments in technology. The rich harvest of that period in the field of the physical sciences led to immense results in the succeeding period, especially in the second half of the eighteenth century, by its use in technology. The great success of the new computers based on the purely outwardly comprehended Physics of Galileo, the profound physical and astronomical understanding of Kepler, the magnificent theoretical and experimental results of the mathematical and physical researches of such men as Leibniz, Huygens and Newton brought mathematical methods into the foreground as a supreme instrument for acquiring knowledge of our world. The great philosophers of the period, above all Descartes, Spinoza and Leibniz availed themselves of it in obtaining a 'clear and definite' picture of Reality.

The impact of the dynamic element characteristic of the Baroque period, on the Renaissance world, which was defined by static form, was marked no less by the creations of an art pulsating with power and tension than by the new dynamics of Galileo, the teaching of Harvey on the circulation of the blood, the analytical geometry of Fermat and Descartes, the dynamic conception of bodies of Leibniz and the differential calculus discovered by Newton and by Leibniz. The seventeenth century more and more replaced the conception of the spontaneity of Nature by laws of mechanics, and thus for the first time attained the conception of Natural Law. At the same time it was recognized that technical achievement must also be in harmony with natural law. In the seventeenth century, to be sure, natural law was still regarded as established by divine decree. Not until the eighteenth century did this conception also become secularized. The idea of Natural Law was matched in the political domain by the desire for all-embracing laws which should admit no individual exceptions. This was important for the formation of a practicable patent law in the realm of invention, which in turn exercised considerable influence on the development of technology.

From the burgeoning experimental research of the Baroque period there appeared a series of scientific instruments and apparatus of which we need name only the microscope, the telescope, the barometer, the calculating machine, the air-pump, the pendulum clock and the thermometer. The introduction of such a body of instruments, which in those days were still constructed by the discoverer himself or under his supervision, betokened a far-reaching introduction of technology into research. A further gift to both pure and applied science from the spirit of dynamics and of continuity peculiar to the Baroque was the new differential calculus which proved, in its application to physical and later to technological problems, to be an immensely useful intellectual tool.

The first really important scientific societies, such as the Royal Society of London and the Paris *Académie des Sciences*, and the earliest learned journals took an active part in the scientific research of the second half of the seventeenth century. There was in fact a general tendency toward co-operation, which was likewise manifested in the formation during this period of trading companies, mostly by private initiative. Experimental research into Nature and technical achievement, at least as regards important discoveries which were in keeping with the contemporary effort toward Utility, played a prominent part both in the scientific Corporations and in the learned journals.

In the Baroque period, as distinguished from the Renaissance, and especially in Germany, there appeared a greater wealth of fantasy and a stronger inclination toward the strange and the mysterious. The flowering of fantasy appeared sometimes also in technology, especially in the inclination towards complicated projects and manifold technical toys which were especially popular in court circles. But by the side of this wealth of ideas, fantasy and inclination toward the supernatural, Baroque man was also characterized by rational sense and a practical economic attitude.

As regards the application of devices and materials, the technology of the seventeenth century man still followed entirely in the path of the preceding period. In simple machines such as water wheels, windmills, treadmills and capstans, use was still made of the elemental forces of water, wind and the muscles of man and beast. Wood was to an overwhelming extent the material with which these simple machines were built. When great technical undertakings were to be carried out as part of the military, court or mercantile demands of the absolute rulers of those days, the desired goal was usually sought by mere multiplication of these simple technological methods and by better organization, though in the second half of the seventeenth century there appeared the first attempts toward the development of really new technological appliances and especially completely changed methods of imparting motion.

The new mechanized cosmology in general, as well as the mercantile tendencies of the seventeenth century, led to mechanization of human work. In the more highly developed manufactures which still in general worked with the traditional simple technical methods, increase of production was sought by division of labour and by better organization. Except in the realm of pure art, the work of the individual lost its import in manufacture; the essential factor was the quantity produced. Manufacture, especially when it was undertaken not by the state but by an individual, meant an invasion of the old static economy, with its strictly limited possibilities of profit, by the dynamic of a progressive and unrestricted struggle for gain.

One important factor in the intensification of research into Nature during the second half of the seventeenth century, increasing the effort to gain control over Nature by technical means and by the great mobilization of economic resources, is to be found in the practical ethics of Calvinism with its clear concern with the world around us. Countries whose peoples were wholly or partly Calvinist, especially the Northern Netherlands, England with its Puritan Free Churches which embraced especially the middle classes, and France until 1685 with its Huguenot community active in science and industry, were all the scenes of important scientific, technical and industrial development. Moreover, the above-mentioned scientific societies and learned journals of the last third of the seventeenth century had an overwhelming proportion of Calvinist interest. It was thus religious forces which were active within the realm of an especially '*active life*'. We shall return to this connection (cf. p. 191). In the eighteenth century the religious impulse retreated, indeed, in the face of increased secularization, just as the trend towards metaphysics which characterized the Baroque period though combined with a strong attraction to the Real, the Economic and the Rational, nevertheless had to yield to a considered utilitarianism and empiric rationalism.

TECHNOLOGY AND PATENTS

Of supreme importance to the development of technology was the formation of an ordered patent law. As early as the Middle Ages we find records here and there of 'privileges' for discoveries, to assure to the discoverer protection from imitation. But a regular system of patent rights was a later development, as for example in Venice at the end of the fifteenth century. During the sixteenth century especially in the Netherlands and in the Electorate of Saxony, definite formulae were evolved for the security of the rights of discoverers, for which the most important qualifications required of the invention were novelty and utility. Charters of patent rights extending over the whole Empire were also issued in the sixteenth century. In England from the end of the sixteenth century the House of Commons was engaged in a struggle against the evil practice of the Crown in granting Privileges at will for trade in all sorts of goods, particularly those providing everyday needs. An action was brought by E. Darcy, who held a royal Privilege dating from 1598 for the import, manufacture and sale of playing cards, against Thomas Allen who had violated the privilege. Allen's advocate Fuller, in a remarkable speech in the Royal Courts of Justice in London in 1602, emphatically maintained that Privileges were valid only and solely for the person who had himself made a new discovery useful to the State. The struggle against the Crown on the question of the wrongful bestowal of Privilege shows us the English citizen already active in defence of his rights. Following is Fuller's historically important speech at the Royal Courts of Justice in London in 1602.

Commonwealths were not made for kings, but kings for commonwealths; and the law, the inheritance of all, binds both the queen and subject. If there were no law, there would be neither king nor inheritance, and to outrun the law is to produce confusion. The queen is not entitled by her patent to do wrong, and her prerogative is no warrant to injure the subject. Letters patent were void if they tended to change the law, or hurt any man's inheritance, or granted anything *contra commune jus* (against common law), or what tended to any general charge of the subject.

By the law of God every man should live by his labour, and therefore, were an act of parliament to prohibit a man from living by the labour to which he was brought up, it would be void by the law of God; and how much more, must letters patent be void.

Arts and skill of manual occupations rise not from the queen, but from the labour and industry of men, and by the gifts of God to them tending to the good of the commonwealth and of the king; and it is the duty of a king, says Bracton, to reject no person, but to make everyone profitable to the commonwealth. . . .

All patents concerning the king and his subjects are to receive exposition and allowance how far they *are* lawful and how far *not* by the judges, and that the judges in the exposition of the Letters Patent are to be guided, not by the precise words of the grant, but by the laws of the realm and laws of God, and according to the ancient allowance thereof. . . .

I will show you how the judges have heretofore allowed of monopoly patents, which is, that where any man by his own charge and industry, and by his own cost and invention, doth bring any *new trade* into the realm, or any *engine tending to the furtherance of a trade*, that *never was used before*, and that for the *good of the realm*, that in that case the king may grant unto him a monopoly patent for some reasonable time, until the *subject may learn the same*, in consideration of the good that he doth bring by his invention to the commonwealth, otherwise not. . . .[87]

By the Statute of Monopolies under James I in 1624, the question of the grant of Privileges came under general rules, and it was established that the Crown might grant Patents only to the first inventors of new manufactures or processes useful to the State.

. . . Letters Patent and grants of Privilege for the term of fourteen years or under, *hereafter to be made*, of the *sole working or making of any manner of new manufacture* within *this realm*, *to the true and first inventor* and *inventors of such manufactures*, which others, at the time of making such Letters Patent and grants, shall *not use*, so also *they be not contrary to the law*, *nor mischievous to the State*, by raising prices of commodites at home, or hurt of trade, *or generally inconvenient*, the said fourteen years to be accounted from the date of the first Letters Patent or grant of such Privilege. . . .[88]

The grant of a *Privilege* to an inventor was still an act of grace on the part of the Crown; not until the end of the eighteenth century was a legal right to a Patent for his discovery conceded to the inventor. Nevertheless the early uniform regulation of the question of Privileges encouraged the development of technology in England. Goethe could with truth say of Englishmen: 'We regard discovery and invention as a splendid personally gained possession, and we pride ourselves on this. But the clever Englishman transforms it by a Patent into real possessions, and thereby avoids all annoying disputes concerning the honour due'.[18] And elsewhere he emphasizes: 'The Englishman is free to use that which he has discovered until it leads to new discovery and fresh activity. One may well ask why are they in every respect in advance of us?'[19]

SCIENCE AND TECHNOLOGY

In the seventeenth century the realm of Nature was estimated as highly as the human world, whereas in the Middle Ages it occupied only the lowest grade. Natural science freed itself more and more completely from theology. Nature was frankly regarded as a field for human activity, although even in this field religious constraint was at first still effective. Already at the beginning of the century, however, Bacon was seeking to lay the theoretical foundations for a natural science whose specific aim was the development by technical means of control over Nature.

For the matter in hand is no mere felicity of speculation, but the real business and fortunes of the human race, and all power of operation. For man is but the servant and interpreter of Nature: what he does and what he knows is only what he has observed of Nature's order in fact or in thought; beyond this he knows nothing and can do nothing. For the chain of causes cannot by any force be loosed or broken, nor can Nature be commanded except by being obeyed. And so those twin objects, human Knowledge and human Power, do really meet in one; and it is from ignorance of causes that operation fails.

. . . .

Now the true and lawful goal of the sciences is none other than this: that human life be endowed with new discoveries and powers.

Now the empire of man over things depends wholly on the arts and sciences. For we cannot command Nature except by obeying her.

Again, if men have thought so much of some one particular discovery as to regard him as more than man who has been able by some benefit to make the whole human race his debtor, how much higher a thing to discover that by means of which all things else shall be discovered with ease![89]

In his Utopia *New Atlantis* Francis Bacon ventured a view into the future of extended human dominion based on diligent empirical research.

Description of Solomon's House

The end of our foundation is the knowledge of causes and secret motions of things, and the enlarging of the bounds of human empire, to the effecting of all things possible. . . .

We also have engine-houses, where are prepared engines for all sorts of motions. There we imitate and practise to make swifter motions than any you have, either out of your muskets or any engine that you have; and to make them and multiply them more easily and with small force, by wheels and other means; and to make them stronger and more violent than yours are, exceeding your greatest cannons and basilisks. We represent also ordnance and instruments of war, and engines of all kinds; and likewise new mixtures and compositions of gunpowder, wildfires burning in water, and unquenchable; also fireworks of all variety, both for pleasure and use. We imitate also flights of birds; we have some degrees of flying in the air; we have ships and boats for going under water, and brooking of seas; also swimming-girdles and supporters. We have diverse curious clocks, and other like motions of return, and some perpetual motions. We imitate also motions of living creatures by images of men, beasts, birds, fishes, and serpents: we have also a great number of other various motions, strange for quality, fineness, and subtility.[90]

Technology became important in the thought of many other leading philosophers and scientists of the seventeenth century besides Bacon. Galileo, the founder of the new dynamics, proceeded in his works to a great extent from technical problems. His work was influenced by the tradition of the Florentine workshops of the experimenting Masters of the Renaissance. In Pisa and in Florence he pursued various studies in physics which exhibit a combination of experimental research and its technical application. And later, in Padua where he taught in the University from 1592, he was occupied with a large number of technical problems such as fortifications, the technique of water-supply, the mechanism of simple machines, widely usable proportional compasses, and the testing of materials. Galileo's famous *Discourses* of 1638, in which he set forth his new dynamics, begin in the busy Arsenal of Venice, and first demonstrate that it is impossible to proceed straightway to complete a large machine even though each one of its parts be of a size which is a given multiple of the size of the corresponding part of a smaller machine.

SALV. The constant activity which you Venetians display in your famous arsenal suggests to the studious mind a large field for investigation, especially that part of the work which involves mechanics; for in this department all types of instruments and machines are constantly being constructed by many artisans, among whom there must be some who, partly by inherited experience and partly by their own observations, have become highly expert and clever in explanation.

SAGR You are quite right. Indeed, I myself, being curious by nature, frequently visit this place for the mere pleasure of observing the work of those who, on account of their superiority over other artisans, we call 'first rank men'. Conference with them has often helped me in the investigation of certain effects including not only those which are striking, but also those which are recondite and almost incredible. At times also I have been put to confusion and driven to despair of ever explaining something for which I could not account, but which my senses told me to be true. And notwithstanding the fact that what the old man told us a little while ago is proverbial and commonly accepted, yet it seemed to me altogether false, like many another saying which is current among the ignorant; for I think they introduce these expressions in order to give the appearance of knowing something about matters which they do not understand.

SALV. You refer, perhaps, to that last remark of his when we asked the reason why they employed stocks, scaffolding and bracing of larger dimensions for launching a big vessel than they do for a small one; and he answered that they did this in order to avoid the danger of the ship parting under its own heavy weight [*vasta mole*], a danger to which small boats are not subject?

SAGR. Yes, that is what I mean; and I refer especially to his last assertion which I have always regarded as a false, though current, opinion; namely, that in speaking of these and other similar machines one cannot argue from the small to the large, because many devices which succeed on a small scale do not work on a large scale. Now, since mechanics has its foundation in geometry, where mere size cuts no figure, I do not see that the properties of circles, triangles, cylinders, cones and other solid figures will change with their size. If, therefore, a large machine be constructed in such a way that its parts bear to one another the same ratio as in a smaller one, and if the smaller is sufficiently strong for the purpose for which it was designed, I do not see why the larger also should not be able to withstand any severe and destructive tests to which it may be subject. . . .

SALV. Even if the imperfections did not exist and matter were absolutely perfect, unalterable and free from all accidental variations, still the mere fact that it is matter makes the larger machine, built of the same material and in the same proportion as the smaller, correspond with exactness to the smaller in every respect except that it will not be so strong or so resistant against violent treatment; the larger the machine, the greater its weakness. Since I assume matter to be unchangeable and always the same, it is clear that we are no less able to treat this constant and invariable property in a rigid manner than if it belonged to simple and pure mathematics. Therefore, Sagredo, you would do well to change the opinion which you, and perhaps also many other students of mechanics, have entertained concerning the ability of machines and structures to resist external disturbances, thinking that when they are built of the same material and maintain the same ratio between parts, they are able equally, or rather proportionally, to resist or yield to such external disturbances and blows. For we can demonstrate by geometry that the large machine is not proportionately stronger than the small. Finally, we may say that, for every machine and structure, whether artificial or natural, there is set a necessary limit beyond which

15. Amsterdam dredger, 1734. Engraving from L. van Natrus, J. Polly and C. van Vuuren, *Groot volkomen Moolenboek*, part I. Amsterdam 1734. The illustration shows a bucket chain dredger driven by a horse capstan. It shows the marked ability of the Dutch millwrights to portray technical details correctly. The dredger was invented in Holland at the end of the 16th century. Since the beginning of the 17th century it has been known with capstan drive

16. Smelting copper at Falun in Sweden, 1781. Painting by P. Hilleström

neither art nor Nature can pass; it is here understood, of course, that the material is the same and the proportion preserved.[91]

In the second Dialogue of his *Discourses* Galileo includes a basic deduction concerning the tensile strength of a beam fixed at one end. He was here attempting to deal with technical problems quantitatively by a scientific method.

Let us imagine a solid prism ABCD fastened into a wall at the end AB, and supporting a weight E at the other end (Fig. 46): understand also that the wall is vertical and that the prism or cylinder is fastened at right angles to the wall. It is clear that, if the cylinder breaks, fracture will occur at the point B where the edge of the mortise acts as a fulcrum for the lever BC, to which the force is applied; the thickness of the solid BA is the other arm of the lever along which is located the resistance. This resistance opposes the separation of the part BD, lying outside the wall, from that portion lying inside. From the preceding, it follows that the magnitude [*momento*] of the force applied at C bears to the magnitude [*momento*] of the resistance, found in the thickness of the prism, i.e., in the attachment of the base BA to its contiguous parts, the same ratio which the length CB bears to half the length BA; if now we define absolute resistance to fracture as that offered to a longitudinal pull (in which case the stretching force acts in the same direction as that through which the body is moved), then it follows that the absolute resistance of the prism BD is to the breaking load placed at the end of the lever BC in the same ratio as the length BC is to the half of AB in the case of a prism, or the semidiameter in the case of a cylinder. This is our first proposition. Observe that in what has here been said the weight of the solid BD itself has been left out of consideration, or rather, the prism has been assumed to be devoid of weight. But if the weight of the prism is to be taken account of in conjunction with the weight E, we must add to the weight E one half that of the prism BD: so that if, for example, the latter weighs two pounds and the weight E is ten pounds we must treat the weight E as if it were eleven pounds.

SIMP. Why not twelve?

SALV. The weight E, my dear Simplicio, hanging at the extreme end C acts upon the lever BC with its full moment of ten pounds: so also would the solid BD if suspended as the same point exert its full moment of two pounds; but, as you know, this solid is uniformly

FIG. 46. *Ultimate strength of a beam.*
Woodcut from: Galileo, *Discorsi e dimostrazioni matematiche* (Leyden 1638)

distributed throughout its entire length, BC, so that the parts which lie near the end B are less effective than those more remote.

Accordingly if we strike a balance between the two, the weight of the entire prism may be considered as concentrated at its center of gravity which lies midway of the lever BC.[92]

Galileo's new scientific knowledge concerning the tensile strength of materials, which was fundamental for the establishment of a rational technique, served as a point of departure for the important researches of Coulomb in 1773 and of Napier in 1826.

From a letter written to Mersenne in 1638 by Descartes, critically discussing the passage of Galileo from which we have quoted above, we learn of the interest which technical problems held for this great philosopher.

Reverend Father 11 October 1638.

I will begin this letter with my remarks on Galileo's book. In general I consider that he philosophizes far better than the average, since he renounces as far as possible scholastic errors and endeavours to consider physical matters mathematically. Therein I entirely agree with him, and I believe there is no other method of discovering truth. . . .

He considers why large machines, though of the same shape and material as smaller ones, are weaker than they; and why a child is hurt less than a large man in falling and a cat less than a horse etc. There seems to me no difficulty in this, nor any reason to base on it a new science. For it is obvious that, in order that the force and resistance of a larger machine shall be in the correct relationship to that of a smaller machine of the same shape, the former must be of a harder and less breakable material since the volume and weight are greater; and there will be the same difference between a large and a small object of the same material as between two of the same size of which one is built of a much lighter material and is therefore firmer than the other. . . .

He compares the force necessary to break off a branch transversely with that which is needed to tear it from above downwards; and he asserts that in the transverse position, it is like a lever whose fulcrum lies in the midpoint of its thickness, which is by no means the case and he gives no proof of it. . . .

I am, my reverend Father,
Your very sincere and obedient servant
DESCARTES[93]

Descartes evolved in his *Discourse on Method* of 1637 the foundations of a rational mathematical-mechanical view of Nature. The scientific results gained by these methods were, however, of importance to him for the sake of their application for the control of Nature.

. . . Although my speculations greatly pleased me, I believed that others also had speculations, which perhaps pleased them more. But as soon as I had acquired some general notions concerning physics, and when, beginning to test them in diverse difficult particulars, I remarked whither they might lead, and how they differed from the principles in present use, I believed that I could not keep them so concealed without greatly sinning against the law which obliges us to procure as much as lieth in us the general good of all men, for they have shown me that it is possible to arrive at knowledge which is very useful in life, and that instead of the speculative philosophy which is taught in the schools, a practical philosophy may be found, by means of which, knowing the power and the action of fire, water, air, stars, heavens, and all the other bodies which environ us, as distinctly as we know the various trades and crafts of our artisans, we might in the same way be able to put them to all the uses to which they are proper, and thus make ourselves, as it were, masters and possessors of Nature. This is to be desired, not only for the invention of an infinitude of artifices which would allow us to enjoy without trouble the fruits of the earth and all its commodities, but principally for the conservation of health, which is without doubt the first good, and the foundation of all the other good things of this life.[94]

Descartes' rigorous method of scientific judgement was, like most seventeenth century scientific philosophy, influenced by Calvinism, distinguished by its consistent and rigorous manner of life and its insistence on useful activity, just as Cartesianism in turn took root particularly in Calvinist Holland. The general impulse of Calvinism at that period, as of Cartesianism, was to utility, to technical efficiency which would render mankind 'Lords and Possessors of Nature', a conception fundamental to their thought.

All too often there arose practical difficulties in the construction of machines useful for technical work and of serviceable apparatus and instruments for purposes of scientific research. The realisation of constructive ideas encountered many obstacles in unsuitable material and inadequate working facilities. Thus Guericke and his craftsmen laboured to fit a piston into a tightly fitting cylindrical vessel. And Pascal complained how hard it was to construct the reckoning machine that he had designed. We reproduce here the Letter of Dedication with which Pascal at the age of twenty-two presented his Reckoning machine in 1645 to the Chancellor Séguier (Plate 9).*

Monseigneur,

If the public finds some use for the invention I have just made for the execution of all sorts of arithmetical processes in a manner no less novel than convenient, they are more indebted to your Excellency

* Chancellor Pierre Séguier (1588–1672) one of the principal founders of the Académie Royale des Sciences.

than to my modest efforts; since I can boast only of having conceived it, while its actual construction is due entirely to the honour bestowed on me by your commands. For the delays and difficulties encountered in the usual methods suggested to me the desirability of a more speedy and easier aid to sustain me in the vast calculations with which I have been occupied for several years concerning several matters in connexion with the post with which you have been pleased to honour my father in the service of His Majesty in Upper Normandy. I employed in research for this purpose all the knowledge of mathematics acquired through my own taste and my early studies, and after profound meditation I realized that it was not impossible to discover such aid. Enlightenment from Geometry, Physics and Mechanics supplied the design and convinced me that it would be infallible in practical use if some workman could construct the instrument of which I had conceived the model. But at this point I encountered obstacles no smaller than those which I sought to escape and for which I was seeking a remedy. Since I had not the aptitude to handle metal and hammer as I did the pen and compasses, while the artisans understood better the practise of their craft than the sciences on which it is founded, I was fain to relinquish the whole enterprise, which yielded me only great exhaustion with no successful outcome. But, Monseigneur, Your Excellency renewed my exhausted courage, and graciously spoke of the simple sketch that my friends had presented to you in terms which made me see it in quite a new light, and with the renewed energy that your praise aroused in me, I was stimulated to fresh efforts and, ceasing from all other occupations, I thought only of the construction of this little machine which I have ventured, Monseigneur, to present to you, having adjusted it so that it performs by itself and with no mental labour the operations of every section of arithmetic in the manner that I had originally conceived.[95]

The reckoning machine of Pascal which performed addition and subtraction was followed in 1671–94 by the reckoning machine of Leibniz for multiplication. These discoveries of the seventeenth century sprang from the endeavour to relieve man, by dint of machinery, not only from physical labour but also from simple mental activity. The great electronic reckoning machines of today, with their manifold potentialities, are the successors of the simple apparatus of Pascal and of Leibniz.

The seventeenth century extended the ideas of the new mechanics also to the world of living beings. Santorre Santorio undertook in 1614 to estimate with scales the insensible perspiration, that is to say, the exhalation from human skin. Harvey, the discoverer of the circulation of the blood, endeavoured, by the application of dynamics to determine the motion of the blood as a process taking place quantitatively in time. The heart was compared to a hydraulic machine. Borelli devoted himself between 1680 and 1685 to the application of the principles of mechanics to the movement of organisms. The arm was regarded as a lever, the heart as a pump. In 1680 Borelli urged that a knowledge of Nature's 'devices' in various animal organisms could be of use in regard to problems of human technology. Here again we observe the seventeenth century conception of the technical application of scientific knowledge for general use.

If the question be posed whether a man can fly by his own strength, it must be considered whether the strength of the human breast muscles (which can be estimated by their thickness) is sufficiently great, so great as to exceed by ten thousandfold the weight of the whole man, including that of the extremely large wings attached to his arm. It is clear that the power of the human breast muscle is far too small for flight; for in birds the mass and the weight of the muscles causing the wings to beat amounts to not less than a sixth of the whole body weight. Therefore the human breast muscles would likewise need to weigh more than a sixth of his whole body weight in order that they could, by beating the wings attached to the arms, exercise a force ten thousand times that of the body weight. But that is by no means the case, as they do not weigh even a hundredth part of the body weight. For these reasons, either the power of the human muscles would have to be increased or the weight of the human body diminished in order to reach the same proportion as in birds.

So it follows that the invention of Icarus is entirely mythical, because impossible; for neither can human muscles be strengthened, nor the body weight be diminished; and whatever apparatus may be used, and even if it be possible to increase the moment of the power, nevertheless it will never be possible to move the bodies, for the power will never be equal to the resistance. Wherefore a wing set in motion by muscle contraction will never develop sufficient power to bear the heavy weight of a man.

So there remains only one possibility, namely to diminish the weight of the human body. Naturally this is not possible by any change in the body itself, but the desired result can be effected in a fashion similar to that by which a piece of lead can be made to float in water by fastening to it a piece of cork; the heavy lead will then float if the combined bulk of the lead and the cork is exactly equal to that of the water displaced by them, in accord with the law of

Archimedes. This method is used by Nature for fish; she places within them a bladder filled with air by aid of which the fish maintain their equilibrium, so that they can rest in a definite position in the water as though they were themselves a part of the water.

Recently some people have imagined that the weight of the human body could in the same fashion be kept hovering in the air; they would attach to a man a large vessel entirely or almost entirely empty of air, and so large that it can support both its own weight and that of the man hanging on to it.

It will be easily understood that this is a vain hope when it is recalled that the vessel which is to support its own weight and that of the man hanging to it would have to be constructed of copper or brass in order that the air could be completely pumped out from it. Its content would amount to more than 22,000 cubic feet, and its walls would need to be extraordinarily thin. But a vessel so large and so thin could not be constructed, and even if this were successfully accomplished, it would not be possible to maintain it uninjured. Above all, it would be impossible to exhaust the air with an air-pump, still less with mercury, since there is not a sufficient quantity of mercury in the whole world. And even if all these difficulties were overcome, a vessel with such thin walls could not withstand the pressure of the air, and it would either break in pieces or be completely crushed.[96]

The mobility achieved by the use of the apparatus described

No sensible person will deny that the works of Nature are in the highest degree simple, necessary and as economical as possible. Therefore machines devised by mankind will doubtless likewise attain most success if they are as far as possible modelled on works of Nature. If then we desire to rest or to move like a fish in water, we shall be successful only if we make use of the same mechanical devices as Nature has employed for fish. For this purpose we need only to carry the great pump R.S. with us and to bind it round our waist like a sword. The amount of air that it contains will be about 1 cubic foot. The little opening at S. must be closed and sealed and the piston T of the cylindrical part UX must be firmly closed at the side. As it can be pulled out or pushed into the pump tube by the handle Y, it is possible to make the air enclosed therein either denser or more rarefied, which exactly corresponds with the nature of air (Plate 10).

Further we will assume that the man AF with his clothes, girdle, head-covering BGHC, and the pump RS with the air enclosed is specifically lighter than the water; and that the upper part MG of the helmet is somewhat above the surface of the water. If a couple of pieces of lead are then attached, the whole mass of the swimming man will be increased and he will be nearly as heavy as the water, so that ultimately only a very small part G of the helmet will emerge. The piston T is now pushed inward to S, so that the air in the pump is compressed; and soon the space in front, TR which was formerly filled with air, is occupied by water. As a result, the space occupied by the man and his pump is smaller than before, the specific weight will become greater and finally become equal to the specific weight of the water. The man will then be able to remain at rest anywhere in the water. If, however, the piston T further compresses the air in the pump RS, and forces more water into it, then the specific weight becomes greater, and the man falls slowly to the bottom. If on the other hand the piston T is pushed towards E, the (enclosed) air owing to its elasticity expands, the water will be forced out of the space TR, the man becomes lighter than the water, and he rises again until a part of the helmet can be seen above the water.

It is now no longer necessary to explain how a man can creep along the bottom like a crab or how he can by paddling with hands and feet swim like a frog in water.

The submersible boat: its make and its use

After we have shown that the pioneer man can live in a room closed on all sides, if the air contained in it which he breathes is renewed, it will then not seem difficult to us to build a boat completely enclosed which can rest immobile under water (Plate 10). And we are also able if we desire to move it downwards, upwards or sideways. The apparatus will be like that described above for immersion, so that the weight of the water displaced by it may be equal, greater or less than its own, so that it can accordingly either rest like a fish under water, or sink, or rise to the surface. This condition is attained if holes N are pierced through the floor E.F. of the boat A.C.E.G., and goat-skin bags ON are so arranged within the boat that they can be fastened at their openings with small nails to the edges of the holes N, whereby the water from without can pour into the bags and fill them. It must of course be prevented from pouring through little spaces and splits within the boat itself. Under these conditions it is clear that the boat will occupy less space in the water if all the containers ON within the boat are filled with water. It will then be of greater specific weight than the water and will sink to the bottom like a stone. If, however, the water is then (perhaps by means of the pole PO) forced out of the bags again through the openings N, the boat will immediately occupy more space than before, and will become of the same weight as the water and will float in it. But as soon as it becomes yet lighter, it will rise.[97]

PHANTASY AND REALISM

In Baroque man, realism and exaggerated phantasy were often closely linked. Especially in the German Baroque mind there was a duality between a profoundly metaphysical turn of mind and tendencies attuned to considerations of economics and utility. So Johann Kunckel, well known for his work in the technology of glass, in his influential *Ars Vitraria Experimentalis* published in 1676, placed great emphasis on 'Experiment and Reason', and on the economic problem of producing the best possible result with the minimum means and expense. And yet in other places, Kunckel could not emancipate himself from fantasy and alchemical tendencies. Spiritually related to Kunckel were two German chemists and technologists, the elder Glauber and the slightly younger Johann Joachim Becher. Both were eager and unstable, and worked on technical and chemico-technological problems with a view to raising German economic power. Glauber, during the period of misery after the Thirty Years' War, in his (German) works *On Minerals* and *On Germany's Welfare*, eloquently endeavoured to introduce an extensive use of the native raw materials of Germany 'for the honour of God and for the service of the fatherland.' He desired also to make economic use of his highly-praised *salt of art* or *marvellous salt*, the *Glauber salt*, as well as of alchemical transmutation of metals.

If by God's grace we find a little piece, how from the imperfect metal buried all ready for us which can be obtained on all sides without trouble, to extract from it some useful gold and silver. We need have no fear that we shall be hindered or held back by water, or by bad air or by spectre or by other evil accident. But if we only have money we can obtain on all sides common lead, tin, copper and iron. What treasure could not Germany have retained at home during this long-drawn out abominable war had there only been men who understood this mining of metal. Have not our enemies taken from our country great ship-loads of lead and tin? Reflect how much time, labour and expense must have been needed to dig such metals out of the earth, and for what a small price they were then sold to foreign countries; and there is no end to such removals which occur only because we have no-one who understands how to work such metals. It is a veritable disgrace to us Germans that we who formerly surpassed other nations in honesty, faithfulness, bravery, understanding and skill, are now so careless and leave to others the advantage in such matters. But it is no wonder that such things happen when those who govern do not offer honourable natural scientists and experienced chemists their help or protection or encouragement for the good of the country. It should be understood how to distinguish such a one from dishonest and corrupt alchemists who wish to teach others how to make gold and yet neither understand nor know the least thing concerning the nature of metal.

Germany is marvellously gifted by God beyond other lands and kingdoms with all sorts of mines. It lacks only experienced persons who know how to use them properly, and then to find timber and sufficient of all other necessities to use with them. Why are we so bad at it that we must send our copper to France or Spain, and our lead to Holland and Venice, that Spanish green and lead-white may be made from them, which we subsequently have to buy back at a great price? Are not our timber, sand and ashes in Germany just as good for making crystal glass as those of Venice or France? And how many similar things can be better made in Germany than in other kingdoms, and yet are not put to use. Whereas we could sell our superfluity for money to other nations, we let it leave our own country in order to enrich others and to denude ourselves. How much of use Germany could attract to herself from neighbouring kingdoms if only she would do so and knew how to. . . . In fine, if God wishes to punish a land, he removes intelligent people from it, and if he wishes to bless it, he sends them to it. How did Venice erstwhile and recently Amsterdam in Holland become so great and powerful except by the experienced and intelligent persons whom they attracted, the good inventions and skilled manufactures that they thereby learnt and exported in great quantity by their navigation throughout the whole world for which they brought masses of gold and silver to their own fatherland? It is far better to sell to others than to be obliged to purchase from others. What do we lack in Germany that has been richly and overflowingly endowed by God and by Nature for every need, if only we understood or chose to understand? Gluttony and drinking are so widespread that even a person who has only a piece of bread over from one day for the next does not leave off but chases it down his gullet and wastes his time eating uselessly like a glutton. And no one whatsoever works in the least at good crafts, or strives toward virtue and wisdom, but all on the contrary waste their time in idleness and hate and persecute knowledge and crafts useful to the country. Therefore God with justified anger sendeth us ever one punishment and torment after another to warn us if no improvement should follow (which it is hard to see) that an even greater evil (from which God graciously preserve us) cannot long fail to come.[98]

Germany's Welfare

Just as every child is bound by God's solemn command to love his parents with all his heart, to fear them and in all just matters to be faithful unto death and obedient to them; so also and no less is every subject and native of a country deemed under obligation not only to be faithful and obedient in all just matters to his divinely ordained rightful chief and the father of his country, but also as far as possible to help to protect and guard him from his enemies, and to warn and preserve him from misfortune.

In this little tract, I will make known to all the world that I have undertaken to demonstrate throughout this year in my Public and my Private Laboratory to high and low, the most useful and important secrets of which I have treated in part openly and in part obscurely in the four published Parts of my *Welfare of the Fatherland*. Among these useful Secrets, the three most important are:

Firstly, how to prepare a good saltpetre from ordinary material to be found all around us.

Secondly, how with ease and in abundance by the use either of volatile or of fixed saltpetre, most usefully and without smelting to extract gold, silver and copper from the ore.

Thirdly it will be shown that Alchemy is veritable, and no dream, fantasy or delusion as hitherto falsely described by the great ignorant majority. But all common, ordinary metals and minerals, such as lead, iron, copper, tin, and also bismuth, cobalt, zinc, calamine, marcasite or pyrites and all other volatile common mineral ores can be rendered merely by fire and salt fixed and mature, so that with great utility and at little cost, much permanent gold and silver can be extracted from them.

These three pieces of information will in my opinion be worthy to be communicated with humility and will be agreeable to the exalted leaders, patrons and patriots of my dear fatherland.

But in order that the great and incredible usefulness to our fatherland of all these three may be fulfilled, I have thought it well to give a little information concerning them.

And firstly as regards saltpetre. It is sufficiently well known to all the world what a useful and very necessary material it is, and that there can never be a superfluity of it that cannot be applied and employed for a useful purpose; not to mention that it is indispensible for the preparation of gun-powder as a powerful weapon against the enemies of our country. Again, great treasures of gold and silver can frequently by its means be extracted from poor ores that are not worth the expense of melting down; the fatherland thereby is benefited since more gold and silver can never be extracted than can be of use to the fatherland. And even if gunpowder, gold and silver are not needed,—a condition which seems very remote,—yet it is impossible to do without good bread; and corn, wine and fruit-trees grow much more abundantly if the corn is soaked in saltpetre before sowing, and a very little is sprinkled on the roots of vines and of fruit trees. The fruits will then ripen much earlier and have a much pleasanter taste than if stinking cow-manure had been used. If then saltpetre, which is so splendid and indispensible, can be so easily and abundantly extracted from timber and stones, and the land thereby made more fruitful than by any other means, and if good gold and silver in quantity can thereby be extracted from common sand and stones, what more can be desired except a healthy body whereby to enjoy the good gifts of God in peace, rest and health? And such health and long life can be completely bestowed by saltpetre.

Regarding alchemy, or the transmutation of common metals into good gold and silver as here announced I declare it to be no idle dream or ill-founded opinion but a truth well tested by experiment, which I have undertaken to demonstrate carefully and in public this summer with God's help and support, for the benefit of the fatherland.

Thus while the fatherland is full on every side of iron, lead and salt besides other common minerals, yet no use is made nor anything constructed from them. But, if such smelting works are established in many places in Germany, it cannot fail that there shall and must be produced in a few years great treasure thereby.

But in order that I may open wider the eyes of the lover of truth, and show him how all this can happen and indeed be true, I inform him as follows that purified metals, iron, lead, copper and tin are not necessary for this smelting work, and they can be used otherwise. But the common natural antimony can be used or sulphuric lead ores which cannot usually be turned into good saleable lead, and such iron or copper ores as from their nature cannot be smelted; but no good saleable copper or iron need be smelted in this operation. I will show also how this work can be accomplished without lead, iron or copper, but can be achieved with volatile pitch and salt which are to be found in all earth; so that a lack of raw material need never be feared.

But I pray to God that He will grant me sufficient strength, understanding and health that I may be able publicly to demonstrate this summer true, unfalsified and useful alchemy, or transmutation of inferior metals into gold and silver, to the glory of God and the great benefit of the fatherland.[99]

Complete proof that our *Salt of the Art* could be useful and serviceable to all human beings, to great and small, to highly placed and lowly placed persons, as well as to all craftsmen and handworkers down to the humblest peasant.

Firstly it will be proved that if this salt were of use for naught except that corn should be soaked in it before sowing, so many blades would shoot up and there would be such increase of growth through its influence that it would be of use to mankind throughout the world, since no-one can live without bread.

If then the peasant gathers more abundant harvest with less toil and expense than by the usual familiar method, he can pay the annual rent due all the better from his superfluity, while exalted persons who will subsist and rule thereby will also greatly benefit.

And since all craftsmen and artists can attain their purposes more easily thereby, it will also be of use to all, including the great lords.

Theologians have matter that is a marvel of God to show to their audience in this recent century, and have matter and reasons to admonish them to abstain from sin and to enter on a better life.

Jurists and those learned in the law should, through the diverse, great and unheard of marvels performed by means of this salt have been able justly to note that God has not permitted this to happen by chance, but wishes to show us that a great change in the whole world must soon follow. Therefore they must execute good law and justice, and for the sake of the gold which yet will not save them, they must not pervert justice, make the crooked straight and the straight crooked to the great injury and prejudice of their neighbour and to the complete destruction of the salvation of their own souls.

Physicians have means thereby to obtain better medicaments and cause to show more diligence with their patients, nor to torture them for so long, and not to leave their fellow-Christians for so long struggling on the Cross for the sake of despicable perishable gold, since they could help them much sooner if they had good medicaments and did not for the sake of the pennies pay many unnecessary visits.

Also apothecaries and surgeons should understand that they have the means to use much more efficacious medicaments, and to prepare plasters and salves whereby their patients should the more quickly and happily be cured.

Every artisan or craftsman, if he can produce his work more easily by the help of this salt, will then in justice under Christ let his product be available at a lower price than before because he had formerly to achieve his work with more labour and expense.

This will be understood by all sellers and buyers, that is to say that they can buy their wares more cheaply and can sell them to others for a lower price.

And if the peasant, cultivator, gardener and similar person who plants necessary means of livelihood, can produce his harvest more easily, more quickly, more perfect, ripe and more abundantly than before, then he can sell them in corresponding greater abundance, can make a better sale or at a lower price and this way can give pleasure to everyone, young and old, great and small, rich and poor, high and low.

Thereby everyone can see what great use and help our *Salt of the Art* can bring to all mankind.

Who would not be so honourable and pious as to deal in this salt and to use it and apply if for himself and his pious neighbour?[100]

The Baroque was the period during which men with all sorts of projects journeyed from court to court, with plans in which there was often an unbridgeable gulf between the idea and practical technical feasibility. In the realm of mechanical technology, and corresponding to the efforts of the alchemists of the day, there appeared plans of mechanisms often far-fetched among which was not lacking the *Perpetuum Mobile* with its ceaseless movement. Böckler, in 1661, gave a careful description of a *Perpetuum Mobile* with water wheel, water elevating screw and grinding-wheels from a figure which had first appeared in 1629 (Plate 10).

The object of inventing this device is to achieve . . . perpetual motion. For the inventor causes the water to flow from the container A through B on to the water wheel C on the axle of which is set up an endless screw D by means of which the worm-wheel E with its axle engages the wheel F attached to it; furthermore this latter rotates the pinion G by meshing with its pins and thereby it rotates the shaft H; since the upper pinion I causes the cog-wheel L with its lateral cogs and axle-tree M to rotate regularly so that ultimately the cog-wheel R engages by its cogs the pinion S on the elevating screw which it rotates and the water, which flowed down at B, is led upwards once again through Q. In order that something useful may be produced by this invention or machine, the inventor has attached a pair of polishing wheels to the axle.

For this apparatus, it must be known whether or not enough water is available . . . a question to which every master workman is urged to give careful consideration.[101]

TECHNOLOGY AND THE FREE CHURCHES

We have already emphasized (p. 171) that in the seventeenth century the countries of the Reformed churches and sects, in which the teaching of Calvin was influential, took a considerable part in the rise of experimental natural sciences and their application for useful purposes, and partook also in the general economic improvement. It was less the theses of learned theology than the propositions of practical ethics which in the later period of Calvinism, that is to say precisely in the seventeenth century, exercised stimulating influence on the activity in scientific, technological and economic problems. Max Weber in 1904–5 was the first to demonstrate the connexion between Calvinist ethics and the development of economics. Troeltsch, Cunningham, Tawney, Müller-Armack and others have devoted further research to these matters. Besides Lecerf, Merton has forcefully revealed to what a great extent Calvinist ethics promoted scientific research and technical achievement.*

Calvin in his teaching developed to its extreme logical conclusion the doctrine of Predestination. Between God and the World is a chasm which only God can bridge. Man is predestined by God to Salvation or to Damnation. God's decree can be influenced neither by good works, nor by Faith, nor by a Church's intervention for grace. And for the man whose anxious enquiry as to the salvation of his soul could not be answered, there remained only strict obedience to the commands of God and ceaseless strenuous work in this world. Later Calvinism indeed recognized in successful work in this world an outward sign of inward grace. But successful labour was understood as the tireless production of useful works for the welfare of mankind and the glory of God. This is the 'inner asceticism' (Weber) of Calvinist puritanism. This attitude inevitably created a favourite attitude towards scientific research and technological activity. Moreover Calvinism repudiated the Platonic Ideas. That meant that the way was free for natural scientists to experiment (cf. p. 19). In 1663, as Merton has shown,[20] of the sixty-eight members of the Royal Society which laid such emphasis on experimental research, no fewer than forty-two were Puritans, including Samuel Hartlib, Sir William Petty,

* See bibliography.

Robert Boyle, and Thomas Sydenham. And especially in the English mines and foundries, and later in the textile industry, Puritans played a predominant part either as entrepreneurs or as creative technicians.

Concerning scientific research and its application, the Reverend John Wilkins said:

> Our best and most divine knowledge is based on action, and those may be justly counted barren studies, which do not lead to utility as their sole aim.[102]

The study of nature served, as was emphasized by the Puritan Boyle, for the glory of God and the welfare of mankind.

> Perhaps it will be no great venture to suppose, that at least in the creating of the sublunary world, and the more conspicuous stars, two of God's principal ends were the manifestation of his own glory and the good of men . . . it will perhaps not be difficult for you, Pyrophilus, to discern, that those, who labour to deter men from sedulous enquiries into nature, do (though, I grant, designlessly) take a course which tends to defeat God of both those mentioned ends.[103]

In his Testament to the members of the Royal Society, the same view is expressed.

> To the Royal and learned Society for the advancement of experimental knowledge. . . . Wishing them a most happy success in their laudable attempts to discover the true nature of the works of God, and praying that they and all other searchers into physical truths may cordially refer their attainments to the glory of the Author of Nature and the benefit of mankind.[104]

In Richard Baxter's *Christian Directory*, written 1664–5, is a clear description of the practical ethics of Puritanism with its insistence on a strenuous life and tireless labour; this assisted technological production to the greatest degree.

Though God *need* none of our *works*, yet that which is *good materially pleaseth him*, as it endeth to *his glory*, and to our *own* and *others benefit*, which he delighteth in.

Question. *But may I not cast off the world, that I may only think of my salvation?*

Answ. You may cast off all such excess of worldly cares or business as unnecessarily hinder you in spiritual things: But you may not cast off all *bodily employment and mental labour* in which you may serve the common Good. Every one that is a member of Church or Common-wealth must employ their parts to the utmost for the good of the Church and Common-wealth: *Public service* is God's *greatest service*. To neglect this, and say, I will pray and meditate, is as if your servant should refuse your *greatest work*, and tie himself to some lesser easie part: And God hath commanded you some way or other to labour for your daily bread, and not to live as drones on the sweat of others only. . . .

Question. *Must every man do his best to cast off all worldly and external labours, and to retire himself to a contemplative life as the most excellent?*

Answ. No: No man should do so without a *special necessity* or *call*: For there are general precepts on all that are able, that we live to the benefit of others, and prefer the common good, and *as we have opportunity do good to all men*, and love our neighbours as ourselves, and do as we would be done by (which will put us upon much action) and that we labour before we eat. And for a man unnecessarily to cast off all the service of his life, in which he may be profitable to others, is a burying or hiding his Masters talents, and a neglect of charity, and a sinning greatly against the Law of Love. As we have *Bodies*, so they must have their *work*, as well as our *souls*. . . .

Take heed of Idleness, and be wholly taken up in diligent business, of our lawful callings, when you are not exercised in the more immediate service of God. . . .

Labour hard in your Callings that your sleep may be sweet while you are in it: . . .

Weary your bodies in your daily labours: . . .

See that thou have a Calling which will find thee employment for all thy time which God's immediate service spareth. . . .

Quest. 1. *Is Labour necessary to all? Or to whom if not to all?*

Answ. It is necessary (as a *duty*) to all that are *Able* to perform it: But to the unable it is not necessary: As to Infants, and sick persons, or distracted persons that cannot do it, or to prisoners, or any that are restrained or hindered unavoidably by others, or to people that are disabled by age, or by anything that maketh it naturally impossible.

Quest. 2. *What labour is it that is necessary?*

Answ. Some labour that shall employ the faculties of the soul and body, and be profitable (as far as may be) to others and ourselves. But the *same kind* of *Labour* is not necessary for all.

Naturally action is the end of all our *Powers*; and the *Power* were *vain*, but in respect to the *act*.

It is for *Action* that God *maintaineth* us and our *abilities*: *work* is the *moral* as well as the *natural* End of *power*. It is the act by the *power* that is commanded us.

It is *Action* that God is most served and honoured by: not so much by our being able to do good, but by our *doing* it: who will keep a servant who is *able* to work, and will not? Will his *mere ability* answer your expectation?

The *public welfare*, or the good of many, is to be valued above our own. Every man therefore is bound to do all the good he can to others, especially for the Church and Commonwealth: And this is not done by *Idleness*, but by Labour! As the *Bees labour* to replenish their hive, so man being a sociable creature, must labour for the good of the society which he belongs to, in which his own is contained as a part. . . .

. . . Though it is said, Prov. 23:4. *Labour not to be rich*: the meaning is, that you make not Riches your chief end: Riches for our fleshly ends must not ultimately be intended or sought. But in subordination to higher things they may: That is, you may labour in the manner as tendeth most to your success and lawful gain: You are bound to improve all your Master's Talents: But then your *end* must be, that you may be the better provided to do God service, and may do the more good with what you have. If God show you a way in which you may lawfully get more than in another way, (without wrong to your soul, or to any other) if you refuse this, and choose the less gainful way, you cross one of the ends of your Calling, and you refuse to be God's Steward, and to accept his gifts, and use them for him when he requireth it: You may labour to be *Rich for God*, though not for the *flesh* and sin.[105]

Be laborious and diligent in your callings; Both precept and necessity call you unto this; and if you cheerfully serve him in the labour of your hands, with a heavenly and obedient mind, it will be as acceptable to him, as if you had spent all that time in more spiritual exercises; . . .[106]

Spend much more time in *doing your duty*, than in *trying your estate*: Be not *so much* in asking, What shall I do to be saved? . . . Give up your selves to a Holy Heavenly life, and do all the good that you are able in the World: Seek after God as revealed in and by our Redeemer: And in thus doing, 1. Grace will become more notable and discernable. 2. Conscience will be less accusing and condemning, and will easilier believe the reconciledness of God. 3. You may be sure that such labour shall never be lost; and in well doing you may trust your souls with God. 4. Those that are not able in an argumentative way to try their state to any full satisfaction, may get that comfort by *feeling* and *experience*, which others get by ratiocination: For the very exercise of Love to God and man, and of a Heavenly mind, and holy life, hath a sensible pleasure in itself, and delighteth the person who is so employed: As if a man were to take the comfort of his Learning or Wisdom, one way is by the discerning his learning and wisdom, and thence inferring his own felicity: But another way is by *exercising* that learning and wisdom which he hath, in reading and meditating on some excellent Books, and making discoveries of some mysterious excellencies in Arts and Sciences: which delight him more by the very acting, than a bare conclusion of his own Learning in the general, would do: What delight had the inventors of the Sea Chart and Magnetic Attraction, and of Printing, and of Guns, in their Inventions? What pleasure had Galileo in his Telescopes in finding out the inequalities and shady parts of the Moon, the *Medicean* Planets, the adjuncts of *Saturn*, the changes of *Venus*, the Stars of the *Via laciea*, etc. . . .[107]

The Puritan must use his time to the utmost. God calls us to work. Baxter repeats this in ever new words.

To *Redeem Time* is to see that we cast none of it away in vain, but use every minute of it as a most precious thing, and spend it wholly in the way of duty. . . .

Time must be *Redeemed* especially for works of *Public benefit*: For the Church and State: . . .

It is *God* that calleth thee to Labour: And wilt thou stand still or be doing other things, when God expecteth duty from thee?

O where are the brains of those men, and of what metal are their hardened hearts made, that can idle and play away that Time, that little Time, that only Time, which is given them for the everlasting saving of their souls! . . .

Thou wilt not wish at death that thou hadst *never laboured* in thy lawful calling. . . . But thou wilt wish then, if thou understand thy self, that thou hadst never lost one minute's Time, and never known those sinful vanities and temptations which did occasion sin. O spend thy Time as thou wouldst review it!

Consider also how unrecoverable Time is when it's past. Take it now or it's lost for ever. All the men on earth, with all *their power*, and all their *wit*, are not able to recall one minute that is gone. . . .[108]

In this Puritanism, the emphasis is not merely 'Use your time well'. The demand is rather 'Use each Minute.' Everyone became conscious of the passage of each little minute. The public clock with minute-hands is the outward expression of this point of view. The command 'Do not waste a single minute' was at this time still rooted in seventeenth century religion. It was for this very reason that it brought so many forces to activity. In the eighteenth century the religious link was lost. There remained only the cold 'Time is money' (cf. p. 252).

Attention is also drawn to the connection between Calvinism and the Genevan clock industry and the production of clocks and of scientific instruments in seventeenth and eighteenth century England.

The reformed churches and sects of England, which we have grouped together under the general name of Puritans, were not included in the Anglican State church. The adherents of these Free Churches could often not hold public offices. For this reason they were forced into the realm of economic and industrial activity, where they were able, with their special ethics of work to develop extensive activity.

Among the sects most economically active in the seventeenth and eighteenth centuries were the Quakers, who after much persecution were tolerated in England from the year 1689. Quaker families took a great part in the development of English mining and foundry establishments (cf. p. 252). These men were distinguished by a democratic cast of mind and by the urge to ceaseless practical work. Their strictly regulated manner of life was simple. It is remarkable that in the *Apology for True Christian Theology* of the Quaker Barclay, researches in geometry are stressed as an occupation during intervals for recreation.

. . . Seeing there are other innocent divertisements which may sufficiently serve for relaxation of the mind, such as for friends to visit one another; to hear or read history; to speak soberly of the present or past transactions; to follow after gardening; to use geometrical and mathematical experiments, and such other things of this nature. In all which things we are not so to forget God, in whom we both live, and are moved.[109]

Their manifold and successful activity in science and technology inspired the Puritans with a strong belief in Progress. However—and this is the difference from the secularized Idea of Progress of later times—in the seventeenth century the prevalent belief, as Edwards testified in 1694, was not only in technical progress, but also still in a future perfection of men's morality and Christian attitude.

Diligent Researches at home, and Travel into remote Countries have produced new Observations and Remarks, unheard-of Discoveries and Inventions. Thus we surpass all the times that have been before us; and it is highly probable that those that succeed will far surpass all other epochs. . . .

. . . And why may there not be expected a proportionable Improvement in Divine Knowledge, and in Moral and Christian Endowments? . . . Can there be any Reason given why God should not prosper Religion as well as Arts?[110]

GREAT TECHNICAL UNDERTAKINGS

Machine technology in the seventeenth century was still operated, as we have shown, by very simple traditional methods. Especially large and difficult problems such as were sometimes posed at the courts in the Baroque period, were solved by the combination of many simple forces in an enterprise supervised by a single will. At the very beginning of the Baroque, we encounter a remarkable technical achievement which, however, involved admirable organisation rather than advanced technology. The vast obelisk, nearly 75 feet 6 inches in height and weighing 327 tons which had been brought in about A.D. 40 under the Emperor Caligula from Helipolis to Rome was, by order of Pope Sixtus V, in 1586 moved by the outstanding engineer and architect Domenico Fontana from the square behind St Peter's to the great approach leading up to the Church. Only simple capstans with block and pulley were used together with the application of human and animal power. About 800 men and 140 horses were used for the erection of the colossus (Plate 11). Everything had to be well organized, and no-one was allowed to leave his place while the work was in progress, under pain of severest punishment. Fontana, the leader of the whole undertaking, in his planning by no means relied only on instinct; rather he had calculated the weight of the colossus, and the number of capstans needed to raise it. Correct static calculation could naturally not be achieved at that time since the science of statics was not developed until much later. Let us now hear Fontana himself on his great and much admired achievement which he carried through successfully by means of the well-organized co-operation of the power of many individual men and beasts.

We, Sixtus V, hereby confer on Domenico Fontana, architect to the Holy Apostolic Palace, in order that he may the more easily and quickly achieve the removal of the Vatican Obelisk to St. Peter's Square, full power and authority during this removal to make use of any and every craftsman and labourer as well as their tools, and if necessary to force them to lend or sell any of them to him, for which he will duly satisfy them with a suitable reward. Moreover he may use all planks, beams and timbers of any sort whatever that are found in suitable places for these purposes to whomsoever they may belong; but he shall nevertheless pay the owners a suitable price valued by two arbiters who shall be chosen by the parties. Moreover he may fell and lop any trees which do in any manner belong to the church of St. Peter, to the Chapter or to its Canons, especially also such as are possessed by the Cemetery, the Hospital of San Sprito in Sassia or the Apostolic Chamber, without any compensation. He can transport these through any place and can pasture therein the animals that he needs for this work without suffering any molestation thereby, but he must pay compensation which will be estimated by experts who will be chosen for this purpose. He may purchase and carry away the afore-mentioned and any necessary objects from anybody without paying duty or tax. Without license or other document, he may take all sorts of nourishment for his own use and for his servants and animals from Rome or from other towns and neighbouring districts. He may take and carry off capstans, ropes and pulleys wherever he finds them, even though they may become broken; nevertheless he must promise to repair them and to bring them back intact and he must pay a just hire for them. Similarly he may use all instruments and objects belonging to the building of St. Peter's and may give orders to the servants and officials thereof that within an appropriate space of time they shall make a clear space on the Square around the obelisk and shall prepare everything necessary for this purpose. In case of need, he may demolish the houses next to the obelisk though the form of compensation to be paid must be firmly settled beforehand.

In short, we give to the here-named Domenico Fontana full authority to do, arrange and demand everything else that may be required for the afore-mentioned purpose; furthermore, he, his agents, servants and household staff may everywhere and at every time bear every sort of arms except those which are forbidden. And we command all magistrates and officials of all the Papal States that in all the aforesaid matters they shall afford help and support to the said Domenico Fontana. All others, however who are in any respect subjects of the Apostolic See, whatever be their rank and station, we command, under pain of our displeasure and a fine of 500 Ducats or more as we may determine, that they shall not dare to obstruct this work or in any wise to molest the aforesaid Domenico, his Agents or his workers, but without delay or any excuse, shall assist, obey, support and aid him.

Given at Rome, at St. Mark's on the 5th October, 1585. . . .

Determination of the Weight of the Obelisk

Before I prepared for the undertaking of the removal, I wished to ascertain the weight of this obelisk that is nearly 75 feet 6 inches high. I therefore caused to be hewn out a palmo [8·7 inches] in a cubical block of the same stone; I found that this block weighed 86 pounds . . . I deduced that the obelisk weighed 963,537 $\frac{35}{48}$ pounds. . . .

I reflected accordingly that a capstan with good ropes and pulleys will raise about 20,000 lb. and that therefore 40 capstans would raise 800,000 lb. For the remainder (of 163,537 lb.) I proposed to use five levers of strong timber, each 42·65 feet long; so that I should have not only a sufficiency but an excess of power. Moreover, according to my dispositions, more light machines could always be added, if the first should prove inadequate.

When my invention was published, it transpired that nearly all the experts doubted whether so many capstans could be brought into co-operation so that they could work with combined force in order to raise so great a weight. They said that the capstans could not all pull evenly, that the one with greatest load would break, thus causing confusion that would put the whole machinery out of gear. I, however, though I had never combined so many different sources of power, nor seen anything similar, nor could be certain by means of any comparison with another, nevertheless felt certain that I could do it; because I knew that four horses, harnessed to one of those ropes as I had arranged, however hard they might strain, they would never be able to break it; but that if one capstan had too great a proportion of the load to bear, it would no longer be able to haul, nor would it be able, as alleged, to break the rope; the other capstans would meanwhile be turned, until each had again taken its rightful share of the burden. Then the first, which had been too heavily loaded, would also begin again and the power of all would be combined. Furthermore in addition I had arranged that after every three or four revolutions, the capstans should be halted and that if the men then felt the ropes and found that one was too greatly strained, they should relax it. . . .

All these arrangements were not new to me and I avoided thereby all dangers and ensured that no rope would break. As it was necessary to build a wooden scaffolding and to create space to set up the aforesaid 40 capstans, it was clear that the Square in question was somewhat too narrow, and that it was necessary to demolish a few houses and to level the ground.

The Scaffolding to Raise the Obelisk

In order to set up the scaffolding, eight wooden pillars or posts were erected, four on one side and four on the opposite side of the obelisk, each about [1·08 m] 3·55 feet distant from the next. Each pillar consisted in section of four timbers, each of which was [490 cm] 19·3 inches thick, so that each pillar was nearly one metre thick. The timbers were attached in such a manner that the mortise-joints should not coincide. They were clamped together at several points by iron bolts and straps, so that they could easily be taken apart again. . . . Around these eight pillars were installed 48 struts. . . . This scaffolding was so firm that the largest building could have been erected on it. But at its summit it was further held by four guy-ropes that rose obliquely from the ground to which they were anchored, and which were tightened by tackles. On the bearers (above on the pillars) were laid five strong timbers, each 21·3 feet long and more than 25½ inches thick, in each direction. Upon which 40 tackles were hung between bearers. These were operated by 40 capstans. . . .

Then the obelisk was covered by double rush-mats, in order that it should not be injured. Above these were laid planks, 2 inches thick. Above these, on either side of the obelisk three iron bars 4¼ inches wide and half that amount in thickness whose lower ends were bent under the obelisk, since it stood on bronze blocks. These iron rods were over two-thirds as high as the obelisk, and were constructed of several pieces attached to one another by hinged joints. They were encircled by nine hoops of the same iron about equally spaced along the length. . . .

The iron work of this covering weighed 24,394·5 lb.; and the planks, ropes and pulleys weighed about the same, so that the obelisk with all this armature weighed over 767,000 lb. While the armature was being set up, the Square was levelled, the capstans installed and the pulleys attached to them. And in order that those who were entrusted with the supervision of the scaffolding might detect which of the capstans lagged behind or was too far ahead, I had a number marked on each capstan and likewise on its guide-rollers and tackles, so that when necessary a sign could be given from the top of the scaffolding as to which capstan should be slackened or tightened, so that the supervisors of the capstans could obey these orders at any moment without the least confusion. . . .

After all the capstans had been marked, each was worked in turn by three or four horses in order to balance the pulling power of the different horses; and after every three or four turns was revised, until they were all hauling equally. This object was achieved on 28 April 1586.*

As very many people crowded to watch so remarkable an enterprise, and to avoid disorder, the streets leading across the Square were barred, and an announcement was made that on the day fixed for the Obelisk to be raised, no-one except the workmen might pass the barriers. Any other person who forced his way in would be punished by death. Furthermore, under pain of severe penalty, no-one might delay the workers, nor speak, nor dispute, nor make any noise in order that the prompt execution of the orders of the officials should not be hindered. To ensure the immediate fulfilment of this decree, the Chief of the Sheriff's Officers should be stationed with his Corps within the enclosure, so that the utmost quiet prevailed among the crowd, partly on account of the novelty of the work and partly on account of the threatened punishments. . . .

On 30th April, two hours before daybreak, two Masses were celebrated in the Church of the Holy Ghost; in order that God, for whose glory and that of the Holy Cross this remarkable undertaking was to be carried out, should grant His grace and should permit it to succeed. And, that He might grant the prayers of all, the workers, foremen and carters engaged on this great work, having by my command been to Confession on the previous day, went all together to partake of the Communion. Also our Lord [the Pope] had on the previous day given me his blessing and advised what I should do. After all had received Communion and had heard the appropriate sermons, he stepped out of the church into the enclosure, and all the workers were ordered to their places. Each capstan had two overseers whose instructions stated that each time the signal of a trumpeter

* The original has 1585, which must be a misprint as the Papal Privilege runs from October 5, 1585.

was heard—whom I placed in a raised position, visible to all, the capstans would start work, and it was their duty to keep a sharp eye open that the work was rightly done; but when the sound was heard of a bell which was hung high up on the scaffolding, a halt had immediately to be called. Within the enclosure, at the end of the Square, stood the Chief of the carters with 20 strong horses in reserve and 20 men at their service. Moreover I had eight to ten reliable men scattered in the Square who walked around and took care that no disorder arose during the work. I had also instructed a detachment of 12 men to carry hither and thither as necessary reserve ropes, capstans and pulleys. These were kept in an elevated position in front of the store-house, whence they could at every signal or command carry out the orders given them, so that no capstan overseer needed to leave his place. But at each capstan I placed both men and horses to work it, that the men might the more intelligently follow the orders of the overseer, since horses alone sometimes either remain still or move too quickly. Under the scaffolding were placed twelve carpenters, who had continuously to drive wooden and iron wedges under the obelisk on the one hand to help to raise it and on the other hand continuously to support it, so that it should never hang free.

These carpenters wore iron helmets on their heads, to protect them in case anything fell from the scaffolding. I assigned 30 men to keep under observation the scaffolding, the capstans and the ropes. On the three levers to the west . . ., I set 35 men ready for service and at the two opposite levers 18 men with a little hand-worked capstan.

After a Paternoster and Ave Maria had been recited by all, I gave the sign to the trumpeter; and as soon as his signal sounded, the five levers and the 40 capstans with 907 men and 75 horses began their work. At the first movement, it seemed as though the earth was shaking, and the scaffolding cracked loudly, because all the wooden members were crushed together under the weight; and the obelisk which had leaned more than 17 inches towards the Choir of St. Peter's assumed a vertical position. . . .

The obelisk was then raised, 23¾ inches in twelve movements, which was sufficient to push in the skids and to take away the metal blocks on which the obelisk had stood. At this height it was held and strong wooden bearers and wooden and iron wedges were driven under the four corners of the obelisk. This happened at 10 p.m. of the same day and the signal was then given with a few mortars on the scaffolding, and the whole artillery expressed its joy with loud thunder. And according to the order, dinner was carried round in baskets to every capstan, that no-one should leave his post. . . .

The obelisk while being raised was, as already described, continuously underpinned by the carpenters with wedges, as though it stood on a pedestal. When this was completed, they proceeded to remove the blocks, of which only two were laid on the surface of the pedestal. Each weighed 440 pounds. One of them was immediately taken to His Holiness the Pope, who manifested great joy thereat. . . .

While the metal blocks were removed from the pedestal the skid was placed on the rollers. The skid was narrower than the foot of the obelisk so that it could be pushed between the supporting timbers under its corners.

The obelisk had now to be laid down which owing to the amount of the movement and the length of the stone was a harder task than the first. For this purpose the pulleys and ropes were differently arranged, so that the western side on which the obelisk was to repose on the skid should remain free. . . .

The accomplishment of these preliminary works required eight days; and on Wednesday the 7th of May 1586 in good time in the morning the whole prepartion was accomplished. At the foot of the obelisk the four pulleys were fastened, and the capstans to serve them stood behind the Sacristry on the West. These began to haul at an early hour in the morning and to draw toward them the foot of the obelisk which rested on a skid which ran on rollers, while the other firmly attached capstans slackened their ropes. . . .

When it was half way down it began of its own accord to slide backwards on the rollers; it was therefore no longer necessary to haul it in this direction, but on the contrary to attach a pulley in the opposite direction to the foot, in order to regulate it according to the desire of the foreman. At 10.0 p.m. it lay firmly held on the skid which had been drawn under it without anyone having been hurt. His Holiness heard of this with the greatest satisfaction and the whole people were so happy about it that the architect was led home with drums and flourish of trumpets. . . .

As the obelisk had to be transported from this point 300 ells to its new position and as it was found on levelling that the new position was 28 feet 6 inches lower than the square where it had formerly stood (that is to say at the same height as the surface of the old base) a level embankment was made (i.e. an embankment with horizontal summit) from the old to the new position; the earth for the purpose was taken from Monte Vaticano behind the buildings of St. Peter's. It was made 71 feet wide at the base, 35 feet 6 inches at the summit and 26 feet in height. Around the site of the scaffolding it was made 67 feet wider at the top, and at the foot about 89 feet wider. It was covered with timbers which were supported by posts and struts; and in many places timbers were laid across it, in order that it should at no point yield to the great pressure. . . .

While all this was being carried out, a layer of worked limestone was placed on the foundation which had already been prepared on the approach to St. Peter's and was to support the obelisk. . . . The pedestal, of bonded white marble blocks was then again placed in the centre. . . .

Raising and Setting of the Obelisk (Plate 11)

On the 10th September 1586, when all was in place, before daybreak, two Masses were celebrated in the Church of the Priory Palace, and as at the laying down of the obelisk everyone who took part in the work went to Communion and prayed to God for a successful outcome. Then every man was placed in position. By daybreak all was in order, and the work began of the 40 capstans, the 140 horses and the 800 men with the same trumpet and bell signals for work and for standstill as before. While the top of the obelisk rose up, its foot was pulled by four capstans, placed on the opposite side so that the ropes which raised the top remained always vertical. The weight to be raised became progressively less, the higher the top rose and the further under it the base was drawn. When the obelisk was half raised, work was stopped and the obelisk was shored up in order that the workers might have their midday meal. After the meal each one devoted himself again zealously to the work. The obelisk was set up in 52 stages, and in many respects it was a beautiful spectacle. Countless people had assembled, and many, in order not to lose their places for the show, remained without a meal until evening. Others made platforms for the people who were streaming in, and earned much money thereby. By sunset, the obelisk was upright; but the skid which had been drawn under it while it was raised, remained beneath it. Immediately the signal was given by mortars on the scaffolding and was answered by guns, and the whole town was filled with joy. All the Roman drummers and trumpeters again hastened to the architect's house and echoed their applause. When the happy news was proclaimed from the scaffolding, His Holiness was giving audience as he had come from Monte Cavallo to St. Peter's in order to receive the French ambassador in public Consistory. Here the news that the obelisk had been raised was brought to His Holiness, and gave him great joy.

The seven following days were occupied in resetting the capstans and fastening the pulleys to the four sides of the obelisk in order to be able to adjust it . . . on the day fixed for the removal of the skid, a beginning was made to work the capstans and pull down the levers so that the obelisk rose slightly; and immediately, its pedestal being broader than the skid, it was underpinned by the carpenters with wedges, so that the skid could be drawn away. Then the bronze blocks were set in their places, and those with tenons were set in lead. Then on the same day the capstans were again operated and the levers pulled downwards, and one wedge after another was struck away, letting down the obelisk gradually so that by the same evening it had come to rest on the blocks; but it was then too late to adjust it. On the following day it was set vertically. . . . Thereupon they proceeded to remove the scaffolding from the obelisk. On the 27th September it was free, and His Holiness commanded that a procession should be arranged to bless the obelisk and to dedicate the golden cross upon it.[111]

Among the most remarkable technical achievements of the seventeenth century were the great new methods of ship-building in Holland and England; and in France, besides extensive fortification work, Louis XIV's great canal du midi, nearly 210 miles in length, uniting the two seas, which Voltaire reckoned to be the most remarkable technical achievement of the period of the *Roi Soleil*, as well as a gigantic pumping installation at Marly which supplied water to the fountains in the gardens of Versailles.

The Marly installation (Plate 12) was built at enormous cost between 1681 and 1685 and was a typical achievement of unbounded princely power in the Baroque period. Fourteen large water wheels, each more than 39·38 feet in diameter, driven by the Seine, worked 221 pumps which raised the water by stages through cast-iron pipes altogether nearly 531·5 feet. Below worked 64 pumps to drive the water nearly 160 feet up to an intermediate reservoir. From this point it was raised a further 185 feet by 79 pumps to a second intermediate reservoir. From this point 78 pumps raised the water yet 187 feet higher. The pumps at the two elevated intermediate reservoirs were driven by the water-wheels in the Seine by means of well-constructed connecting rod systems, that is by the same apparatus for transmission of power that we have already observed in German mining in the sixteenth century (cf. p. 150). The output of this enormous establishment, whose cost of maintenance was exceptionally high, amounted to only 80 horsepower in terms of water actually raised. This corresponds to the output of the engine of a modern 3½ ton lorry. In his *Architecture hydraulique* of 1739, Bélidor described the installation.

Description of the Machines at Marly

It seems as though no machine has ever been built concerning which there has been so much report as that at Marly. It can well be placed among those works which were typical of the magnificence of Louis the Great. And in fact only this monarch could have forced such a river as the Seine out of its natural course in order to reach the summit of a mountain as high as that up which it now rises. Poets have caused their heroes to accomplish marvels by the help of their gods; but this great king, without recourse to fantasy, found in his treasury and in the skill of those who sought to add to the greatness of his glory everything which he needed to carry out his great plans. The position which he himself chose in Marly forest for a castle to be built may be regarded as one of the most beautiful in the world. A lovely situation and most charming countryside was endowed by Nature with everything that could be desired except only water. But how could one do without this in such a place which it was desired to decorate richly with everything that the imagination can conceive to have graced those magical places, so magnificently described for us by the Romans, to give the utmost in pleasure and gaiety? A less mighty Prince would have been diverted from his plan by such a difficulty but he wished to show that he could successfully complete the greatest undertaking. He immediately commanded that the most skilful people should be found in France and in foreign countries who would be attracted by the special benefits with which he rewarded good service and would compete for the honour of serving him.

Since in those times it was sufficient to possess some good ability and skill in order to gain the ear of the ministers, he found one, named Rennequin from Liége, a man with splendid natural aptitude for machinery who did not lack courage for so great an undertaking as to bring water in plenty to Marly and Versailles as though it flowed there from springs. The machinery that he built for this purpose started work in the year 1685. It is said that it cost more than 8 millions. . . . The machinery lies between Marly and a village on a branch of the Seine. At this point the river is held partly by the installation itself, partly by a weir or dam. In order, however, to avoid creating an obstacle to navigation, a canal was dug two miles above Marly, along which ships can pass. 30 or 35 fathoms above the machinery an ice grating has been placed in order to prevent blocks of ice or logs from being carried along by the stream and causing damage. The better to protect the sluices which converge on the water wheels, they are provided with a special grating or lattice of beams, which holds back anything which has come through the ice grating.

The engine consists of 14 water wheels which all combine to bring into action certain water pumps, whereby the water is forced up to the tower at the top of the mountain, where it is all collected, and led through several pipes to an aqueduct built on arches and from there it flows to the reservoirs which receive it. [112]

The Leipzig mechanician Jakob Leupold in his *Theatrum Machinarum hydraulicarum* of 1725 gives technical details of the pumping apparatus.

First it must be known that all . . . wheels are arranged in the same way; and that what one does, they all do and that whoever understands the structure and working power of one, understands also the others.

The whole engine is really a force pump, for the inventor understood well that water in a cylinder or pipe cannot be brought all at once to such a height, and that neither cylinder, valves nor pipes could possibly stand up to such work (just as in our mines we do not raise the water all at once in a single pipe or cylinder, but raise it through various intermediate stages); therefore each wheel has two cranks or offset pins; the one on the one side, drives six piston rods in six cylinders to pump the water of the Seine halfway up the mountain into a reservoir or water-tower. The cranks on the other side of the axle drive a system of overland connecting rods which run to the afore-mentioned tower and raise the water by pressure pumps; and from the tower it flows to the level desired. . . .

The most important thing to learn about these machines . . . is that the architect, has taken a cylinder of only 4 inches bore and given the piston 4 feet stroke; therefore he needed 6 cylinders for a single wheel. Many might think that if he had used at least one 6 inch cylinder, he would have needed only about three of them and thereby would have been able to save money and labour. . . . But if it is considered to what a height the water has to be forced through such cylinders, the reason will be easy to find and understand, that it has been planned with good forethought and also from necessity, for if the water was forced up 200 feet by such a piston, only four Leipzig inches in diameter—and the Parisian one is much bigger—then the load would be 8 cwt. which with all six cylinders would amount to 48 cwt. If a cylinder were used only 4 inches wider, that is to say 8 inches in diameter, that would result in a pressure of 32 cwt.; not to mention the result if it were desired to make a 12 inch cylinder which is by no means the largest known, since they have been found 13 to 16 inches in diameter.

Moreover it seems remarkable that he made the suction pipes fairly large in proportion to the cylinders; but in order that the water should not fall back, he has bent those pipes below . . . and it is known that this method serves as well as a valve.[113]

This vast and costly establishment at Marly, and its relatively small output shows us clearly the limits of the old power-machine technology. To seek a more efficient and reliable prime mover than the traditional wind and water wheels was in fact to become the urgent task of the technology of that period.

THE STRUGGLE FOR A NEW PRIME MOVER

Many heads were occupied from the end of the seventeenth century with the question of water supply, and especially with the removal of mine water which, as the shafts were driven deeper and deeper, could no longer be mastered by the engines of the period. In this connection we need name only Huygens, Leibniz, Papin, Savery and Newcomen.

Leibniz, not only a philosopher and mathematician, but also active in the realm of natural science and technology, endeavoured in 1681 in the Harz Mountains where there was not sufficient water-power, to raise the water out of mines with the help of a wind-driven engine which he constructed. Calvör describes it:

When a shaft is sunk, water at once collects in it and penetrates into all the passages, so that it is necessary to remove it if any work is to take place in the shaft; and not only are water-works required, but also deep adits are needed to carry away the water which has been raised. . . .

Many craftsmen have promised to provide for the mines such machines to deal with the floods, needing no water or less than that required for the devices hitherto used; but when it came to a test . . . none worked in a practical way and for all sorts of reasons no test at all was made of most of the suggestions. . . .

In the spring of 1678, the Privy Councillor, Minister for mines and also tithe-collector of Clausthal, Peter Harzingk suggested that water could be forced out of the mines by wind-mills, in order to economize water by use of the wind, and by such an alternation to keep the devices constantly in action. He exhibited a model that he had made for the purpose. . . . So in August 1679 Duke John Frederick resolved to build a wind-mill, and at the same time informed the Mining Office that the inventor was His Highness, Councillor Godfrey William Leibniz who for that purpose would come to Clausthal. We may deduce that the model had been anonymously sent to the Court and Mining Councillor Harzingk himself, in order first to hear the opinion of the Mining Office concerning it. . . .

The world famous Leibniz stated in Hanover that it would be possible to remedy the flooding in the mines by a combination of wind and water power; so that a notable additional quantity of ore would thereby be produced with great advantage to the mine. For that purpose he was prepared at his own expense to erect a windmill at a mine chosen as serviceable for adequate proof of the usefulness of his invention, and to test it for a year; in evaluating its results one could presume that the same would apply for use in other mines whether old or new, deep or shallow. . . .

The Councillor therefore proceeded to build the test windmill device at the Catharina in order to raise the water directly from the mine. . . .

When the Councillor's carpenters, whose master was a miller named Hans Linse, started the completed wind-operated device for the first time in a strong wind in 1681, the shutters which had been made in the sails to open in a strong wind fell out; and it was necessary to brake the wings by holding them back with chains; from which, as well as for other reasons, it was thought that the devices were not based on a specific completed model, but that the inventor was still daily seeking by all sorts of changes to produce what he had hoped and promised. When the damage had been repaired this apparatus made a good impression at Martinmas 1682, when it worked for an hour and completely raised eleven consignments. Then a beam broke to which the control rods were attached; because at that time there was not yet a staple attached to it, as was the case later.

When this wind device was brought back into use by the insertion of a staple it worked once more, but a wind storm . . . in 1683 twisted the whole roof, and the horizontal shaft together with all the wings were shattered and finally thrown off. After this all was made afresh and the device worked again when there was sufficient wind. . . .

From the above . . . it may be seen that the device was of some use in a good wind; thus once during 2 or 3 consecutive days it raised 14 consignments to the House of Israel water course but did not raise them to the adits. In a light wind it remained motionless, and in a strong wind a wing or connecting rod or something on the large spur wheel was often broken. . . .

Above all, it had demonstrated that a wind device could raise water, though not so great a quantity as a water device; but that they were not yet able to make it so reliable that it would not break down too often when raising a heavy burden; perhaps this was also partly to be blamed on the fact that its construction had not been carried out under the supervision of mining officials, and that an engineer had not arranged the parts in the proper proportion and order; and my Lord the Councillor often complained of the inaccurate work of his often refractory workers, and that people constantly broke into the upper part of the windmills by force, took away things that had been provided, turned the sails, and did much damage. . . .

It is therefore most astonishing that this great man did not weary of this engineering work which had cost him so much money, time, fatigue, travel, correspondence and controversy; but in spite of the difficulties he had encountered, he continued to propose new machines, as I have in part described. . . . Despite his deep insight, he was nevertheless ignorant, at least at first, of the Harz mines, of the condition of the shafts, the machines, and of the work they had to perform, as well as of the difficulties to be overcome; and he had to rely on other people, who were often refractory and unreliable in their work; for he was not permanently in the Harz, and could not be present when hindrances occurred in course of the work.[114]

Leibniz described as follows the design for a machine using wind power with variable gear by means of a conical chain wheel.

Raising Water by Means of Wind Power

Windmills enabling water to be brought up from deep mines have this drawback, that in a strong wind the gear is driven too fast so that a part is easily broken; while in a light wind there is not sufficient power and only long strokes are wanted wherein the connecting rod hangs more or less close to the centre shaft of the wing, whereby the stroke is made smaller or larger; then the piston in the cylinder of the pumps goes all too slowly and loses the water again. And in order to remedy this I have at last devised this method which is in my opinion the most perfect that has been proposed (Fig. 47).

A. is the wing of the windmill. An iron chain E.F.G.H.L.E. runs on the axle, that is to say it is led from the axle E. beneath the roller F. and thence on to the round frame G.H. from there back onto the roller L. and then comes again to E. The frame consists of a wheel Q. and about 10 ferrules from which beech rods pass upwards together towards H.G. and there are fastened to the vertical shaft. If the chain is placed higher or lower around the frame as at P. the vertical shaft revolves quickly or slowly and for that purpose countershaft M.N. must be moveable so that the chain can be tightened or slackened.

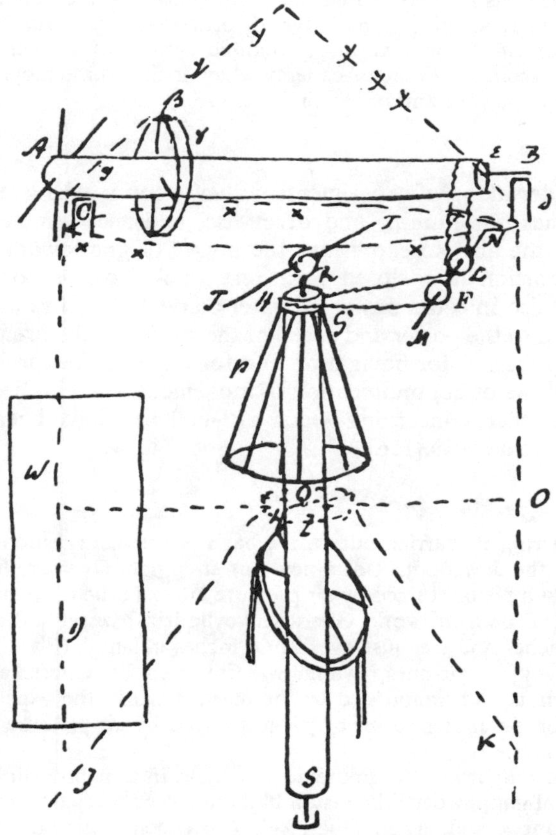

FIG. 47. *Leibniz's design for a windmill, to raise water from mine shafts,* 1678. Freehand sketch by Leibniz.—From: Leibniz, *Nachgelassene Schriften physikalischen, mechanischen und technischen Inhalts*

On the vertical shaft R.S. is a sloping collar or ellipse 1.2, which alternately raises and depresses two pistons 1.3 and 2.4, during rotation.[115]

The efforts of Leibniz by the above described installations to control the floods in the mines were, as we have heard unsuccessful. But in France a way had already been found which soon led to a first primitive engine in which the elementary power of animal or human muscle or of wind or falling water was no longer used but was replaced by a different method.

As early as 1666 Huygens had urgently proposed to the Minister Colbert the founder of the mercantilist economic system of France, practical suggestions from the French Academy which included attempts to use the power of gunpowder and of steam.

The undertaking of experiments with a vacuum exhausted by help of machines (air pumps) and otherwise, and ascertainment of the weight of the air. Research into the power of gunpowder of which a small portion is enclosed in a very thick iron or copper case. Research also into the power of water converted by fire into steam. Research into the power and speed of the wind and the use which can be derived from it for navigation and for engines. Research into the force of blows or communication of movement when bodies encounter one another, concerning which as I believe I have been the first to give the true laws.[116]

The experiments carried out on the basis of these suggestions were the prelude to the development of a new power-engine. Guericke had shown in 1661 that a piston, forced by air pressure into an evacuated cylinder can be utilized to perform work. Guericke's cylindrical vessel had a diameter of 15·35 inches and was just over 21·64 inches in length (Plate 13); it was evacuated by an air-pump which was invented by Guericke. In 1664 Caspar Schott had announced to the learned world the experiments of Guericke on obtaining power by pistons forced by air pressure through a metal cylinder.

Huygens sought in 1673 to create a vacuum in a metal cylinder by the explosion of gunpowder. The piston had then to be pushed through by air pressure, and could, as in Guericke's experiment, furnish power. The successful experiments with the little gunpowder engines moved Huygens to ideas for the future concerning the application of power engines of this sort, once they were developed in the right way. He thought, with a true technical instinct, of the propulsion of vehicles, ships and aeroplanes. Let us hear Huygens himself:

The force of gunpowder has hitherto served only for violent action . . ., and although it has long been wished to restrain its excessive speed and force for the attainments of different aims, no one, so far as I know has up to now met with success. . . .

About three months ago there came into my mind the suggestions I will now make for this purpose; since that time I have worked on the discovery in order to perfect it, while I made various attempts and innumerable experiments of which the success has finally so satisfied me that I ultimately dared even when they had only been carried out in models to deduce that the affair would, on a large scale be equally successful or even better, for reasons which will be recognized once the machine has been explained (Fig. 48). The violent action of the powder is by this discovery restricted to a movement which limits itself as does that of a great weight. And not only can it serve all purposes to which weight is applied but also in most cases when man or animal power is needed, such that it could be applied to raise great stones for building, to erect obelisks, to raise water for fountains or to work mills to grind grain, when there is not sufficient space or facilities to use horses. And this motor has the advantage that it requires no expenditure on maintenance while it is not in use.

It can also be used as a very powerful projector of such a nature that it would be possible by this means to construct weapons which would discharge cannon balls, great arrows and bomb shells with perhaps, as great force as from cannons or mortars. Even according to my approximate estimation this motor saves a great part of the powder that is now used. And, unlike the artillery of today these engines would be easy to transport, because in this discovery lightness is combined with power.

This last characteristic is very important and by this means permits the discovery of new kinds of vehicles on land and water.

And although it may sound contradictory it seems not impossible to devise some vehicle to move through the air; for the great obstacle to the art of flight has hitherto been the difficulty of building very light machines which could develop really powerful motion. But I admit that it will still require a good deal of science and inventiveness to achieve such an undertaking.

It remains only to estimate how far the power of gunpowder can

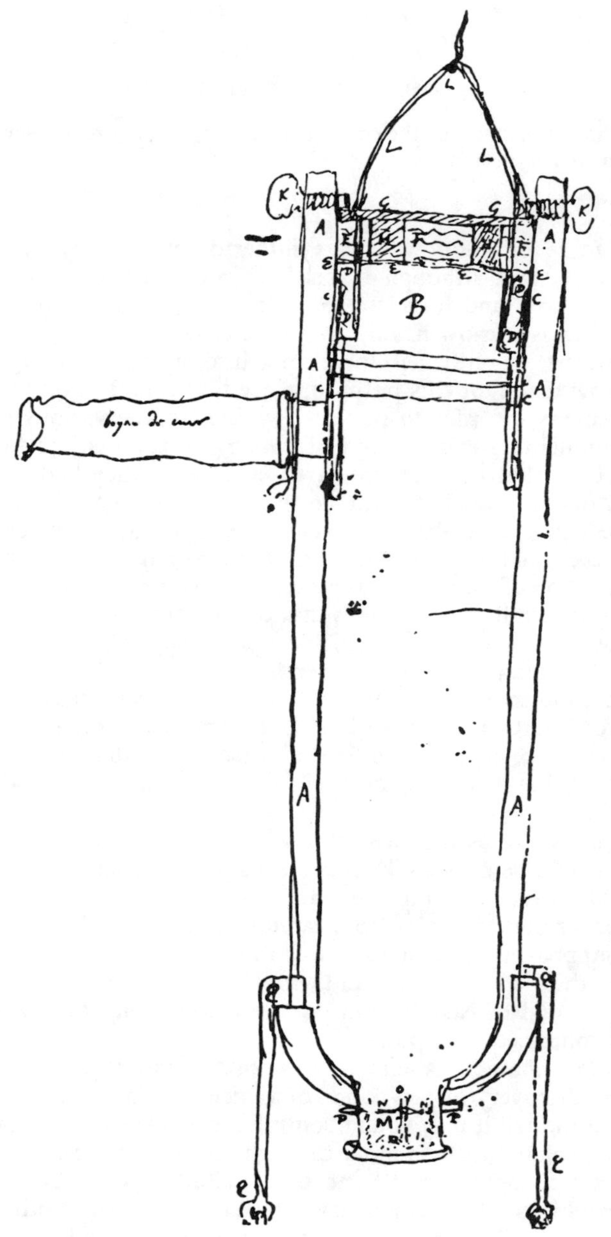

FIG. 48. *Gunpowder machine by Chr. Huygens*, 1673.
Freehand sketch by Huygens. AA: thickness of cylinder; B: hollow in piston; below: screw which holds the powder; top left: valve tube.—From: Huygens, *Œuvres complètes*

be increased by this discovery. I find, by calculation based on my experiments, that 1 lb. of powder can supply power to raise a weight of 3,000 lb. at least 30 feet; whereby it is possible to estimate the power of the new motor which I regard as greater than can be obtained from the usual use of gunpowder. . . . The discovery can serve all sorts of needs where it is required to combine great power with lightness, as in flying, which can no longer be rejected as impossible, although its realization will still require much more work.[117]

In a letter of September 22, 1673, to his brother Lodewijk, Huygens sketched with great clarity the method by which his machine worked.

A few days ago I let the gentlemen of our Academy and also Monsieur Colbert see a summary of a discovery which was considered very good and from which I should hope for great results, if I were sure that it would be as successful on a large as on a small scale. It concerns a new propulsive power by means of gun-powder and of air pressure. Here is the description of it (Fig. 49).

AB is a tube, well polished within and of unvarying bore, a piston D is in the upper part of the tube, and can move within it but cannot emerge from the top, because a stop is fastened there which prevents it. Within the base of the tube is screwed a little capsule for which leather is used to make it completely tight. At the points EE of the tube are openings, and hoses of moist leather EF are attached to these. Before the capsule C is fastened, a little gunpowder is placed in it with a tiny piece of tinder. After this has been lit at one end, the capsule is fastened. The fire then ignites the powder which flames right up through the tube, filling it and expelling the air through the hoses (or valves) EF. which are soon closed by the pressure of the outer air and pressed flat against the openings, which are furnished with gratings in order that the leather hoses shall not be drawn into the tube. Now while this tube (cylinder) remains in this manner empty or nearly empty, the air presses most powerfully on the piston D and forces it down through the tube, taking with it the rope DK and at the same time the weight G or anything else that is attached to it.

The strength of this force can easily be calculated by means of our knowledge of the amount of air pressure on a given surface. And if the tube is 1 foot in diameter, the pressure of the air on the piston is 1,800 lb. and correspondingly for other sizes of the surface. But that

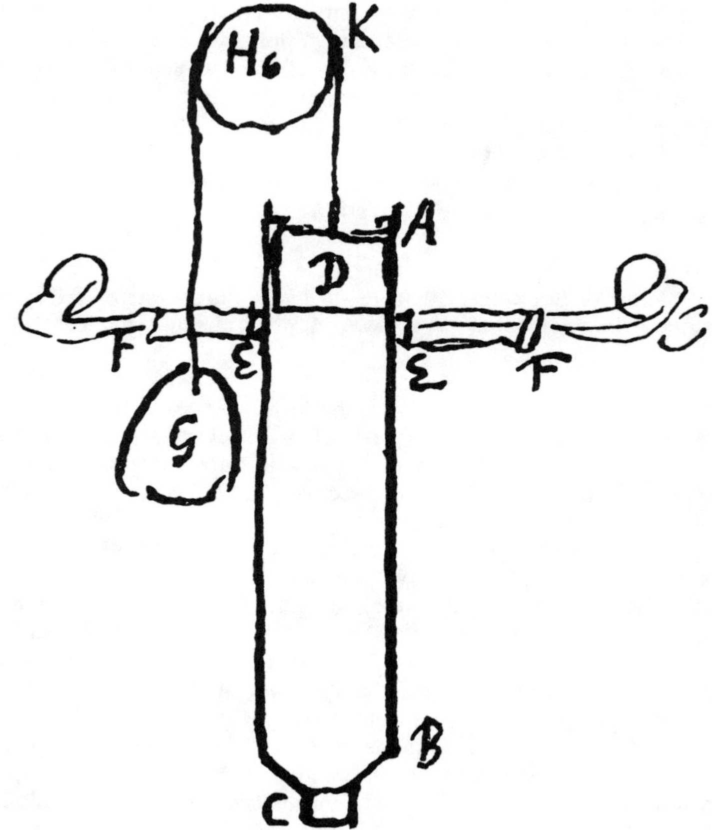

FIG. 49. *Gunpowder machine by Chr. Huygens*, 1673.
Freehand sketch by Huygens in a letter to his brother Lodewijk dated 22.9.1673.
From: Huygens, *Œuvres complètes*

is only the case if the tube has been completely evacuated of air; a little, however, always remains in the tube. If the diameter of the tube is $2\frac{1}{2}$ inches, and its length 2 feet, it would be evacuated of air by between ·1763 and ·2116 oz. (*Avoir du poids*) of gunpowder; but about one sixth of the air would remain in the tube. If the diameter is one foot and the length four feet, it can be evacuated of air by $\frac{3}{32}$ oz. (*Avoirdupoids*) but still nearly half the air will remain in the tube which very greatly lessens the effectiveness. But I think that this fault arises in part because the openings are much too small to let out the air, which must be definitely settled by further experiments. Meanwhile with this one foot tube that has only half the air evacuated, I have been able to show surprising results by drawing up both men and weights hanging on to the rope H.G. If air could really be removed, it would be quite another matter, and since the tube does not need to be very strong, since it is convex to the pressure of the outer air, and as it could therefore be made very light, it would not be impossible to build a machine in this manner (I do not dare to say for flight) but that would at least raise itself in the air and carry with it anybody who was bold enough to get in it.[118]

The possibility of a new and peaceable application of gunpowder found a warm welcome. In an anonymous article which referred to a letter of Huygens concerning gunpowder engines,[21] the chief of the establishment at Zödtenburg enthusiastically celebrated the new discovery, in which he, in true Calvinist spirit recognized a technical work for the greater glory of God and for the benefit of mankind.

Ad majorem Dei gloriam

Profound meditation is urged upon researchers concerning the propulsive power of gunpowder; and their perceptive intelligence is challenged to realize the possibility of diverting the vast power of gunpowder to healthier applications than those hitherto known.

Only irreverent thinking can deny such an incontrovertible truth as that in the sight of God only wholesome applications and uses appertain to all that is created and can be manufactured therefrom; for everything exists for the benefit of mankind, if only they had a serious desire for it. Nevertheless it is only too well known that the corrupted and misled spirit of man is concerned in countless cases not with salutary application, but allows itself to become passionately concerned with applying all its acumen to discovering how to misuse things, which according to the intention of the Creator should be useful to mankind, so that they should have every cause to praise him on their account.

Such misuse had induced some men to describe the discoverer of gunpowder simply as a sorcerer in monk's garb instructed by satan; for owing to its violent power it seems impossible to achieve aught but explosions, loss of life and destruction. Similarly doubtless in the immemorial past, people thought of the propulsive power of flowing water and of wind, before they had been made serviceable for useful purposes for mankind by wise and industrious mechanical craftsmen who used first simple wheels and later toothed wheels.

The above mentioned view of the satanic origin of gunpowder should therefore be set aside and replaced by the following:

1. The inventor of gunpowder whoever he was was a capable chemist.

2. The skilful achievements of chemistry are hated neither by God nor by Nature, nor are they directed against God's will; since they can instantly convert active poisons into blessed and healing potions.

3. It is possible by means of some form of control to force the aforesaid propulsive power of gunpowder, however sudden and violent it may now be, into ordered channels, so that it should be adaptable depending on the arrangements made for driving an ordinary mill or for performing other work; and this aim may be obtained if earnest prayer for divine support is combined with enthusiastic pyro-mechanical labours, and if mind and hand are ceaselessly busied with this work; and above all if the afore-mentioned fundamental demonstration [of a pious use of the gunpowder] and the praise of the Almighty Creator are kept in mind, —rather than its direct tangible use, which divine Providence will in its wisdom confer on the present century or on future ones.

More than two and a half years have elapsed since a number of researchers were publicly charged with the said problem concerning gunpowder, namely that it could and should be used for other purposes than hitherto. And the researchers should thereby be urged to the bold attempt gradually to give up their enthusiastic researches which were placing the misapplied use of the violent power of gunpowder . . . in a very favourable light. And at the same time they were to devote a part of their efforts, for the glory of the Creator, to discovering a new use for gunpowder, which had been latent in it from the beginning but had gone unheeded, because all who have hitherto occupied themselves with its application have stood under the spell of that terrible misconception that gunpowder was useful only for an idle and wasteful display of its flashes and flying sparks, or else to wound, to kill, to explode, to burst, to ruin, in short to unhinge the whole world. It is but too easy to suppose that our researchers, under the influence of this prejudice, have given either no thought or not sufficiently serious thought to a useful application of gunpowder, especially as no prospect of an important or obvious application attracted them in this direction. Only one man has been found who, with a view to a possible advantageous application, has freed himself from the above mentioned preconceived opinion . . . and has honoured the attempt with his attention, in a French letter of the 24 May 1686.

'I have received the problem communicated by you concerning a new application of gunpowder. In my opinion it may certainly be hoped to attain this end. Seven or eight years ago I showed to Monsieur Colbert an engine which I had had built for this very purpose, and which was illustrated in the *Proceedings* of our Academy. It worked as follows; a tiny quantity of gunpowder, about a thimbleful, was able to raise some 1,600 lb. five feet high; and not with such violence as is usual, but with moderate and steady power. Four or five servants, whom Monsieur Colbert ordered to pull the rope attached to the engine, were quite easily lifted up into the air. Nevertheless there was a certain difficulty in constantly reproducing this power.'

The writer of this letter communicated two quite unusual and really incomparable discoveries;

1. One or two drachmas of gunpowder, a thimbleful, will raise a weight of 1,600 lb. five feet;

and furthermore,

2. This occurs without the usual violence, but with moderate and steady power.

The first of these discoveries arouses admiration, but is in conformity with principles already accepted. The effectiveness of the powder could naturally be increased either by addition of more powder or by improvement of the piston. Nevertheless the second discovery would appear to extend beyond these known principles, and must be the more highly valued since it approaches the miraculous. . . .

Doubtless therefore all those who are plagued by curiosity to see this simple and really useful experiment could take the trouble to make such a machine or another which is suitable to impel any weight chosen as desired. They should allow themselves to be helped therein by men who are familiar not only with the use and misuse of gunpowder but also with the art of mechanics. Especially they will need for this, the magnificent work of the late Monsieur Bondel. *The Art of Shooting with iron balls filled with powder*, a work that would perhaps be more correctly entitled *The Art of Thoroughly Understanding the Nature and Characteristics of Natural and Violent Motion.* In it will be found many demonstrations directed toward the aim here mentioned. And there can no longer be any doubt that the fact that a tiny quantity of powder can raise 1,600 lb. so high, can, some day, be put to general use as soon as an inventor turns his attention to solving the many difficulties, especially those which obstruct the repetition of regular action. It remains only to add that at this point besides the description and exhibition, a drawing of the machine itself could very easily have been shown, by means of which its manner of working could have been shown; did not the ease with which this can be manifested and understood seem to make this quite superfluous, especially since its effectiveness has already been more than sufficiently demonstrated. Moreover, gifted investigators have had sufficient reason to believe positively that the making of such experiments, all too rare, each of which needs special consideration, may ultimately lead to a fruitful contribution useful to everyone. Meantime, the decision rests with God alone. He will according to his merciful judgment at the right time make it evident that all creation is appointed for the welfare and service of mankind. It is therefore the duty of man not only to believe this truth, but to work with all his power that he may use and enjoy everything with acknowledgement and gratitude. Praised therefore be the most holy name of Him through whose goodness the first stage of this apparent impossibility (namely the useful application of gunpowder) has been overcome; Praised, say I, be his name to all eternity! Amen.[119]

Huygens' gunpowder machine was however not capable of practical development. The difficulty of producing a good vacuum and 'bringing perpetually fresh power to bear' could not be overcome. Experimenting with gunpowder was dangerous. Then in 1690 it occurred to Huygens' assistant Papin to evacuate the cylinder by condensation of steam (Fig. 50). From a letter of his to Count Philipp Ludwig von Sinzendorff in the early 1690's, we learn to understand clearly Papin's atmospheric engine and the mining difficulties that were to be overcome by his discovery.

Monseigneur,

I have with deep humility been conscious of the honour which your Excellency has conferred on me when you graciously wrote to me from Bohemia inviting me to visit, at your expense, a mine which had become unworkable owing to the extent of underground flooding . . . I would very greatly desire to evince to your Excellency the strength of my zeal as well as to render to you my most humble service, if we only had peace; but unfortunately we see the peace disturbed in our neighbourhood. And the uncertainty of the incidents of war reminds me that I must not, at such a time, leave my family for so long a period.

I do not doubt, Monseigneur, that your Excellency's mine can be dried out by help of one or other of the engines described in my letter to His Excellency Count von Solms. But as Your Excellency's mine is far removed from the rivers, I must confess that it would involve much more work and expense to apply these machines there. This has led me to perfect an invention that I consider very advantageous for this sort of work and which I described in August, 1690, in the Leipzig *Acta Eruditorum* (Fig. 50).

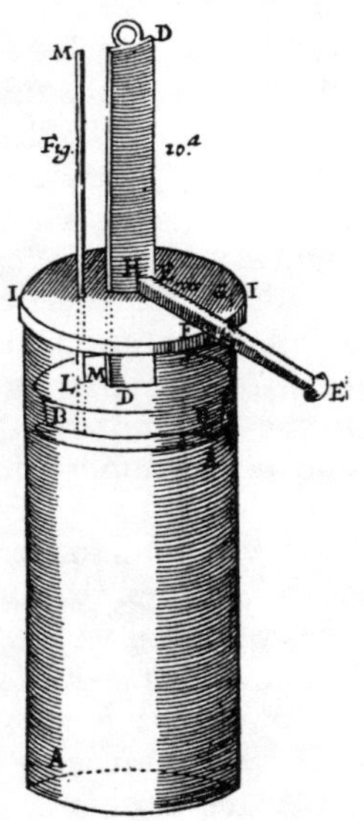

FIG. 50. *Papin's atmospheric steam engine,* 1690.

Engraving from: Papin, *Recueil de . . . quelques nouvelles machines* (Kassel 1695)

As water has the property that it is converted by fire into steam . . . and can then be easily condensed again by cold, I thought that it should not be difficult to build engines in which, by means of moderate heat and the use of only a little water, that complete vacuum could be produced which had been sought in vain by the use of gunpowder. And among various constructions which could be devised for the purpose, the following seemed to me to be the best. AA is a tube, of the same diameter from end to end, and well closed below; BB is a piston fitted into the tube; DD is a haft fastened on to the piston; EE is an iron rod which can turn on an axis at F; G is a spring which presses on the iron rod EE in such a manner that it engages in the notch H, as soon as the piston with its haft is sufficiently raised for the said notch H to appear above the lid II. L is a little hole in the piston through which the air from the lower part of the tube AA can escape when the piston is first pushed into the tube. Now in order to use this apparatus, a little water is poured into the tube AA, to the height of 3 or 4 lines; the piston is then introduced and pushed to the bottom so that the water rises into the little hole L which can then be closed with the rod MM. Now the cover II is placed over it, and this also has holes, which are necessary in order that it can be fixed without trouble. After a moderate fire has been kindled under the tube AA, this latter will warm up very quickly as it is made of a very thin sheet of metal; the water contained therein changes to steam and exercises such strong pressure that, overcoming the atmospheric pressure, it drives the piston upwards until the incision H appears above the lid II, and the iron rod EE will be clicked into it by the spring G, which will not occur without some noise. Then the fire must immediately be removed, and the steam in this light tube will quickly be condensed again into water by the cold and will leave the tube completely evacuated of air. Then all that is needed is to move the bar EE until the incision H is freed, so that the piston can descend freely. And in fact, the piston, driven by the whole pressure of the atmosphere, immediately drives downwards and, with a power dependent on the diameter of the cylinder, carries out the movement desired. . . .

I have ascertained by experiment that the piston, raised by the heat to the upper part of the cylinder, will immediately return again to the bottom, and this happens several times in succession so that one might suppose that there was no air at all exerting pressure from below or offering any obstacle to the descent of the piston. Now my tube, whose diameter is only 2½ inches is nevertheless able to raise 60 lb. the whole distance through which the piston falls. And the tube does not even weigh 5 oz. . . .

I have also proved that one minute is sufficient time for a moderate fire to force the piston to the top of the tube. . . .

If consideration is given to the magnitude of the force that can be generated by this means, and the small cost of the wood that has to be used, it must certainly be admitted that this method is far preferable to the use of gunpowder. . . .

It would lead us too far afield to discuss how this discovery could be applied to extract water from the mines, to throw bombs, to sail against the wind or the other similar applications which might arise. . . .

As these tubes would not conveniently put into motion the usual oars, rotary oars would have to be used. . . . It would be easy to set axles in motion by means of our tube (cylinder), having the oars fixed at the extremities. It would then only be necessary to furnish the hafts of the pistons (piston rods) with teeth, in order to set in motion small toothed wheels fixed to the axles of the oars. And provided that there were three or four tubes, which would work on one and the same axle, then these could cause a perpetual uninterrupted motion of the axle. . . . The clock maker is obliged every day to fix on to axles, toothed wheels which unfailingly rotate with the axle if they move in one direction; but which can freely rotate in the opposite direction without imparting any motion to the axle. This axle can therefore have a direction of rotation which is contrary to that of the aforesaid wheels. The greatest difficulty lies in the erection of a workshop in which, without great trouble, tubes (cylinders) could be made which need to be light, of considerable diameter and of regular shape throughout their length. . . .

In case it is agreeable to your Excellency to make use of this discovery I can definitely assure you that I know now a very good method of making without trouble tubes that shall be large, light and undeviating in width; and that I should regard it as an honour if I were able to lay before you evidence of my zeal. And with respect and homage I am, Monseigneur, your Excellency's most humble and obedient servant.

Denis Papin.

Doctor of Medicine, Professor of Mathematics in the University of Marburg, and Fellow of the Royal Society of London.[120]

For Papin as for Guericke difficulties impeded the production of accurate cylinders, in which the piston moved tightly.

He also ascertained by experiments how strong the parts of a steam engine must be in order to resist the pressure of steam.

Papin, who as a Huguenot had had to flee from France, reported in 1698 from Cassel to the philosopher Leibniz on a steam pump with a boiler separated from the cylinder; in which the steam itself drove the piston during the working stroke.

Cassell, 25th June 1698

. . . The method in which I now use fire to raise water rests always on the principle of the evaporation of water. But I now use a much easier method than that which I published. And furthermore besides using suction, I use also the force of the pressure which water exerts on other bodies when it expands. These effects are not limited, as in the case of suction. So I am convinced that this discovery if used in the proper fashion will be most useful. . . . I personally almost think that this discovery can be used for many other purposes besides raising water. I have made a little model of a carriage which is moved forward by this power. And in my furnace it shows the expected result. But I think that the unevennesses and bends in large roads will make the full use of this discovery very difficult for country vehicles; but in regard to travel by water I cherish the hope of reaching this aim in time if I could find more support than is now the case. . . .[121]

Some years later Papin wrote to Leibniz:

Cassell, 24 Dec. 1704

. . . I shall perhaps be told, Sir, that it is better to use the power of the rivers rather than human power to raise water. To which I reply that if a river is available as desired it would be better to use it than to make use of the fire machine. But there are many cases when rivers are completely lacking or are so distant that the maintenance of the machine would cost almost as much as if the water were raised by human power. I have never seen the machines at Marly; but shrewd people have assured me that their maintenance costs annualy 100,000 francs without counting the expense of installation. Furthermore it does not produce a great deal of water.

I say further that our machine can be applied to many other purposes for which the current of the rivers cannot be used. I think also, Sir, there is no reason to doubt that his Highness could not do me the honour to use the plan that I propose without injury to his reputation. I rely on you in this matter.[122]

Papin in 1706 complained of his difficulties in attempting to raise water by steam pump, in writing to Leibniz who was immensely interested in all these questions.

Cassell, 19 Aug. 1706

. . . Strong tubes of cast iron were available, so it was thought that it would be best and quickest to use them. I personally emphasized that the cement which was smeared over the joints between the tubes would not offer the necessary resistance. But the others continued with it. And when it came to the test it was seen that the water in fact leaked through all these joints. And at the lowest joint there was such a strong stream of water that His Highness said at once that this experiment could not be successful. But I very humbly asked him to wait a little as I thought that the machines would supply sufficent water to let it reach the top although so much had been lost in so many places. And actually, as the experiment proceeded we saw the water flowing out of the highest point of the tubes four or five times.

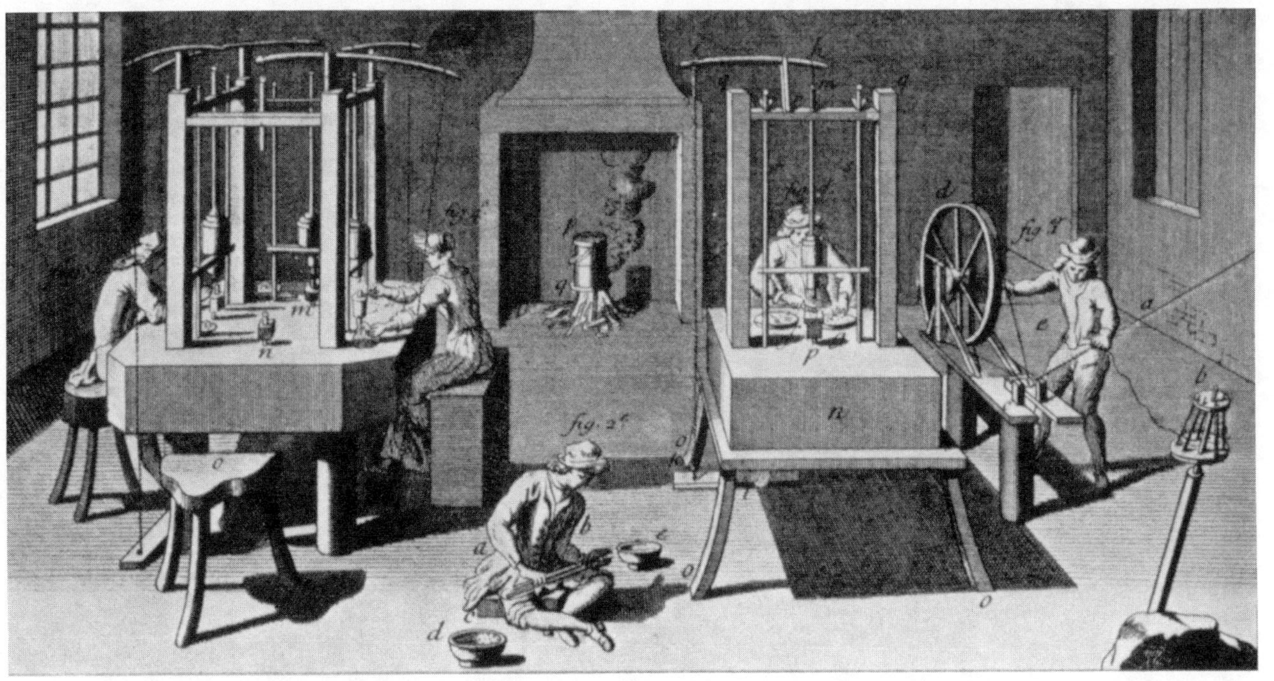

17. Pin manufacture, 1762. Engraving from Réaumur, *L'art de l'épinglier*. Figures 1 and 2 forming the heads, figures 3 and 5 pressing the heads onto the pins

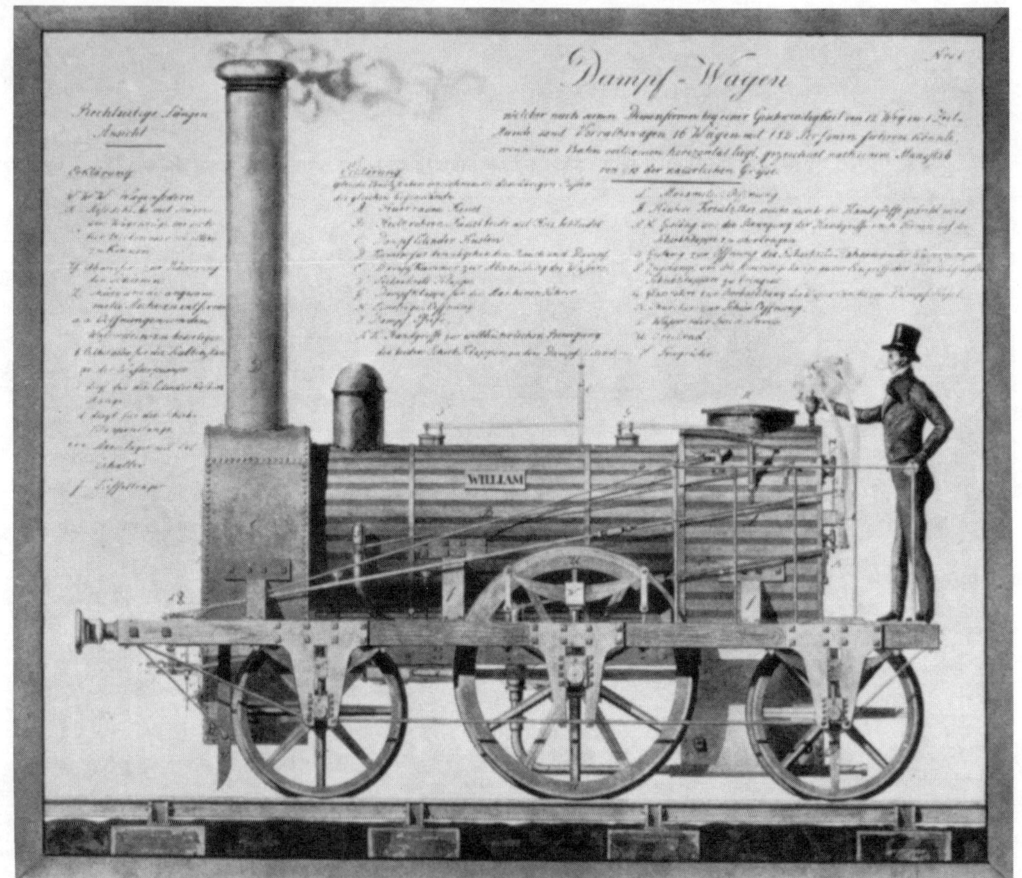

18. English locomotive of Stephenson's successful construction 'Patentee', 1837. Drawing by von Seeger

They wanted then to renew the cement but as it was very warm quite a lot of it entered the pipes. It fell onto the valve and prevented it from closing when they wished to make a second experiment. So his Highness ordered that tubes of copper plate should be made. As these tubes will be very well soldered together they will not be subject to the same drawbacks as those of cast iron. . . . Meantime the absence of His Highness caused much delay; for, at the moment, the workmen only labour when they have nothing else to do. There is complete uncertainty as to the time at which his Highness will return. But most people think it will not be before Michaelmas. The top of the house to which we have forced the water to rise, my Lord, has been measured at only 70 feet.[123]

In the beginning of 1707 Papin had sent to Leibniz his work which had just appeared, *The New Art of Raising Water efficiently with the aid of Fire*, in which he described the construction perfected in 1706 of a high pressure steam pump (Fig. 51). The philosopher thanked him for the gift in a letter written from Berlin on February 4, 1707. His suggestions betray a good technical mind.

Berlin, 4 Feb. 1707

I have just been honoured by your letter as well as by the delightful present of the excellent book which you have published on how to use the evaporation of water by heat in machines. I am happy to see in the preface that I have incidentally been in part the cause of this beautiful work. . . .

I imagine that there will soon be a great consumption of water in the retorts of your fire machines, and that it will be necessary to think out a way of replacing it, a sort of stopcock could accomplish this (Fig. 51).

I have an idea that perhaps will not displease you, namely the useful application of the steam while it is still hot when it leaves the pump (i.e. the cylinder) as the piston is pushed up. . . . In order to make good use of the otherwise superfluous heat (of the steam). . . . I would lay around the vessel Q.N. a sort of mantle or case, partly filled with compressed air; and within this case I would let the steam enter in such a way that before it streams powerfully into the open air it would be between the case and the vessel. And while it warms this vessel it would as a result contribute toward the work of the compressed air contained therein.

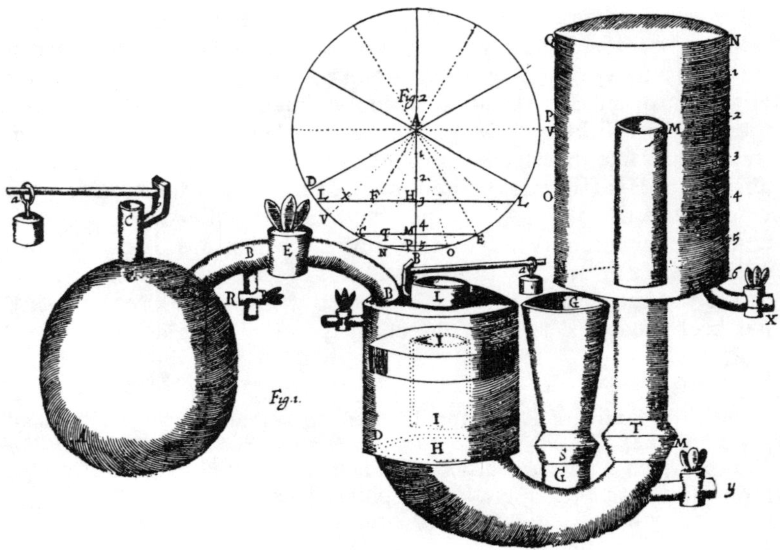

FIG. 51. *Papin's direct-acting high pressure steam pump (without condenser)*, 1706.

Left, the copper boiler (51 cm. dia.) with safety valve; in the centre, the cylinder (D) with the piston (F); right, the compressed air chamber.—Engraving from: Papin, *Ars nova ad aquam ignis adminiculo efficacissime elevandam* (Frankfort-on-Main 1707)

I will not speak of the superfluous heat of the furnace and the smoke which emerges from it which can be similarly useful, among other ways by heating the water of the funnel G. and of the tube H. .

Furthermore I have no doubt that you could, if you so desired, easily arrange that the stopcocks E. and N. (on the upper left of the cylinder) should be opened and shut by the machine alternately, without having to use a man for this. . . .[124]

Meanwhile Thomas Savery had invented in England in 1698 a steam pump in which alternating action took place between the direct work of the steam and the indirect force caused by the creation of a vacuum. He called his little work published in 1702 *The Miner's Friend* and his machines were to help the miner in his struggle against flood water. But they could not fulfil this promise as the delivery height in the mines was too great. In some country residences, however, it was used for water supply. Since

1711–12 Thomas Newcomen had carried out practical work in mines, and had built large atmospheric steam engines, probably as a continuation of Papin's experiment. They had a boiler separated from the cylinder and a large balancing beam. At one end hung the piston, at the other a counter weight was fastened which was attached to the pump-rod. Newcomen's machine worked. It was used in numerous mines. But it was not practical owing to its great loss of energy. Enormous masses of coal were devoured. Only when the thermal properties of steam were more exactly understood, could new methods be devised. Here came the work of Watt which belongs to the period of rationalism which will be reviewed in the following Part.

PART V

THE AGE OF RATIONALISM

PART V

INTRODUCTION

FROM the first quarter of the eighteenth century, the spiritual trend that we have sketched as typically Baroque began to slacken. The leaning toward the mysterious, the feeling for metaphysics, were lost. Religious ties were further loosened. Furthermore, as against the previous period, an emphatically one-sided rationalism appeared. We speak of a period of rationalism; but the rationalism of the eighteenth century was no longer that of the great systems of the Baroque period. Attention was directed much more to single facts. An empirical rationalism was therefore the sign of the times. The effort was made to penetrate by means of Reason both religious problems and the traditional procedure of technical work. Development in this respect was uneven in the various countries. In France, rationalism gained especially wide adherence, while England continued to tend rather to pure empiricism. Rationalism made its way also into Germany. But the philosophy of Leibniz, which offered scope for metaphysical impulses too, still stood guard for a long time against one-sidedness, though Christian Wolff, who made a systematic synthesis of the view of Leibniz, himself tended rather to pure rationalism. Kant succeeded in both clarifying and deepening rationalist philosophy. And the tendencies both of German classicism and subsequently of German romanticism offered a protection against the undue growth of rationalism. Where the door was opened wide to western influences, as in the circle of Frederick the Great, these were greatly modified, and the use of reason was transferred from the realm of thought and knowledge to that of ethics and duty. But in spite of all the variations and divergent developments, the period was characterized by universal emphasis on the power of Reason.

Quantitative natural science, based on the combination of experiment and reason began its triumphal march. This opened a new period in the history of technology too. The effort was now made to apply the concepts of science to technological activity, hitherto based predominantly on traditional practice and experience. In this way a systematic rational technology, built on scientific discoveries, began its career in the cultural history of man. It is nevertheless remarkable that not until the second half of the eighteenth century did great and beneficial technical creation take place, while the first half of the century was devoted rather to the collection, classification and scientific development of technical knowledge already available.

The natural sciences, which had greatly developed in the Baroque period, made further strides in the eighteenth century. The integral calculus was systematically improved. Mechanics by the application of the new calculus was elevated to an exact science, which was ultimately able to condense many phenomena into a few fundamental formulae. The law of heat led ultimately, once a serviceable instrument for the measurement of temperature had been invented, to a clear distinction between a degree of temperature and an amount of heat, and thus also to the understanding of specific heat, immensely fruitful for further development. This made possible for the first time the development of a rational heat engine.

Chemistry, which first became an independent science in the Baroque period, and attained at the end of that period, in the Phlogistonic teaching, a theory which, in spite of all deficiencies, succeeded in providing a common point of view for a number of detached phenomena, was greatly stimulated in the eighteenth century by a series of important discoveries concerning gases. These discoveries of the phlogistonic period were closely connected with work on the laws of heat. The connection between calorific study, phlogistonic theory and researches on gas is natural if we reflect that phlogistonic theory regarded phlogiston as a hypothetical heat substance or even as a subtle gaseous matter which in fact was ultimately identified with hydrogen. Thus we understand that the chemist Black, inheriting the phlogistonic theory, occupied himself intensively with the chemistry of gases and with the laws of heat. But this calorific study greatly influenced Watt and the development of the steam engine. Only toward the end of the century, when in the face of new experimental data, the phlogistonic theory could no longer answer requirements, there appeared in the seventies a changed logical evaluation of chemical phenomena, opposed to that of the phlogistonic period. Attention was now above all directed to the relative weights of the matter involved in processes: Chemistry, though later than physics, thus flowed into the stream of rational natural science whose aim was Measureability.

All the above-mentioned scientific successes sooner or later had favourable effects on technology, which in the eighteenth century began to assume the character of applied science. This change was marked, for example, in that the construction of machinery was treated more than formerly, as a part of mathematics, whereas in the Renaissance it had been regarded rather as an appendage to architecture. It is true that the Renaissance period had also shown an urge toward scientific understanding of technological activity. But by and large, this was the work of empiricists who were independently seeking a scientific explanation of technical work. The scientific corporations remained fairly aloof. But in the rationalist century, it was, at least on the Continent, mostly the men of science who sought rational explanations of technical activity. Since the Baroque period, rationalist philosophy had attained to the Chairs in the great Schools, and had sought to render the influence of mathematical thought influential in all departments of intellectual and material culture.

The period from the middle of the seventeenth to well beyond the middle of the eighteenth century saw the blossoming of the idea of mercantile economy. The state strove to attract as much money as possible into the country. An attempt to gain this end was sought in State support of trade and manufacture and by increased exports. The entry of finished products was restricted by protective duties. Private factories were favoured by special Privileges (Plate 14); and much enterprise was undertaken by the State itself. In England, however, conditions were different. In the seventeenth century, as already explained in our Section on the Baroque period (p. 191). Non-conformist Church circles became prominent, which in the pursuit of an 'asceticism' of work on earth zealously advocated technical progress and sought, in opposition to the State, gradually to wrest for themselves all those freedoms which helped towards the realization of their unbridled activities. These pioneers of technical and industrial progress in England were more or less in opposition to the State. There thus developed in England technical and industrial activity, strongly based on private enterprise, and independent of the State. Technology developed at first entirely in a practical direction. The most able from among the ranks of the craftsmen were able to rise, with much less hindrance than on the Continent where handicraft remained much more limited to the narrow traditional circles established by the guilds and by the boundaries of rank. Consequently we find that on the Continent it was the State which was concerned to see technology and industry develop.

England soon attained a considerable advantage over the Continent in the realm of technology. The general favourable political situation, the natural wealth of indigenous raw materials, and protective measures favouring inventions, besides the restless spirit of enterprise of Puritanism were the main-springs of this exceptional development. At the head of the great technical achievements of the second half of the eighteenth century, stood the steam engine constructed by the greatly gifted technical mind and scientific spirit of Watt. It was this which first made possible an increase in the production both of ore and of coal beyond the wildest dreams of contemporaries. The introduction of coke into the blast furnace process, the discovery of crucible-cast steel, of cylinder-blowers and of the puddling process opened the way for iron as the most important raw material. Improved machine tools and new productive machines, especially in the department of textile technology, together with the steam-engine as a source of propulsive power soon led to a complete revolution in industrial conditions. Manual home industry, whether independent or for employers, yielded inevitably more and more, especially in textile work, to mechanized production in central workshops. The increased need for chemicals due to the rapid improvement in textile technique and especially to the wider use of cotton gave a powerful stimulus to chemical technology, which developed in the second half of the century the earliest processes for the production of sulphuric acid and of sodium carbonate on a large scale, besides the lead-chamber and Leblanc-processes. These important chemical substances are indeed economically almost as essential as coal.

The increase in the course of the eighteenth century in the output of ore and coal, and the rise in production in manufactures and factories, as well as the demands of military technology influenced the development of roads and methods of transport. The construction of roads, bridges and canals, based no longer solely on practical experience but also on scientific knowledge, became more active. In English mining districts, tracks were laid for the transport of coal and ores. Vehicles drawn by horses were driven along wooden, or from the last third of the century along cast-iron rails. These railways had developed on the principles of the late medieval German mining technique which influenced English mining. The combination of iron rails and steam vehicles—both originating as early as the eighteenth century—led in the beginning of the nineteenth century to the first steam railway.

Watt's first steam engine started work in a foundry in 1776. In 1787 Watt's engine was introduced into the cotton-spinning industry. By the end of the century, the number used in this industry had risen to eighty-four. The production of raw iron in England was quadrupled in the short period between 1788 and 1806. So the ground was prepared in the second half of the eighteenth century for a comprehensive industrialisation which developed widely during the nineteenth century. The general mechanization by means of new machines, at first run by water-power, and later the introduction of the steam engine as the source of power were not without violent social repercussions which will be discussed later (cf. pp. 295–8).

MASTER CRAFTSMEN AND MILLWRIGHTS

We first encounter in the first third of the eighteenth century the striving toward a more strictly rational appreciation of technical production as compared with the seventeenth century, and most impressively in the skilled Leipzig mechanic, machinery constructor and mining commissioner Jakob Leupold, who most fortunately combined a scientific mind with practical skill. He wrote a comprehensive work[22] on the construction of apparatus and the machine technology of his time; it was intended especially to serve the mechanic and master-craftsman who were the actual mechanical engineers of the period, and beyond this to increase the economic prosperity of Germany by increasing 'crafts, mines and manufactures'. Leupold outlined the tasks of the engineer as follows:

As in earliest days were our mechanics, such today are our engineers, who may be required not only to tear down a fortress, and then to build it up again, but also to produce all sorts of engines based on mechanical principles, and with equal ease to defend or to annihilate a fortress. Moreover he must invent many sorts of useful engines to lighten work; and he must often render a thing feasible that has appeared to be impossible.[125]

Leupold inveighed emphatically against boastful project-mongers, self-styled inventors, 'perpetual-mobilists' and producers of miraculous works—these last often as ignorant as they are deceitful, who were incapable of calculating in their designs 'Power, load and time'. So the objective and rational spirit of Leupold turned against this mere paper project making such as characterized many seventeenth century books on machines, with their often complicated but ultimately useless mechanisms.

Some people desire to obtain great force with little power, and that in a limited time. In a few words; they seek that which for unnumbered years many people with terrible expense, anxiety and effort have succeeded in conceiving but, before bringing into being have to their great distress lost again, namely the *Perpetuum mobile*. For whoever seeks from power more work than is produced so far by our calculus or theory of mechanics, is in fact seeking in vain the *Perpetuum mobile* and will not find it . . . therefore my intention and effort is directed only and solely to teaching that person who wishes to be made a mechanic. Nor will he have to trouble himself further than to know fundamentally what his machine should in theory be able to do and wherein it has hitherto failed in practice—that is to say he must know what sort of theory is most important or in what manner he can arrive at it, namely by avoidance or elimination of friction and by the right application of force. . . .[126]

If people are examined and considered who boastfully present themselves as omniscient, they are most often those who have not properly mastered the elementary theoretical principles of mechanics and who wish nevertheless to appear as great practitioners. Some have indeed a certain idea of the theory, and are not untalented in practice either; only in their conceit they hurry too much and therefore omit to consider beforehand all the circumstances which could arise in their suggested improvement and could cause them inconvenience. So heedless indeed are they sometimes that they have not even troubled to see the work which they aspire to improve, far less are they aware what in fact the machine should or does already achieve. . . . But if this operator had made himself acquainted with the affair beforehand and had seen how much had already been achieved without his imagined skill, then he would not thus have prostituted himself and would not have given displeasure to so many high officials and colleagues, yes even to the lord of the land himself. . . .

Since then it is an utter impossibility to induce a simple machine to exert a whit more effect than another machine, or an ordinary lever with a sharp fulcrum, still less is it possible to achieve results from complicated machines equipped with many wheels, screws, etc., for these as a rule deal with greater loads. And as they have far more pins, teeth, gears and planes which grind, drag and resist, and cause a much greater impediment and resistance to the force, they achieve far less than the theory would indicate.

This resistance and in general the friction called by Germans and master craftsmen stagnation or constraint, is the chief and main cause why one machine does not accomplish so much as another, or as theory and calculation would suggest.

But the other main cause is unskilful application of power. For he who has the skill to eliminate friction in his machines and to apply the power rightly according to the requirements will be able to answer the other point of the question: whether the machines can be improved, in the affirmative and he will be known as a skilled craftsman.

That friction is the great robber of power can be seen on a goods vehicle; for whether it were loaded with fifty or even a hundred cwt. a man would be able to push it with one finger along a completely even and flat horizontal plane; but that a number of strong horses can hardly manage this is the result only of friction caused by the wheels on their axles, and also with their felloes and nails along the rough pavement or roads.

Whoever therefore wishes to accomplish with his machines as much as they should in theory will as far as possible abolish all friction, which may be achieved as follows:

1. If the machine goes fast.
2. It must not be too heavily laden.
3. Few of its parts and pieces must move in their bearings, or rub, slip, slide or resist one another.
4. All such parts must be hard, slippery, round and smooth and must not lack adequate lubrication.

That friction increases more than proportionately to the load, has been proved on a machine or crane with iron wheels and mechanism.

For I have made an experiment and have hung a weight of fifty pounds on to the shaft, and one pound was then sufficient as counterweight; and if it had to move it was necessary to add one ounce to the Force applied. But after I had hung a burden of three hundred pounds, according to calculation six pounds six ounces should have moved this burden; but twelve pounds were hardly sufficient, so much had the friction increased even in a machine in which everything was round, smooth and well lubricated. What then might be expected with wooden, rough and unfinished machines?[127]

The best machines and to be chosen above all others are those which consist of the fewest parts or which are the simplest, which produce least friction; which are not too heavily loaded, and where power can be conveniently applied without any waste.[128]

Leupold devoted himself also, in the true engineering spirit, to the important problems of manufacture and showed how springs, screws, toothed wheels and pistons as well as suitable lubricants should be made. He often treated individual mechanisms and parts in detail, as for example the various sorts of valves and pistons. This constituted the important step in machine construction, already taken to some extent by Leonardo da Vinci and by Cardan, 'of seeing the general in the particular' (as Reuleaux[23] calls it) and of regarding machines not as hitherto solely as complete units, but rather isolating the several mechanisms and treating them separately, and with only secondary consideration of their special application. Leupold exercised great influence. James Watt was among his readers, and he learnt German for the purpose from a German-Swiss living in Glasgow. A plan of 1725 to improve the mining machinery in the electorate of Saxony is further evidence of Leupold's effort to perfect machinery by considering it scientifically.

That a complete understanding of machine construction is most necessary, in fact indispensible in the mining industry has long been agreed.

Many indeed consider that it is so advanced that little or no improvement is possible, or that the failings are such that there are no further means of improvement; wherefore either the old methods must be retained, or quite new inventions must be devised. But all this is without foundation so long as we do not really understand our machines so that we can base our calculation both on theory and on practice, which has hitherto scarcely been the case. How can anyone say: The machine is not performing as it should; How can anyone assert with a clear conscience that he will improve an old machine or produce a new one that will do much more than its predecessor, when he does not know whether the old engine is not already doing all that God has empowered it to do, and as much as theory would expect from it? . . .

Therefore, as His Majesty the King of Poland and His Highness the elector of Saxony has been graciously pleased to command me to supervise and to improve the organization of the engineering and machines of all the mines, I have considered with all care how I could without loss of time arrange and complete such an establishment as shall be fundamental, universal and durable; so that not I alone, but all others also, overseers, carpentry and mining foremen, indeed everyone who needs or at least wishes to know such things will be able to acquire a complete and fundamental grasp of them. . . . But for such attainments, the following preparation is necessary:

1. A clear list must be made of all devices and machines which are in mines or foundries, in clear outline, in all working positions, their parts and component pieces must be drawn accurately to scale, and as far as possible clearly described and calculated by Theory.

2. The power of each engine must be calculated as accurately as possible and the work of which it is capable; or what this engine is now doing and what according to principles or theory it should be doing. From this it will be clearly seen to what extent the engine is doing its work and whether or how far improvement is possible.

3. At the same time, all the mechanical and physical foundations and causes of both performance and non-performance should be explained clearly by experiments and sketches, together with the calculations, both geometrical and mechanical, or however else may be clearest; and instructions given in what respect and how it is hoped to effect an improvement.

4. The faults should be set forth with their causes, and what complaints the people with engineering knowledge had already been able to make about them; also the remedies that have already been applied and how they have worked.

5. I have invented devices for measuring power, especially water-power, so that by means of an accurate clock with a second-hand or at least with a minute-hand, and of certain tables and rules, anyone should be able to calculate that in one minute or in one second, so much water will flow through. . . .

6. I will also do so with engines on outlying mines. . . .

7. All this shall appear in two special books or *Theatra Machinarum* printed at my expense. . . .

8. I will produce diverse machines, inventing them entirely anew, in order to investigate the power of falling water, both under-shot and over-shot, and to instruct as to the best method; for hitherto the greatest ignorance has been evinced in this matter which is, however, one of the most important items; for the whole question of improving machinery depends on (1) the right application of force, (2) the elimination of friction and the failure to attain simplicity. . . .

9. I will give faithful guidance and teaching by experiments and on the engines at every mining-town or district where I find mining-crafts and persons who have need of the principles governing mechanics and their engines, and who are desirous of knowledge. . . .

10. Thereupon it will be quite easy to begin improvement, and to discuss with all who understand mine-work and are experienced engineers the improvements proposed—*pro* and *contra*—and thus seriously to set to work. . . .[129]

Among the most experienced technical engineers of the period before the general adoption of steam-engines were not only the master craftsmen on whom lay especially the responsibility for the construction and maintenance of the mining engines, but also the millwrights. The calling of millwright which like that of the engineer had developed from the rank of carpenter—since wood was at first the main material for building machines—had already to some extent passed beyond the narrow boundaries of the guilds. These millwrights were in general familiar with many branches of craftsmanship and also commanded a certain amount of theoretical knowledge. The Dutch books on mills of the first half of the eighteenth century bear witness to the very soundly based technical ability of these men (Plate 15). The English engineer Fairbairn who himself was of this origin gave an excellent description of the manifold capabilities of these builders of mills and millwrights.

The millwright of former days was to a great extent the sole representative of mechanical art, and was looked upon as the authority in all the applications of wind and water, under whatever conditions they were to be used, as a motive power for the purposes of manufacture. He was the engineer of the district in which he lived, a kind of jack-of-all-trades who could with equal facility work at the lathe, the anvil, or the carpenter's bench. In country districts, far removed from towns, he had to exercise all these professions, and he thus gained the character of an ingenious, roving, rollicking blade, able to turn his hand to anything, and, like other wandering tribes in days of old, went about the country from mill to mill, with the old song of '*kettles to mend*' reapplied to the more important fractures of machinery.

Thus the millwright of the last century was an itinerant engineer and mechanic of high reputation. He could handle the axe, the hammer, and the plane with equal skill and precision; he could turn, bore, or forge with the ease and despatch of one brought up to these trades, and he could set out and cut in the furrows of a millstone with an accuracy equal or superior to that of the miller himself. These various duties he was called upon to exercise, and seldom in vain, as in the practice of his profession he had mainly to depend upon his own resources. Generally, he was a fair arithmetician, knew something of geometry, levelling, and mensuration, and in some cases possessed a very competent knowledge of practical mathematics. He could calculate the velocities, strength, and power of machines: could draw in plan and section, and could construct buildings, conduits, or watercourses, in all the forms and under all the conditions required in his professional practice; he could build bridges, cut canals, and perform a variety of work now done by civil engineers. Such was the character and condition of the men who designed and carried out most of the mechanical work of this country, up to the middle and end of the last century. Living in a more primitive state of society than ourselves, there probably never existed a more useful and independent class of men than the country millwrights. The whole mechanical knowledge of the country was centred amongst them.[130]

TRADE AND FINANCE

As we have already emphasized (p. 233), the eighteenth century was greatly influenced by mercantile tendencies. The Swedish engineer Christopher Polhem 'a skilled mechanic who understands both theory and practice and investigates the efficiency of all machines from first principles',[24] exerted himself to make his country technically independent of England. His *Patriotic Testament* of 1746 gives a picture of this effort (e.g. Plate 16).

We have perceived with pleasure from several books that have appeared recently how one and another, each according to his fashion, has endeavoured to demonstrate the necessity for manufacture and factories for the salvation of the state from its present dangers; it is, however, astonishing that so few have advocated the

19. Steam-driven soap factory, 1850. Lithograph from *Album des célébrités industrielles*. Vol. 2. Paris 1865

20. The gigantic Corliss Steam Engine at the International Exhibition in Philadelphia, 1876. Cylinder bore 102 cm. Output 1400 H.P. Woodcut from *The Scientific American*, Vol. 34, 1876, p. 351

improvement of our own raw materials, as though unlimited trade through all sorts of work in iron, steel, copper and brass, did not deserve to be classed as manufactures as much as many other things.

On the contrary there is more tendency to regard it as stupidity and lack of understanding if anyone opens his mouth on this subject; the impossibilities and dangerous results are described, until I fear, our trade will no longer be able to escape foreign domination. But how long will this continue? Is it not now time to gain fresh courage, while some powers remain to us and almost the whole of Europe except ourselves is involved in war.

The domestic economy of other countries may serve us as a model. Every nation under the sun seeks first to build up its own affairs because this is the safest road to riches. Nowhere outside this country are there found iron foundries where the best and greatest possible quantity of iron is not worked by manufacture. As a result, for every two cubic metres of coal the country receives a profit of fifty to a hundred dollars of copper coins; but we gain for every two cubic metres hardly seven dollars of copper coins so long as foreigners will pay no more than 24 dollars for 2·95 cwt. of iron; whereas to obtain 2·95 cwt., twelve cubic metres of coal is required for mining and smelting. I do not mention wages for the process from extracting the iron from the mines to smelting it into ingots, nor the value of the iron ore itself, for all of which we receive nothing so long as we deliver the iron to the foreigners for so low a price. According to the most accurate calculations, we are destroying our mines and our forests as posterity will realize, and our descendants will bemoan with sorrow.

I have then undertaken to show how more profit and advantage can be obtained from but few mines and woods if we, following the foreign fashion, do not sell our iron in rough smelted bars, but in future, at least in part, through all sorts of installations and useful work increase its value before we export it; whereby not only will many thousands of persons obtain a means of livelihood but also our woods will be spared or at least will be applied to greater use and profit for the State; instead of as at present being diverted only to the profit and advantage of the foreigner.

God help us to open our eyes, and preserve me from death before I can rejoice over a happy change in this matter.[131]

Mercantilism developed most markedly in France. Colbert had already presented to the Paris *Academy of Sciences*, founded in 1666, numerous practical proposals which, in the light of the mercantile needs of the time, were to serve the requirements of trade and manufactures. So about 1695 it was decided to undertake under the leadership of the *Academy of Sciences* a comprehensive scientific description of manufacturing processes and of technical contrivances, entering into the most minute detail by word and illustration, in order to bring to light and into general use throughout the country exceptionally rational but little-known processes, and in order to be able to compare individual practices often kept secret; it was thus hoped to permeate all manual production with the scientific spirit for the benefit of the State. In the year 1711 the young Réaumur was commissioned to sift all the material and to urge on the work necessary for this 'summary of the condition of the arts'. It was not until after the death of Réaumur (1757) that in 1761 this great work began to appear as *Descriptions of the Arts and Crafts*. 121 parts with over 1,000 copper plates were published up to 1789. In the Introduction to this work which contributed greatly towards the improvement of French manufacture, the following is remarked:

The Royal Academy of Science in Paris . . . had hardly been founded before it decided to describe gradually all the processes of the mechanical arts; for it was convinced that this undertaking would contribute equally to the establishment and increase both of these mechanical arts and also of science.

Since the crafts were born in gloomy times, in which industry groping always in the dark, could achieve from century to century but slow progress, long before the establishment of learned societies; there can be no objection to the recognition that in those times and places where science was industriously cultivated remarkably speedy progress was achieved. One would soon be convinced of this on considering present conditions of the clock-making industry, of pyrotechnics, of those arts appertaining to navigation, and to the manufacture of instruments for geometry, optics, astronomy and surgery and indeed so many other activities including those which usually occupy scientific academies, and comparing them with the conditions of these crafts a hundred years ago. An immense difference will be observed which is not the result of mere chance but of the efforts made during this period to bring to perfection geometry, mechanics, optics, chemistry, anatomy etc.

What new degrees of perfection of the arts may not be expected if the scholars who have acquired knowledge and experience in the various departments of natural science give themselves the trouble to examine and to explain the often ingenious labours undertaken by the skilled man in his workshop? Thereby they will themselves perceive the needs of a craft, the limits which have held back the craftsman and the difficulties which obstruct him, and the help which can be given to one craft by another, which the worker is seldom in a position to perceive. The surveyor, the engineer, the chemist will give new insight to an intelligent craftsman so that he may overcome obstacles which he has not ventured himself to remove. They will also put him in the way of making new and useful discoveries. At the same time they will learn from him what part of the theory must be most earnestly pursued in order better to explain the practical work, and how to produce certain and reliable rules concerning a number of delicate operations depending at present on a correct eye or a skilled hand, the success of which is but too often uncertain.

This was the intention of the *Academy of Sciences* which continually directed its work towards the Useful, when they imbued their members with the desire to work on the description of the crafts. Since the beginning of this century the Academy has ceased to collect material for this purpose, but the subject is immeasurable, and can only be achieved in the course of a long time.[132]

At about the same time as the *Descriptions* appeared, the great French *Encyclopedia* of the Enlightenment,[25] the product of the initiative of Diderot and d'Alembert. The essence of this work was the combination of the spirit of logical analysis with the spirit of practical fact. The great picture of rational man in a rational world that was here unfolded, laid the first foundation for a general culture. The Encyclopedists endeavoured to use science to form public opinion. The enlightenment of wide circles through the *Encyclopedia* led to a balanced culture, partly elevating and liberating both culturally and socially; but partly levelling down in its effect. The inclusion of technology, indeed to a most remarkable degree contributed to a wider extension of technical knowledge, to the great advantage of practical technical work. We owe the technical passages especially to Diderot. He never tired of seeking out factories and workshops and giving scientifically considered reports of the machines and processes that he saw there. It was through the *Encyclopedia* which was widely circulated that the general public first became conscious of technology. It is true that even before the *Encyclopedia*, natural science and technology had attained no inconsiderable position in the higher general culture of the eighteenth century. The notable results especially in physics in the seventeenth and eighteenth centuries had influenced this period, so favourably inclined to all experimental and intellectual science, so that a certain degree of knowledge of this nature had been absorbed into the so-called general culture of the higher circles. Thus in 1737 F. Algarotti wrote on Newtonian physics for women, and between 1768 and 1772 Euler published letters on natural history to a German princess. But the *Encyclopedia* spread a general knowledge of science and technology to wider circles.

In 1755 in the article entitled 'Encyclopedia', in the fifth volume of this great French *Encyclopedia,* Diderot emphasized that the aim of an Encyclopedia should be 'to collect the knowledge which is scattered over various parts of the world in order to reveal the general system to our contemporaries, and to hand it on to those who will come after us in order that the work of past centuries shall not have been useless and that our successors, because they will be better instructed will also be finer and happier men, whereby we shall not die without having served mankind.'

Descriptions and *Encyclopedia* may be technically very similar but the spiritual soil from which the two works grew was very different. The *Descriptions* originated in the atmosphere of state sponsorship of craftsmanship and industry in the spirit of Colbert's mercantilist Paris Academy. The beginnings of this work go back, as we have remarked, to the end of the seventeenth century. The *Encyclopedia* on the other hand, oriented toward liberal enlightenment, pointed to the future. Réaumur, the director of the Academy who for a long time had prepared the undertaking of the *Descriptions,* and Diderot, the chief exponent of the idea of the *Encyclopedia* were opponents. According to the evidence of Réaumur the Encyclopedists had copied some of their plates from engravings in the *Descriptions.* With biting sarcasm, the ageing Réaumur remarked of Diderot, who was writing the technical passages of the *Encyclopedia,* 'the son of the cutler was atavistic in reserving for himself the passages on the technical crafts'.[26] and he often observed with reference to the *Encyclopedia* that it is dangerous to combine science and politics.

Germany manifested in the eighteenth century, and especially in the Lutheran territories, as has been pointed out by Müller-Armack,[27] a general vigorous development of state economic policy based on a literature composed largely by professors and theologians and in which trade organization also played a part. This so-called *financial* or fiscal literature reflected something of the narrow confines of the small States. Especially Johann Beckmann, Professor of economic science at Göttingen, the originator of trade and technology as Collegiate sciences, worked at the collection and scientific explanation in Germany of the arts of the technician and craftsman. 'Scholarship will help to increase trade' was his motto.[28] University teaching in technology as a part of political science, as we often find in the eighteenth century was to make the future administrators in the princely Councils familiar with questions of trade and manufacture. Already in 1727, Frederick William I had introduced technology into the curriculum of political and fiscal science at the Universities of Halle and of Frankfurt on the Oder.[29] But this attempt, based on economics, to cover technological fields in the universities, was not strong enough to develop. With the decline of the mercantilist period and the rise in the nineteenth century of liberal economic opinion, exclusively legal training became the most important for the higher administrative officials. But technological studies were cultivated in their own colleges.

The French *Descriptions* won the approval of the German mercantilists. In 1762 J. H. G. von Justi began a German translation. Daniel Gottfried Schreber, Johann Conrad Harrepeter and Johann Samuel Halle continued this work. Justi wrote in the Introduction as follows:

It may be said without flattery that one of the main sources of the great flowering of French manufacture was chiefly to be sought in the Paris Academy of Sciences. It is certain that French manufactures are liked by foreigners chiefly for the beauty and durability of their colours; and this perfection of the colours is solely due to the efforts of the Academy. It always occupied some of its members in ceaseless experiments for the improvement of the colours, and these improvements were prescribed by law for the dyers by rules imposed by the government.[133]

As a remarkable example of a treatise in the *Descriptions* we cite the comments on scientific practice by the engineer J. R. Perronet (cf. p. 261) on the manufacture of pins in 1762 (Plate 17). By rational division of labour, a considerable increase of production was attained, although the machines used were still relatively simple. The activity of a single worker was changed to pure mass production. Manual operations began to be assessed according to the time occupied.

In a period of twelve hours, operatives draw 15 lb. of brass through the three first holes of the die plate; but they only manage 10 lb. through the smaller holes, as the wire has lengthened. It may be reckoned that a worker on the average draws 28 lb. of wire, for which he receives about one stiver per lb.; but he must supply the necessary hand tools while the master need contribute only the tartar. Moreover the maintenance of the tools diminishes his earnings by one third. . . .

The work which the wire cleaners have to undertake is usually hard; for the wire must be struck violently for a long time. Usually it is struck three times gently, then once hard. The wire cleaner does not carry on this work for more than an hour, in which time he can clean a coil of brass wire weighing from 25–30 lb. He then draws this wire through the die plate which is for him to some extent a rest; and as it should occupy a day to draw 15 lb. through three holes, he will need not more than half an hour to clean the wire which can be drawn in a single day.

The work of the wire straightener is exhausting for he can straighten in one hour 600 klafters; and as he runs through this length twice, when he goes to the machine, he produces 1,200 klafters or half a mile in an hour. . . .

When a coil of wire which is 5 klafters long is to be cut into pieces each 4 inches and 9 lines in length, this takes 22 minutes. For the straightening of wire of various thicknesses and cutting it up a worker receives 1 stiver for every dozen of pins which consists of 12 thousand; and he has to make a thirteenth dozen in addition to replace those which are spoilt or faulty. In one day he can cut up from 8–10 of these dozens for which he will earn from 8–10 stivers. . . .

The man who makes the points can hold 25 pieces of wire at a time on the pointing ring if the pins are thick; or 40 if they are fine. These are called a grip. . . . A worker can point in one day 15 dozen thousands of thick and thin pins and must add a thirteenth to replace waste. For 12,000 he receives 15 deniers ($\frac{1}{12}$ of 5 centimes), and in this fashion if he is able to work continuously he can earn each day 18 stivers and 9 denier. But the best manufacturers in Laigle sell daily through the bank no more than 7–8 dozen thousand pins which is only half the number that a worker can manufacture and this is just as well; for their chests suffer great injury from the copper dust which they inhale at every breath; and the glass screen protects their eyes indeed from the larger pieces but not from the fine dust. The one who turns the wheel is paid 1 stiver 9 deniers per dozen thousand from which a thirteenth is regularly deducted for waste. This appears higher than the pay received by the man who points the pins, whose work needs more skill and who suffers greatly from dust. But against that the man who turns the wheel has hard and exhausting work, and he must also, since he has not to turn continuously, fold the paper and undertake various other duties. . .

The polisher receives a stiver for every 12,000 pins and must similarly throw in the thirteenth. He makes as many pins as the man who puts on the points, and earns a fifth less than he. . . .

For cutting the shafts of various thicknesses, the worker receives 9 deniers per 12,000 and always throws in a thirteenth. He usually cuts in one hour 3 of these dozen thousand, and if he exerts himself a little he can cut as many as 4 dozen, so that in less than 3 hours

7 or 8 dozen thousand can be made; that is, as many as can be dealt with usually by the best merchants in Laigle in a single day. For this reason a single cutter can execute the work of two or three factories and in this way can earn some 15 stivers daily. . . .

In one minute the man who cuts the pin heads makes about 70 cuts with his scissors. When he has cut 12 times he taps with the broad side of the scissors to make the pegs again lie evenly. There are cutters who, although this rearrangement requires great accuracy and considerable dexterity, have nevertheless the skill to cut the whole set of 12 pegs one after another without needing to disturb the work by arranging the ends of the pegs together. Since the worker with his scissors can execute 70 cuts in a minute, this produces 4,200 in an hour; and as with every cut of the scissors he goes through 12 pegs, therefore such a person can complete 50,400 heads of fine pins in one hour. This would occur if he had to hurry with the work. But usually a worker can cut 30,000 heads mixed large and fine in one hour; nevertheless as the eyes become very tired with the effort he cuts daily only 15 dozen thousand. Such a person receives for turning 12 thousand heads 3 deniers, and for cutting them 9 deniers; and as he is able daily to cut 15 dozen thousand he earns 11 stivers and 3 deniers. . . .

A man can stamp in one minute 20 pin-heads mixed thick and fine; and as he strikes each head 5 or 6 times, the anvil receives from a 100 to 120 blows per minute. A stamper usually prepares 1,000 pins in an hour, and from 10–12 thousand in a day; not counting the thirteenth of waste. The stampers receive two different rates of wage; namely 9 stivers for 12,000 to which a thirteenth thousand must always be thrown in, for stamping thick pins from number 22 to number 14; and eight stivers for pins below these numbers which bring him 7 or 8 stivers per day; but they have themselves to provide the punches and dies, which altogether cost 10 stivers; they must also, if pins of other thicknesses have to be made, have them recut which costs every month about 2 stivers. . . .

A good woman pin sticker can put 4 dozen thousand pins into the papers. She receives 1 stiver for each dozen thousand. Usually they only put in 2 or 3 dozen. Furthermore they must sort the pins themselves so that faulty ones are laid aside. As these persons normally perform these three duties, pricking into the papers, arranging and selecting the pins, they earn 2 stivers and 6 deniers for each 12,000 large and fine pins. The strongest women earn more than 4 stivers a day. Children between 7 and 8 years of age can earn 1 stiver daily simply by pushing the pins into their papers. But these women usually print the sign or trade mark of the merchant on the papers. They get through about 1,000 per hour pressing the paper with the flat of the hand on to the wood-cut which is firmly attached to the table, and transferring the colour on to the wood-cut by means of a thick brush which has been dipped in cinnobar, prepared and mixed with size. . . .

Processes needed for manufacturing 12 thousand No. 6 pins 9 lines long:

	Livres	Stiver	Denier
Wire 1 lb. 9 oz. and 6 scruples	2	9	7
Straightening and cutting in long pieces		1	
Pointing		1	3
Turning the wheel for pointing		1	9
Polishing the points		1	
Turning the wheel for polishing		1	
Cutting the shafts			9
Turning the heads			3
Cutting the pegs			9
Fire to anneal the heads			3
Stamping			8
Tartar to scour the pins		1	
Inserting the pins in the paper		1	
The book of paper costs 6 stivers. 5 oz. 3 scruples are needed for every 12,000 pins No. 6 which amounts to		2	
For maintenance of hand tools and for losses may be reckoned		4	
As a result 12,000 pins No. 6 cost	3	7	3

[134]

THE LIBERAL ECONOMY

As against State patronage of economic undertakings, trade and manufacture under mercantilism, there appeared in English industry, and also in the circles of the French encyclopedists and physiocrats a marked tendency to economic liberalism. This individualist and liberal economic attitude, demanding free competition and free trade as the bases of a healthy economic system, was expressed by Adam Smith in 1776 in his work on *The Wealth of Nations*. We will cite a few passages from Smith's book in which he advocates the division of labour.

To take an example, therefore, from a very trifling manufacture; but one in which the division of labour has been very often taken notice of, the trade of the pin-maker; a workman not educated to this business (which the division of labour has rendered a distinct trade), not acquainted with the use of the machinery employed in it (to the invention of which the same division of labour has probably given occasion), could scarce, perhaps, with his utmost industry, make one pin in a day, and certainly could not make twenty. But in the way in which this business is now carried on, not only the whole work is a peculiar trade, but it is divided into a number of branches, of which the greater part are likewise peculiar trades. One man draws out the wire, another straights it, a third cuts it, a fourth points it, a fifth grinds it at the top for receiving the head; to make the head requires two or three distinct operations; to put it on is a peculiar business, to whiten the pins is another; it is even a trade by itself to put them into the paper; and the important business of making a pin is, in this manner, divided into about eighteen distinct operations, which, in some manufactories, are all performed by distinct hands, though in others the same man will sometimes perform two or three of them. I have seen a small manufactory of this kind where ten men only were employed, and where some of them consequently performed two or three distinct operations. But though they were very poor, and therefore but indifferently accommodated with the necessary machinery, they could, when they exerted themselves, make among them about twelve pounds of pins in a day. There are in a pound upwards of four thousand pins of a middling size. Those ten persons, therefore, could make among them upwards of forty-eight thousand pins in a day. Each person, therefore, making a tenth part of forty-eight thousand pins, might be considered as making four thousand eight hundred pins in a day. But if they had all wrought separately and independently, and without any of them having been educated to this peculiar business, they certainly could not each of them have made twenty, perhaps not one pin in a day; that is, certainly, not the two hundred and fortieth, perhaps not the four thousand eight hundredth part of what they are at present capable of performing, in consequence of a proper division and combination of their different operations.

In every other art and manufacture the effects of the division of labour are similar to what they are in this very trifling one; though, in many of them, the labour can neither be so much subdivided, nor reduced to so great a simplicity of operation. The division of labour, however, so far as it can be introduced, occasions, in every art, a proportionate increase of the productive powers of labour. The separation of different trades and employments from one another, seems to have taken place in consequence of this advantage. This separation, too, is generally carried furthest in those countries which enjoy the highest degree of industry and improvement; what is the work of one man in a rude state of society being generally that of several in an improved one. In every improved society, the farmer is generally nothing but a farmer, the manufacturer nothing but a manufacturer. The labour, too, which is necessary to produce any one complete manufacture is almost always divided among a great number of hands. How many different trades are employed in each branch of the linen and wollen manufactures, from the growers of the flax and the wool to the bleachers and smoothers of the linen, or to the dyers and dressers of the cloth! The nature of agriculture, indeed, does not admit of so many subdivisions of labour, nor of so complete a separation of one business from another, as manufactures. It is impossible to separate so entirely the business of the grazier from that of the corn-farmer, as the trade of the carpenter is commonly separated from that of the smith. The spinner is almost always a distinct person from the weaver; but the ploughman, the harrower, the sower of the seed, and the reaper of the corn, are often the same. The occasions for those different sorts of labour returning with the different seasons of the year, it is impossible that one man should be constantly employed in any one of them. This impossibility of making so complete and entire a separation of all the different branches of labour employed in agriculture, is, perhaps, the reason why the improvement of the productive powers of labour in this art does not always keep pace with their improvement in manufactures. The most opulent nations, indeed, generally excel all their neighbours in agriculture as well as in manufacture; but they are commonly more distinguished by their superiority in the latter than in the former.[135]

This great increase of the quantity of work which, in consequence of the division of labour, the same number of people are capable of performing, is owing to three different circumstances; first, to the increase of dexterity in every particular workman; secondly, to the saving of the time which is commonly lost in passing from one species of work to another; and lastly, to the invention of a great number of machines which facilitate and abridge labour, and enable one man to do the work of many.[136]

It is the great multiplication of the productions of all the different arts, in consequence of the division of labour, which occasions, in a well-governed society, that universal opulence which extends itself to the lowest ranks of the people. Every workman has a great quantity of his own work to dispose of beyond what he himself has occasion for; and every other workman being exactly in the same situation, he is enabled to exchange a great quantity of his own goods for a great quantity, or, what comes to the same thing, for the price of a great quantity of theirs. He supplies them abundantly with what they have occasion for, and they accommodate him as amply with what he has occasion for, and a general plenty diffuses itself through all the different ranks of the society.[137]

The natural effort of every individual to better his own condition, when suffered to exert itself with freedom and security, is so powerful a principle that it is alone, and without any assistance, not only capable of carrying on the society to wealth and prosperity, but of surmounting a hundred impertinent obstructions with which the folly of human laws too often incumbers its operations; though the effect of these obstructions is always more or less either to encroach upon its freedom, or to diminish its security. In Great Britain industry is perfectly secure; and though it is far from being perfectly free, it is as free or freer than in any other part of Europe.[138]

Emancipation from State patronage in favour of free and independent activity evolved very slowly on the Continent. In this respect England, as has been remarked, was far in advance. Linked with this development was the effort of the technically productive middle class to provide itself with institutions for technical training and further education. As an expression of liberal economic opinion, there had been founded already in 1754 in England the *Society for the encouragement of arts, manufactures and commerce*, which was by no means a state establishment, but was an association of men for the purpose of furthering their own independent technical work.

ENGLAND'S LEAD IN TECHNOLOGY

The intense economic and technical activity peculiar to the Puritan mind was also operative in the English colonies in America. The first great American scientist and technician, Benjamin Franklin, who was also a great statesman, exemplifies for us a violent impulse toward economic activity and a frank seeking after gain, though here, in contrast to the

Puritanism of the seventeenth century, the religious motive was more in abeyance.

Remember, that *time* is money. He, that can earn ten shillings a day by his labour, and goes abroad, or sits idle one half of that day, though he spends but sixpence during his diversion or idleness, ought not to reckon *that* the only expense; he has really spent, or rather thrown away, five shillings besides. . . .

. . . In short, the way to wealth, if you desire it, is as plain as the way to market. It depends chiefly on two words, *industry* and *frugality*; that is, waste neither *time* nor *money*, but make the best use of both. Without industry and frugality nothing will do, and with them everything. He, that gets all he can honestly, and saves all he gets (necessary expences excepted) will certainly become *rich*—if that Being who governs the world, to whom all should look for a blessing on their honest endeavours, doth not, in his wise providence, otherwise determine.[139]

We have briefly spoken above of the great technical progress during the eighteenth century in the English motherland (cf. p. 233). The rapid technical development in the second half of the eighteenth century was to a considerable extent caused by the increased activity of iron foundries, which was marked by a series of significant inventions of which the most important was the discovery of a process to smelt serviceable pig-iron from the ore by coke extracted from coal instead of by charcoal from wood that was becoming ever harder to obtain. We owe to Abraham Darby the Elder the first beginnings of this discovery about 1717. His son, Abraham Darby the Younger, took an important part in developing the process. A letter of the year 1775 from Abiah, wife of the younger Darby, to a Quaker friend, gives us a clear picture of this development. It throws light at the same time on the atmosphere in which these austere, industrious Quaker families, devoted to technological progress, who played so important a part as iron-masters in establishing iron foundries, did their work. Not only the Darby family, but also Benjamin Huntsman, Sampson Lloyd, Jeremia Homfray, the Reynolds family and many others who contributed to the establishment of English foundries, were Quakers.

Esteemed Friend, Sunniside.

. . . I cannot help lamenting with thee in thy just observation, 'that it has been universally observed, that the Destroyers of mankind are recorded and remembered, while the Benefactors are unnoticed and forgotten.' This seems owing to the depravity of the mind, which centres in reaping the present advantages, and suffering obscurity to vail the original causes of such benefits; and even the very names of those to whom we are indebted for the important discoveries, to sink into oblivion. Whereas if they were handed down to posterity, gratitude would naturally arise in the commemoration of their ingenuity, and the great advantages injoyed from their indefatigable labours—I now make free to communicate what I have heard my Husband say, and what arises from my own knowledge; also what I am inform'd from a person now living, whose father came here as a workman at the first beginning of these Pit Coal Works.

Then to begin at the original. It was my Husband's Father, whose name he bore (Abraham Darby and who was the first that set on foot the Brass Works at or near Bristol) that attempted to mould and cast Iron pots, &c., in sand instead of Loam (as they were wont to do, which made it a tedious and more expensive process) in which he succeeded. This first attempt was tryed at an Air Furnace in Bristol. About the year 1709 he came into Shropshire to Coalbrookdale, and with other partners took a lease of the works, which only consisted of an old Blast Furnace and some Forges. He here cast Iron Goods in sand out of the Blast Furnace that blow'd with wood charcoal; for it was not yet thought of to blow with Pit Coal. Sometime after he suggested the thought, that it might be practable to smelt the Iron from the ore in the blast Furnace with Pit Coal: Upon this he first try'd with raw coal as it came out of the Mines, but it did not answer. He not discouraged, had the coal coak'd into Cynder, as is done for drying Malt, and it then succeeded to his satisfaction. But he found that only one sort of pit Coal would suit best for the purpose of making good Iron.—These were beneficial discoveries, for the moulding and casting in sand instead of Loam was of great service, both in respect to expence and expedition. And if we may compare little things with great—as the invention of printing was to writing, so was the moulding and casting in Sand to that of Loam. He then erected another Blast Furnace, and enlarged the Works. This discovery soon got abroad and became of great utillity.

This Place and its environs was very barren, little money stiring amongst the Inhabitants. So that I have heard they were Obliged to exchange their small produce one to another instead of money, until he came and got the Works to bear, and made Money Circulate amongst the different parties who were employed by him. Yet notwithstanding the Service he was of to the Country, he had opposers and ill-wishers.

My Husband's Father died early in life; a religious good man, and an Eminent Minister amongst the people call'd Quakers.

My Husband Abraham Darby was but Six years old when his Father died—but he inherited his genius—enlarg'd upon his plan, and made many improvements. One of Consequence to the prosperity of these Works was as they were very short of water that in the Summer or dry Seasons they were obliged to blow very slow, and generally blow out the furnaces once a year, which was attended with great loss. But my Husband proposed the Erecting a Fire Engine to draw up the Water from the lower Works and convey it back into the upper pools, that by continual rotation of the Water the furnaces might be plentifully supplied; which answered Exceeding Well to these Works, and others have followed the Example.

But all this time the making of Barr Iron at Forges from Pit Coal pigs was not thought of. About 26 years ago my Husband conceived this happy thought—that it might be possible to make bar from pit coal pigs. Upon this he Sent some of our pigs to be tryed at the Forges, and that no prejudice might arise against them he did not discover from whence they came, or of what quality they were. And a good account being given of their working, he errected Blast Furnaces for Pig Iron for Forges. Edward Knight Esqr a capitol Iron Master urged my Husband to get a patent, that he might reap the benefit for years of this happy discovery: but he said he would not deprive the public of Such an Acquisition which he was Satisfyed it would be; and so it has proved, for it soon spread, and Many Furnaces both in this Neighbourhood and Several other places have been errected for this purpose.

Had not these discoveries been made the Iron trade of our own produce would have dwindled away, for woods for charcoal became very Scarce and landed Gentlemen rose the prices of cord wood exceeding high—indeed it would not have been to be got. But from pit coal being introduced in its stead the demand for wood charcoal is much lessen'd, and in a few years I apprehend will set the use of that article aside.

Many other improvements he was the author of. One of Service to these Works here they used to carry all their mine and coal upon horses' backs but he got roads made and laid with Sleepers and rails as they have them in the North of England for carrying them to the Rivers, and brings them to the Furnaces in Waggons. And one waggon with three horses will bring as much as twenty horses used to bring on horses' backs. But this laying the roads with wood begot a Scarcity and rose the price of it. So that of late years the laying of the rails of cast Iron was substituted; which altho' expensive, answers well for Ware and Duration. We have in the different Works near twenty miles of this road which cost upwards of Eight hundred pounds a mile. That of Iron Wheels and axletrees for these waggons was I believe my Husband's Invention.

He kept himself confined to the Iron Trade and the Necessary Appendages annex'd thereto. He was just in his dealings—of universal benevolence and charity, living Strictly to the Rectitude of the Divine and Moral Law, held forth by his great Lord and Saviour, had an extraordinary command over his own spirit, which thro' the Assistance of Divine Grace enabled to bear up with fortitude above all opposition: for it may seem very strange, so valuable a man should have Antagonists, yet he had. Those called Gentlemen with an Envious Spirit could not bear to see him prosper; and others covetious; strove to make every advantage by raising their Rents of their collieries and lands in which he wanted to make roads; and endeavour'd to stop the works. But he surmounted all: and died in Peace beloved and Lamented by many.[140]

After John Kay had in 1733, by the invention of the high speed shuttle, accelerated the process of weaving, there was an alternation between the progressive mechanization of spinning and of weaving. A more efficient weaving apparatus led to a shortage of yarn and thereby necessitated an acceleration of the process of spinning. Then a further improvement of the apparatus for spinning necessitated a yet greater mechanization of the loom. We will not discuss the individual discoveries in this sphere, each conditioned by the other. This development, notably accelerated by the permission granted by the English Parliament in 1774 for the use of indigenous pure cotton cloth, soon demanded a better source of power than the human muscle. Water-power for spinning machines, introduced in 1775, was an improvement; but water-power was not always available in sufficient measure. So the demand grew for an effective source of power independent both of weather and of season. Here help was forthcoming from Watt's steam engine with its mechanical power.

Starting from Newcomen's atmospheric steam-engine (Plate 13) which did not work very economically, Watt succeeded in designing an engine which consumed only a quarter of the amount of coal needed by the old Newcomen engine. Watt achieved this, no longer like Newcomen, by atmospheric pressure but by using the pressure of steam, and furthermore by surrounding the cylinder with a further steam jacket and thus preventing the harmful initial condensation of the steam, and finally by introducing a condenser separated from the cylinder. We encounter in Watt's successful efforts to achieve a serviceable steam-engine, two lines of development: the constructive, which derived from Guericke, through Papin, Savery and Newcomen; and the physical, which is characterized by the work of the seventeenth and eighteenth century on the phenomena of heat. Watt's engine, which was to be further perfected by him, was the product of a mind with real genius for construction, and in contrast to that of Newcomen it is the product of technology governed by science, of the application of physical knowledge of the properties of steam ascertained by experiment and measurement.

To convert the inventive thought of this genius into practice was not an easy task. Watt also was plagued by difficulties in the production of accurate cylinders. Here there was help from John Wilkinson's invention of an improved boring machine. Finally the happy union of Watt with a capable business man, Matthew Boulton, contributed to the triumphal march of the new invention. Watt's first steam-engine operated in 1776 in Wilkinson's iron foundry. Soon it was working in mines, in foundries, and from the eighties it was the source of motive power, especially in textile factories, and for productive work in many places. The development of ever new tools and engines of increasing size and capacity was first rendered possible by Watt's steam-engine, which supplied ample power.

Watt's famous first patent for a steam-engine may be given here as one of the most important documents in the history of technology of the eighteenth century.

SPECIFICATION OF PATENT, JANUARY 5TH, 1769, FOR A NEW METHOD OF LESSENING THE CONSUMPTION OF STEAM AND FUEL IN FIRE ENGINES.

TO ALL TO WHOM these presents shall come, I, JAMES WATT, of Glasgow, in Scotland, Merchant, send greeting.

WHEREAS His Most Excellent Majesty King George the Third, by his Letters Patent, under the Great Seal of Great Britain, bearing date the fifth day of January, in the ninth year of his said Majesty's reign, did give and grant unto me, the said JAMES WATT, his special licence, full power, sole privilege and authority, that I, the said JAMES WATT, my executors, administrators, and assigns, should, and lawfully might, during the term of years therein expressed, use, exercise, and vend throughout that part of his Majesty's Kingdom of Great Britain called England, the Dominion of Wales, and Town of Berwick upon Tweed, and also in his Majesty's Colonies and Plantations abroad, my new invented 'METHOD OF LESSENING THE CONSUMPTION OF STEAM AND FUEL IN FIRE ENGINES;' in which said recited Letters Patent is contained a Proviso, obliging me, the said JAMES WATT, by writing under my hand and seal, to cause a particular description of the nature of the said invention to be inrolled in his Majesty's High Court of Chancery, within four calendar months after the date of the said recited Letters Patent, ...

NOW KNOW YE, that in compliance with the said Proviso, and in pursuance of the said Statute, I, the said JAMES WATT, do hereby declare that the following is a particular description of the nature of my said invention and the manner in which the same is to be performed (that is to say): MY METHOD of lessening the consumption of steam, and consequently fuel, in fire engines, consists of the following principles:

FIRST, that vessel in which the powers of steam are to be employed to work the engine, which is called the *Cylinder* in common fire engines, and which I call the *Steam Vessel*, must, during the whole time the engine is at work, be kept as hot as the steam that enters it; first, by inclosing it in a case of wood, or any other materials that transmit heat slowly; secondly, by surrounding it with steam or other heated bodies; and, thirdly, by suffering neither water nor any other substance colder than the steam to enter or touch it during that time.

SECONDLY, in Engines that are to be worked wholly or partially by condensation of steam, the steam is to be condensed in vessels distinct from the steam vessels or cylinders, although occasionally communicating with them: these vessels I call *Condensers*; and, whilst the engines are working, these condensers ought at least to be kept as cold as the air in the neighbourhood of the engines, by application of water, or other cold bodies.

THIRDLY, whatever air, or other elastic vapour, is not condensed by the cold of the condenser, and may impede the working of the engine, is to be drawn out of the steam vessels or condensers by means of pumps, wrought by the engines themselves, or otherwise.

FOURTHLY, I intend in many cases to employ the expansive force of steam to press on the pistons, or whatever may be used instead of them, in the same manner as the pressure of the atmosphere is now employed in common fire engines: in cases where cold water cannot be had in plenty, the engines may be wrought by this force of steam only, by discharging the steam into the open air after it has done its office.

FIFTHLY, where motions round an axis are required, I make the steam vessels in form of hollow rings, or circular channels, with proper inlets and outlets for the steam, mounted on horizontal axles, like the wheels of a water-mill; within them are placed a number of valves, that suffer any body to go round the channel in one direction only: in these steam vessels are placed weights, so fitted to them as entirely to fill up a part or portion of their channels, yet rendered capable of moving freely in them by the means hereinafter mentioned or specified. When the steam is admitted in these engines, between these weights and the valves, it acts equally on both, so as to raise the weight to one side of the wheel, and by the re-action on the valves successively, to give a circular motion to the wheel, the valves opening in the direction in which the weights are pressed, but not in the contrary: as the steam vessel moves round, it is supplied with steam from the boiler, and that which has performed its office may either be discharged by means of condensers, or into the open air.

SIXTHLY, I intend, in some cases, to apply a degree of cold not capable of reducing the steam to water, but of contracting it considerably, so that the engines shall be worked by the alternate expansion and contraction of the steam.

LASTLY, instead of using water to render the piston, or other parts of the engines, air and steam-tight, I employ oils, wax, resinous bodies, fat of animals, quicksilver, and other metals, in their fluid state.

IN WITNESS whereof I have hereunto set my hand and seal this twenty-fifth day of April, in the year of our Lord one thousand seven hundred and sixty-nine.

JAMES WATT.

Sealed and delivered in the presence of

COLL. WILKIE.
GEO. JARDINE.
JOHN ROEBUCK.

BE IT REMEMBERED, that the said JAMES WATT doth not intend that anything in the Fourth Article shall be understood to extend to any engine where the water to be raised enters the steam vessel itself, or any vessel having an open communication with it.

JAMES WATT.
Witnesses. COLL. WILKIE.
GEO. JARDINE.

And be it known that the aforesaid James Watt appeared on the twentyninth day of April in the year of our Lord 1769 at the Chancellory of our Royal Lord, and acknowledged the above description with all that is therein included and described. Wherefore the aforesaid description shall be stamped in accordance with decree of the sixth year of the reign of the deceased King William and Queen Mary of England etc.

Inrolled the twenty-ninth day of April, in the year of our Lord one thousand seven hundred and sixty-nine.[141]

At the same time as the first Watt steam engines moved their parts, J. Smeaton appeared with his improved atmospheric steam engine. Smeaton, without changing anything essential in the method of work of Newcomen's engine, had by experiment and calculation determined the most favourable dimensions for the engine and had had the individual parts made with special care.

In the spirit of a scientifically conducted technology, this famous engineer succeeded in developing the atmospheric engine to the highest possible perfection. But, thanks to Watt's brilliant genius for discovery, his engine required only half as much coal for the same output as Smeaton's atmospheric engine.

England's leading position in technology and industry was undeniable at this period. On the Continent development took place much more slowly. It is true that the first Watt steam engine to be constructed in Germany, a simple acting low pressure engine for water supply, started work in 1785 at the Hettstedt mine, not far from Mansfeld; but in other branches of technology it gained a place in Germany only after the turn of the century.

German visitors to England at the end of the eighteenth century were amazed at the achievement of the English foundries, manufacturers and factories, which were for their time gigantic technical marvels. In 1791 the young twenty-year-old Georg Reichenbach who with J. Utzschneider and Joseph Liebherr was later to found a mathematical-mechanical Institute at Munich, on a journey to England together with the Bavarian engineer inspector Joseph von Baader in order to study English engines, saw in the factory of Boulton and Watt in Soho, Watt's double acting steam engine for rotary motion. Reichenbach secretly executed a sketch (Fig. 52) and wrote concerning it as follows in his diary:

Today after our five o'clock meal we travelled by Post from London to Birmingham, and on the 10th July, 1791, we arrived here at mid-day. On the same day we drove to Soho to visit Mr. Boulton; there Herr Baader explained to him the object with which I had come, namely in order to study the mechanism of the Watt fire-engine; but he did not seem very pleased at this for his character is very secretive. We had supper with him and at 10 o'clock on the same evening we drove to Birmingham. The next morning we visited Boulton again and were successful in seeing Mr. Watt. There we stayed the whole day and I was able to see some fire engines. In the evening at 12 o'clock Herr von Baader left Birmingham for Wigan and the iron foundry there. On the following day (12th June) I left my inn to go to Soho but unluckily I lost my way for four hours as I could not ask anybody the way; when I at last found it and went to my right quarter I was truly perturbed at my unfortunate position regarding learning anything and at having been separated from all my friends—but I endeavoured as speedily as possible to reconcile myself to this unpleasant position; and I soon observed it had its advantageous side, for I was able by giving a few small tips to obtain the opportunity, despite the secrecy of Mr. Watt and Mr. Boulton, thoroughly to study the mechanism of the fire—or steam engine. I worked at my drawings for six weeks, for I had to maintain secrecy not only against Mr. Boulton but also against all the workers who were there. For this reason, this work cost me indescribable labour, for not only could I ask no questions of anybody but also might not for fear of arousing suspicion; so on the other hand I was only allowed to look at them at certain times. . . .

This is the arrangement of the Watt engine (Fig. 52) . . . The steam boiler A must bear to the content of the cylinder B the relation

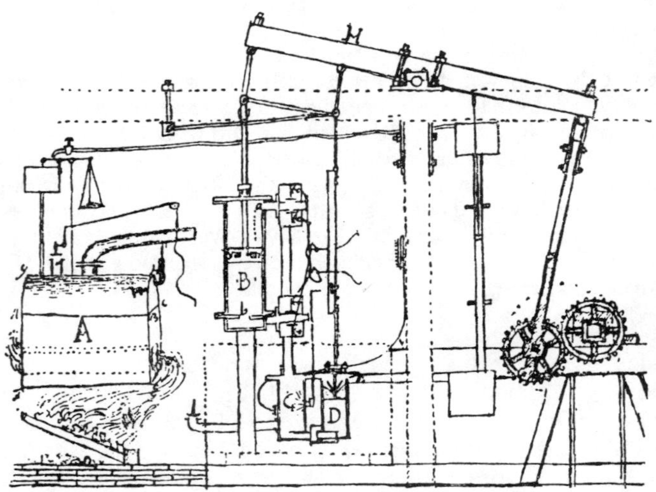

FIG. 52. *Watt's double-acting steam engine with rotary drive*, 1791. Sketch by G. Reichenbach.—From: *Reichenbach's englisches Tagebuch* 1791. MS. in the collection of manuscripts in the German Museum, Munich. Length of the beam, about 4·75 m.

170,035 to 11,793, or to put it more shortly 14·418 to 1 or even more briefly, almost 14½ to 1. . . .

The engine made 24 strokes per minute; the consumption of steam per second without loss is 5·04 cubic feet. The area of the valves is seven square inches. The speed of the steam flowing through the valve is 104 feet per second. The fire and its circulation round the boiler covered a surface of 80 square feet. Accordingly 16 square feet of surface exposed to the fire and its circulation gives 1 cubic foot of steam per second. . . .

Mr. Watt had brought the durability and usefulness of the engine to the very highest pitch and he gives his steam a pressure of 8 feet of water above atmosphere; that is to say the pressure of this steam against a complete vacuum would be equal to the weight of a column of water of 41 feet, by English measure.[142]

TECHNOLOGY AND SCIENCE

There was a great effort in the eighteenth century to apply science to technological production. Thus the French engineer B. F. de Bélidor sought as early as the second third of the eighteenth century, as Christian Wolff expressed it 'to apply the theory of mechanics to machines',[30] more than hitherto. In Bélidor, Wolff saw the realization of his own efforts and those of his contemporaries to improve craftsmanship by the use of mathematical science and thereby to advance human happiness. Bélidor was a pioneer in the application of the recently discovered integral calculus to the solution of hydrotechnical problems, that is to say for tasks 'which concerned wholely and solely practical execution'.[31] Bélidor assisted in establishing the renown of applied sciences in France, which was further enhanced during the last third of the eighteenth century by C. A. de Coulomb, J. R. Perronet and M. R. de Prony, to mention only three names. England's course lay rather in the direction of purely practical experimentation. But we must not overlook the great importance for Watt's steam-engine of scientific experiments on the theory of heat.

The application of science to technology all too often encountered opposition from purely practical men, especially since sometimes it did not attain the desired success. Pope Benedict XIV in 1742–3 ordered a static examination of the dome of St. Peter's, which showed signs of damage. A Commission of three mathematicians among other things gave advice as to the cause of the damage, and the methods to cure it. They endeavoured to solve the problem mathematically by noteworthy mechanical considerations which would be regarded today as inadequate. Many voices indeed were raised against 'mathematical' methods. G. Polenia reports, among other things, the objections of an unknown critic.*

* The remarks of the unknown critic are placed here in inverted commas.

'Consideration of the damage which was observed on the dome of St Peters, the causes of this damage and the appropriate means to remedy them by no means constitutes one of those occasions which make mathematical theory more necessary than experience. . . .

. . . This assertion can also be confirmed by a consideration of the suggestions made by the three mathematicians in their report to the architects. Their opinion is proved to be . . . false. . . . Since St Peter's dome was originally projected, designed and constructed without mathematicians and above all without the mechanics, so much cultivated today, it will also be possible to reconstruct it without first requiring mathematicians and mathematics.' And he (the unknown author) shows that Buonarroti was no master of mathematics, and yet was able to design the dome of the Vatican.

'I yield to no one in proper esteem which I credit to beautiful, pure and clear sighted science. And with esteem I unite love; otherwise I would not have devoted to the study of this science through many years as much time as other preoccupations allowed me. But just because I greatly esteem this science, its misuse fills me with anger; particularly since I know that it is the usual misfortune of the best things to be most violently abused.'[143]

Frederick the Great in a letter to Voltaire in 1778 most amusingly ridiculed a water-raising plant made from calculations by Euler which were based on mathematical principles, but did not function.

25th Jan. 1778

. . . The English have built ships with the most advantageous section in Newton's opinion, but their admirals have assured me that these ships did not sail nearly so well as those built according to the rules of experience. I wanted to make a fountain in my garden. Euler calculated the output of the wheels which should have raised the water into a reservoir, from which it was to flow again through canals and again mount on high in the fountains at Sans Souci. My lifting-gear was carried out according to mathematical calculations but could not raise a drop of water to fifty paces from the reservoir. Vanity of vanities! Vanity of mathematics![144]

Witness to the successful application to technical problems of science based on experiment and calculation is borne by the works of Coulomb in the last quarter of the eighteenth century. The editor of the 2nd edition of Coulomb's *Théorie des Machines Simples* composed in 1781, wrote as follows:

Everything is coherent in the development of the human spirit. When science makes progress, crafts and industry also improve. This has been observed in France, especially during the last thirty years or so. And Coulomb made an extraordinary contribution to this improvement by his treatises in which he described his researches. In them may be found numerous experiments which are described with all the necessary details to carry conviction of the accuracy of the results obtained, which are amply portrayed by analytical treatment. This latter requires mathematical knowledge in order to be understood. But craftsmen who are little versed in mathematical science can nevertheless study the whole experimental part with profit.

It is just by the study of the works of great men that we develop our own knowledge. He who neglects this study and regards it as of little use will not succeed in acquiring the theory of the craft that he practices. And even if a few rise by their own resources above their fellow-workers by talent, fortunate gifts and favourable circumstances, they have only succeeded in finding a theory which, so to speak, leads them unconsciously after much seeking. All these inconveniences are avoided by those craftsmen who have accustomed themselves to read and study the works of the experts who have written concerning their crafts. It is just for this class of craftsman that the works of Coulomb . . . can be of great use. There they will find described the ingenious means invented by the author to give to his experiments the greatest attainable accuracy. They will also recognize there the care which he exercised to avoid, in the course of experiments, those accidents which the men working under him might have encountered.[145]

Most notably, Coulomb sought (1776) very ably to apply mathematical expressions of Maxima and Minima to the problems of structural statics. Thus he was able to calculate correctly the relative strength of a horizontal flexible beam of rectangular cross-section and built into a wall at one end, which Galileo had not succeeded in doing. We give as examples of Coulomb's illuminating work passages on the determination of the pressure of earth against revetment-walls (Fig. 53) and the calculation for vaults.

On Earth Pressure

We will turn to the determination of earth pressure (Fig. 53) against revetment-walls which hold the soil. . . . We will assume a closed right-angled triangle as the plane of the angle of shear of which one of the two sides containing the right angle will be perpendicular, and the hypotenuse will coincide with an inclined plane down which it seeks to slip. If now this triangular prism (of earth) urged by its own weight is held by a horizontal force consisting of cohesion and friction applied along the whole length of the hypotenuse, we shall be able easily to determine the reaction of this horizontal force by the principles of statics. We notice further that the Earth-mass which is here regarded as homogeneous may sever in shearing, not merely along a straight line, but along some curve. It follows that in order to support the pressure of a surface against a vertical wall

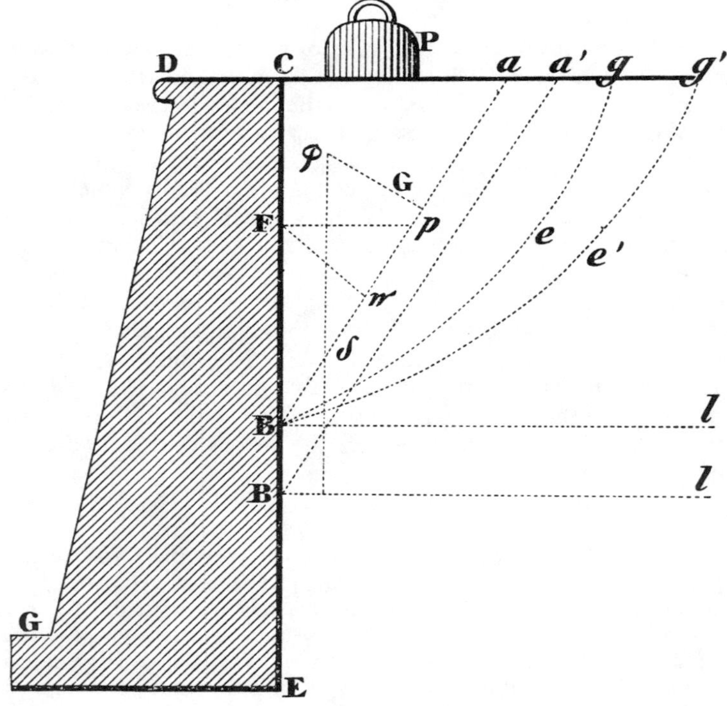

FIG. 53. *Coulomb's theory of earth pressure*, 1776.
Drawing from: Mém. de mathém. prés. à l'Acad. des Sciences, Paris. (1776)

we must seek from all the planes which can be inclined to the indefinite vertical wall and which are forced diagonally downward by their weight and restrained by friction and cohesion, that one in which the horizontal force required for balance is at a maximum. For it is clear that any other figure which would require a smaller horizontal force for balance could not shear the homogeneous mass of earth. As experience shows us, an approximately straight line is the line of cleavage of the earth-mass which would move the wall, it is in practice sufficient, from among all the triangles which press on a vertical wall to choose that one which requires the greatest horizontal force to restrain it. As soon as this force is determined the dimensions of the revetment-wall can easily be deduced from it.

On Vaults

Whatever be the number and direction of the forces exercised on a vault which is built according to the preceding assumptions (in which no account is taken of friction and cohesion in the joints of the stones of the vault) the shape of this vault is, as I have proved, like that of a catenary which would be pulled downward by the same forces. . . . The formulae which have been found, which do not allow for friction and the cohesion of the joints, cannot, however, be used in practice. All geometricians who have been concerned with this matter are convinced of this. So in order to achieve results that can be used, calculation must be based on assumptions that are close to Nature. These assumptions consist usually of the division of the vault into several parts after which the conditions of equilibrium of these parts must be sought. But as this division is fairly arbitrary I have sought, by laws of maxima and minima to determine which would be the actual points of fracture in vaults that are too weak. . . I have as far as possible endeavoured to describe clearly the principles I have used so that even a craftsman of little education will understand them and be able to use them.[146]

French technology, which was distinguished by the application of scientific principles to the solution of practical problems, found at the end of the century in the *Polytechnic* at Paris a worthy institution for study and cultivation. This institution of higher education in the field of the scientific study of technology owed its foundation in 1794–5 primarily to the military needs of France in the Revolutionary Wars. The chemist A. F. de Fourcroy, who was especially helpful in spreading the teaching of Lavoisier, outlined briefly in 1794 the numerous divisions of the engineering profession for which the *Polytechnic* was to supply the scientific foundation. There were then attached to the *Polytechnic* numerous specialised colleges.

> We need:
>
> 1. Military engineers for the construction and maintenance of our fortified establishments, for attack on and defence of places and military camps, for the construction and maintenance of military buildings such as barracks, Arsenals etc.
> 2. Engineers for bridges and roads [civil engineers] to build and maintain supply lines on land and on water, such as roads, bridges, canals, sluices, harbours, dams, light-houses, and buildings for the navy.
> 3. Surveyors for the production of general and special maps of land and sea.
> 4. Mining engineers for the discovery and use of minerals, for the treatment of metals and for the improvement of foundry processes.
> 5. Marine engineers, for the navy, who will be in charge of the construction of all marine transport, who will provide the ships with the best characteristics for their special services, and who will be responsible for the supply to the harbours of timber and all kinds of materials.[147]

In the structure of the first German Polytechnic, which was founded in Karlsruhe in 1825, there may be recognized a certain amount of dependence on the Paris Polytechnic (p. 317).

A brief glance should be directed at the effort characteristic of Lutheran Pietism in Germany to increase the scientific and technological culture of its youth. The consideration of objects of technology and natural science was, to be sure, intended primarily to serve Pietistic devotion,[32] but also responded to the general inclination of the period toward these subjects. The Secondary School for Modern Subjects founded in Halle in 1708 by Christoph Semler and the Economic and mathematical Modern School of Johannes Julius Hecker founded in Berlin in 1747 flourished in this spiritual soil. The recognition of the cultural value of work brought forth a series of educational works of a thoroughly practical character, which eagerly adopted technical and technological subjects. The authors were primarily educationalists, above all Pietistic theologians. The emphasis, characteristic of Pietism, on the study of nature for the glory of God, and the effort to combine rationalism with empiricism, brought about a certain resemblance between these efforts and those of English Puritanism.[33]

PART VI

THE PERIOD OF INDUSTRIALIZATION

PART VI

INTRODUCTION

THE nineteenth century was a period of rapid industrialization. This process was accompanied by widespread economic and social upheaval. The steam engine as a source of motive power, numerous newly invented production machines, and iron as an important raw material began their triumphal march. The agents of this introduction of technology were the middle class whose minds were filled with the liberalism that had erupted in the French revolution. The old guild organization with its patriarchal conditions disappeared with industrialization. The position of the indigent industrial workers developed from this. Social agitation developed.

Whereas industrizaliation in England was accomplished by free enterprise, in Germany the State had to take part in the development of national industry. The formation of an industrial State in Germany was, as has been clearly shown especially by Franz Schnabel,[34] not after all an extensive cultural problem, and this for two reasons. On the one hand it was necessary to raise the general standard of the people's education, in order to increase spiritual and also material needs and in order to promote the understanding by the working classes of the new machinery; and on the other hand it was necessary to form a body of more highly endowed technologists able to meet the increased scientific and technological demands of industry. This great cultural task, however, required the attention of the State. Moreover, industrial enterprise needed stimulation. The spirit of free enterprise for the benefit of the State could only be expected if individuals were encouraged to share in the community life. A prerequisite of industrial activity for the benefit of the State was that the industrialist should have a larger share than hitherto in the destinies of the State. So industrial development and constitutional aspirations were closely linked.[35] And the same was true of the mass of the people. Here also the co-operation of the people could be demanded only if they were also given a voice, self-respect, and self-government.[36]

The steam engines which characterized the epoch of industrialization extended also to transport. Steamship and steam railway were the new means of transport, without which the great rise of production in the factories would have been impossible. The telegraph using the galvanic current helped the news services to conquer Space and Time. And the new method of illumination by gaslight enabled factories to work day and night.

'There must be light when a worker has to control 840 threads' said Christian Peter Wilhelm Beuth of the cotton-spinning factories in Manchester, illuminated by gaslight (cf. p. 298). Besides the steam engines, other sorts of prime movers were introduced in the last third of the nineteenth century which, however, we will discuss later (cf. p. 339).

The nineteenth century prepared the way to an ever greater extent for scientific technology. In the first half of the century architecture in particular used scientific technological methods—we think of the French building engineers—but so also did the precision mechanical and optical industry; we mention here especially Joseph von Fraunhofer. In Germany scientifically conducted construction of machines began only in the middle of the century. France with her *École Polytechnique* was in this respect more advanced. The application of technology in the nineteenth century in its turn exercised an influence on scientific research. Technology provided research with tools which permitted entirely new methods.

The middle class of the nineteenth century, in spite of considerable adverse criticism and in spite of the burning social question arising especially from the speedy rise of technology, was animated by a firm belief in the irresistible advance of science and technology which would continue indefinitely, of its own accord. This faith in the power of progress enabled them to undertake the boldest technical projects.

With industrialization, began a great increase in population. Between 1800 and 1940 the population of the Latin countries doubled; in German countries it trebled. It is easy to see the cause of this increase of population in the rapid development of economics, technology and hygiene. We, however, should like to agree with E. Wagemann[37] that these external influences were not demographically decisive. It seems probable here that certain internal factors may have been at work. In Slav countries where technical progress was much smaller, we have to record between 1800 and 1940 a fourfold increase of population. And population figures in the last century and a half have risen as much in Asia as in the regions of Western culture. Rapidly increasing industrial production may therefore itself have been rendered possible at least in part by a rapid rise of population.

Individualism and liberalism in the period of the first development of industry gave way later to new alliances which took place among the workers in their own organizations, and among the employers in various sorts of combines such as syndicates, cartels, and trusts. In the development of joint-stock companies a democratization of capitalism took place.

THE WIDENING USE OF THE STEAM-ENGINE

Watt's double acting steam-engine with rotary motion (Fig. 52) was the chief factor in the speedy development of industry. We have mentioned that the steam-engine invaded the cotton-spinning industry in 1787. In Germany between 1794 and 1825 the engineer August Friedrich Wilhelm

Holtzhausen was foremost in constructing first the atmospheric, then the single and finally the double acting Watt steam-engine. Holtzhausen constructed his engines in Gleiwitz in Upper Silesia. The first steam-engine in Westphalia was also built in Silesia, in 1801. They stimulated the able Westphalian cabinet-maker and mechanician Franz Dinnendahl to build steam-engines himself. Dinnendahl gives us in his autobiography a vivid description of the great difficulties encountered in producing these engines —in the first place an atmospheric engine, and then those 'according to the new principle' of Watt, and especially in the manufacture of their various parts.

In the year 1801, before the districts of Essen and Werden had united with Prussia, I built the first heat engine on the old principle at the works of the aforesaid Wohlgemuth Mine in Werden. The whole personnel of the District Department of Mines, especially Mr Crone, and even foreign mining authorities who had had the opportunity of seeing steam-engines, all doubted that I should be able to accomplish such a work. Some swore that it was impossible, and others prophesied my downfall since, because as an ordinary craftsman I was getting on well, I was undertaking work beyond my sphere. Certainly it was a weighty undertaking, especially since in those parts there was not even a smith competent to make a proper screw, let alone other parts of the engine such as valve gear, piston rods and boiler-work or who understood drilling and turning. I myself understood cabinet-making and carpentry; but I now had to undertake smith's work without ever having learnt it. Nevertheless I forged almost the whole engine with my own hand, including even the boiler; so that for 1 to 1½ years I was engaged almost entirely on smith's work, and therefore myself replaced the lack of such craftsmen. But also lacking in this district were well equipped sheet works and expert sheet-workers, wherefore nearly all the plates of the first boiler were imperfect and brittle. Equally imperfect were those parts of the engines which had to be produced by foundries, such as cylinders, steam pipes, pumps for the shafts, pistons etc. This obstacle was also overcome by the ingenuity of Mr Jacoby, owner of the Sterkrade foundries in the region of Dislaken, near Wesel, to whom I communicated my ideas; as a result I succeeded in getting this foundry to supply all the parts necessary for an engine—at first indeed imperfectly, but now as well finished as possible. The boring of the cylinder presented me with fresh obstacles; but again I did not allow myself to be dismayed and I manufactured a drilling-machine without ever having seen one.

So after indescribable obstacles that would perhaps have deterred many other persons in my position, I was able at length to get so far that the first machine on the old principle was complete. . . .

During this time, I heard that engines were being built on a new principle by Boulton and Watt; I had never either read of them nor seen them, but I pondered day and night as to whether and how I could accomplish this. At length I learnt that such a machine had been built at the salt works at Königsborn near Unna. I accordingly went there, inspected it, and had hardly considered it for an hour before I was so well acquainted with it that I felt able to build such a machine myself.

Through the first engine that I had built here on the original principle, I had gained the confidence of the public who in fact knew little or nothing of any difference between an engine on the old principle and one on the new. In the meanwhile, shortly after the construction of the first engine, I had been invited through the recommendation of Frau von Scherp zu Baldeney to build a steam engine for a lead mine in the neighbourhood of Aachen. I went there, made a contract with the works and undertook the erection of a 32 inch machine on the old principle for 5,000 Imperial dollars. But as I had to construct the pit plant and the building at my own expense, I had stipulated at least 3,000 Imperial dollars too little which actually put me in no little difficulty, especially as I was anxious to attempt to construct this engine on the new principle. I had no means as I had built the first machine, which was 20 inches, for 2,400 Imperial Marks and thus had earned nothing from it. So I had my doubts as to whether I should be able to put the matter through. But my courage did not desert me even now; rather were all these circumstances a spur to exert myself yet further. I myself with my assistants and a brother worked day and night until the engine was completed. As a precaution, I had arranged this engine according to the old and the new principle; and in order that I should not in the very beginning suffer loss of esteem, I first set it in motion on the old principle. It went splendidly, and worked excellently from the start. Confidence in me increased daily, and after this was firmly established I ventured to operate the machine on the new principle. I screwed up the cylinder at the top, set the air-pumps in motion and the steam piping in order, and ran the engine on the new principle. I cannot describe the joy that it gave me when I saw that the engine still carried out its work. This was in the year 1803–4, at the same time as the Essen and Werden districts became Prussian. . . .

I was thereupon commissioned by the worthy local Mines Office to build a 40 inch engine near Essen on the so-called Röttgersbank. . .

The difficulties which I had to combat in building this machine gave to the Regional Officers sufficient material to take their revenge on me, since they believed they had previously been down-graded by Director Cappel because he preferred my suggestions to theirs not only as regards the preparatory arrangements but also on many other points. In every possible fashion they belittled me both at the works and before the public. This happened for example when the engine was not ready in the 18 months within which I had contracted to build it. The fact that I had to cast the first cylinder 5 times before it was sufficiently perfect, since such a massive piece of work had never previously been cast at the foundry and it was first too hard, then too gritty, first too small, then too large; that I had to make it in three pieces as the smelting furnace could not hold so great a mass of iron as was needed for the whole cylinder, and that therefore without any fault of mine, more than 11 months had been lost, was not taken into account.[148]

In addition to the 40-inch engine to prevent flooding, Dinnendahl also built a 15-inch double-acting engine for hauling coal (Fig. 54). We give below the last section of Dinnendahl's estimate of expenses for this engine.

	Recapitulation (*sic!*) of the estimate of expenses	Imperial dollars.	Stivers
I.	Cast iron	1050	
II.	Forgings	339	28
III.	Building wages and stones	77	30
IV.	Woodwork	377	17
V.	Materials	157	30
1.	Carpenter's wages	76	28
2.	Assembling the engine	320	
3.	For spur wheel and shafting . . .	112	
4.	For my layout and supervision . .	260	
5.	For unforeseen expenses	100	
	ESSEN, NOV. 1807. Total expenditure .	2868	13
	Dinnendahl		[149]

The steam engine was first used in Germany chiefly for mines and foundries. It advanced there as a motive power in factories, but slowly as compared with England. The first engine used in Germany for power was installed in 1800 in the royal porcelain factory in Berlin. From about the 1820's the steam-engine was used increasingly in the continental factories (Plate 19).

When Sadi Carnot wrote his important work on *The Motive Power of Heat* in 1824, he took as his point of departure the steam-engine of his period, that greatly admired source of power which began to transform the whole of technology. But in spite of all the success of the steam-engine he was struck by the lack of a serviceable theory of this 'universal motor'. Carnot sought now to establish the general conditions under which mechanical power can be obtained from heat. By thinking out the reversible cycle he reached an understanding of the ideal maximum of work, which is solely dependent on the amount of heat conveyed and the temperatures between which this transfer takes place, but not on the source of power. Carnot's cyclic process as the general principle of every heat engine was of the greatest importance for the further development of heat engines. The evolution of thermodynamics associated with Carnot laid the foundation for the new scientific treatment of thermal motors, which especially

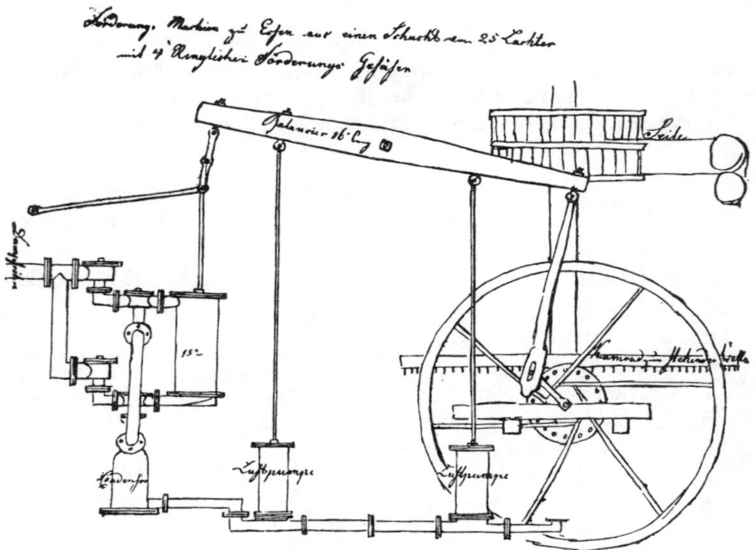

FIG. 54. *Dinnendahl's* 15-*inch, double-acting steam-engine for hauling coal,*

for a coal mine on the Röttgersbank near Essen, 1807. Freehand drawing from Dinnendahl's sketch book. From the manuscript collection of the German Museum, Munich. In 1807/08 Dinnendahl built a 40-inch pumping engine and the above 15-inch coal hauling engine.

occupied Gustav Zeuner from 1860. We will now give a few general passages from Carnot's valuable publication of 1824 on *The Motive Power of Heat.*

Reflections on Motor Power of Heat and on Engines

Paris 1824.

The study of these machines is of the greatest interest, for their importance is enormous and their use increases day by day. They seem destined to effect a great revolution in the civilized world. Already the heat engine works our mines, propels our ships, deepens our harbours and rivers, forges iron, fashions wood, grinds our corn, spins and weaves our fabrics, pulls the heaviest loads etc. It seems as though one day it will be the general motive power which will overpass animal power, waterfalls, and wind currents. It has the advantage over the first mentioned motor-power of cheapness and the inestimable advantage as against the others of being always and everywhere applicable and never ceasing work.

Once the improvements in heat engines are sufficiently advanced to make their installation and fuel cheap, it would enable industry to advance to a degree which it is difficult to predict.

For it will not only be possible to replace the motive powers now used by a more powerful and convenient motive power which can be obtained and conveyed everywhere; but it will create a rapid development in the industries in which it is used,—indeed it may well establish new industries.

The greatest service which the heat engine has rendered to England is without doubt the fresh stimulus given to the exploitation of her coal mines, which were languishing and in danger of collapse as a result of the ever growing difficulties of controlling and removing water. In the second place must be mentioned the services rendered to the iron industry both in respect of the widespread replacement of wood which was just beginning to be exhausted, and on account of power machinery of all kinds, whose application was simplified or made possible by heat engines.

Iron and fire are recognized as the sustenance and support of mechanical industry. Perhaps there is no single factory in England whose existence is not founded on the use of these elements and in which they are not abundantly applied. If England today were to be deprived of her steam engines, she would also be robbed of coal and iron; all her sources of wealth would be cut off and all her means of development would be destroyed; this would mean the ruin of this vast Power. The destruction of her fleet, which is regarded as her most certain protection, would perhaps be less fatal for England.

Rapid and safe navigation by means of steamships may be considered an entirely new achievement of heat engines. This has already made possible the establishment of speedy and regular communication across the inlets and great rivers of the old and new continent. It has enabled wild territories to be traversed which previously could hardly be penetrated; it has also permitted the fruits of civilization to be carried to regions of the earth which would otherwise have had to wait for many years. Steam-ship navigation in a sense brings far distant nations close to one another. It unites the peoples of the world as though all inhabited the same country. Is not in fact the reduction in the length, effort, uncertainty and dangers of travel the equivalent of a significant shortening of distances? . . .

In spite of the numerous works on heat engines, and in spite of the satisfactory position they have now achieved, their theory has advanced but little, and the attempts to improve them have almost been conducted by chance. . . .

The phenomenon of the generation of motion by heat has not been regarded from a sufficiently general point of view. It has only been investigated in engines whose principles have not allowed the full development of which it is capable. In such engines it appears, so to speak, limited and incomplete, so that it is difficult to recognize its principles and to study its laws.

In order to consider the principle of the generation of motion by heat in the most general way, it is necessary to imagine it separated from all mechanism and every especial application; it must be considered as applied not only to steam engines, but to every conceivable heat engine whatever may be the material used, and in whatever manner it is operated.[150]

The steam engine effected a specially profound transformation in the realm of transportation. In 1807 Robert Fulton built the first serviceable steamboat *Clermont* which plied between New York and Albany. The distance of 149·129 miles (240 km) was covered in 32 hours. Already in 1819 the first larger steamship *The Savannah* which indeed still used sails as well, crossed the Atlantic in 29½ days. Steamship navigation increased rapidly after 1840. Great ship yards were founded, the screw propeller was introduced and iron took its place as a material for ship building. Transoceanic trade was immensely stimulated. The continents approached nearer to one another.

In the early days of steamship navigation, the engineering works of

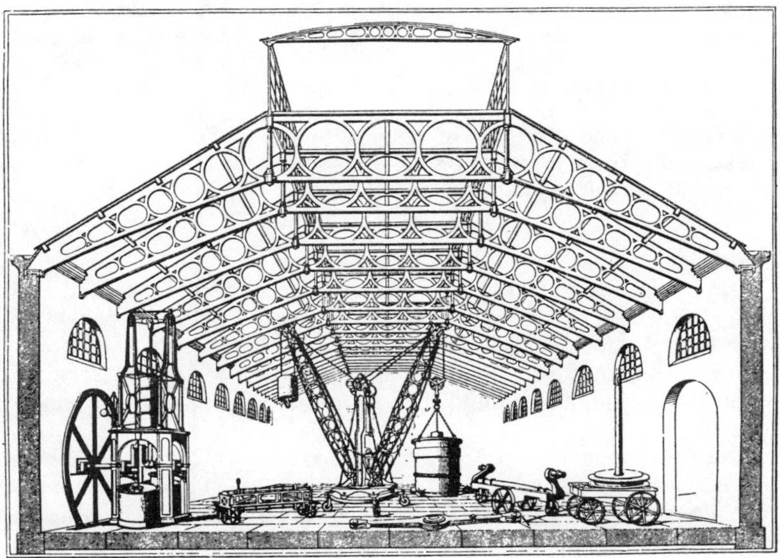

FIG. 55. *Marine steam-engine works of Maudslay, Sons and Field in London, 1834*

Woodcut by Bury from: St. Flachat, *L'Industrie. Exposition de 1834* (Paris 1834)

'Maudslay and Field' became supreme through their accurately made marine engines (Fig. 55). Machine tools of masterly design permitted the manufacture of unusually accurate component parts (cf. p. 283).

Parallel with the steamship as a new means of transport was the steam railway. This first made possible the extensive transport of goods which was required by the growing number of factories. There were already, in English foundries, in the last third of the eighteenth century, iron railroad tracks for horse-drawn vehicles. Previously there had been wooden rails as in German late medieval mines; but the lack of wood and the increased supply of iron gave rise to iron rails. The combination of iron rails and steam cars led to the steam locomotive running on iron rails. The brilliant Englishman Richard Trevithick who had already in 1798 constructed a serviceable high pressure steam-engine, in 1803–4 built the first locomotive in the world to run on iron rails. But not until 1829 did George Stephenson build a steam locomotive that was capable of development. By 1830 he was able to open the first passenger steam railway in the world, the line from the Port of Liverpool to the cotton city of Manchester, a distance of some 30 miles (Plate 18). The lay-out of this railway was from the technical point of view a difficult undertaking, for it had to traverse 3½ miles of swamp, and required the erection of sixty-three bridges and viaducts, the excavation of nearly 2 miles of cuttings, and the boring of a tunnel 1¼ miles long. In spite of all obstacles, the idea of railways spread, as we have stated, with extraordinary speed. Railway lines were the arteries of industry.

In Germany one of the first of the great founders of industry, Friedrich Harkort, as early as 1825 was occupied with the establishment of railways for the stimulation of industry and trade.

Railroads

The prosperity of a country is remarkably increased by rapid and cheap transport of goods and this is ensured by canals, navigable streams and good roads.

The state should for this reason not regard road tolls as a source of financial gain, but should levy only the cost of first-rate maintenance.

Railways appear to offer greater advantages than the means hitherto adopted.

In England over 150,000,000 Prussian dollars have been assigned to this purpose; a proof that the undertakings had strong public opinion behind them.

In Germany, too, people are at least beginning to speak of such things, to which the following remarks may perhaps furnish some contributions.

It is not my intention to enter into details of the question; a few general outlines will suffice for the present.

The Western railway from London to Falmouth will have a length of 400 English miles.

A new railway 32 miles long is being proposed from Manchester to Liverpool, although there is a connection by water.

Experiments which took place for that purpose in Killingworth proved that an engine of 8 horse-power would move a weight of 48 tons along the level at a speed of 7 miles per hour.

Let us imagine such a level stretch from Elberfeld to Düsseldorf; then 1,000 cwt. would be transported from the one place to the other in 2½ hours with a use of 5 bushels of coal for the whole journey.

An 8 horse-power engine would convey 1,000 bushels of coal from Steehle to the Rhine in 3 hours, that is to say would bring financial ruin to Ruhr shipping.

All the Ruhr mining companies would gain from a railway the inestimable advantage of a speedy and regular market with great economies in freight.

Within 10 hours 1,000 cwt. could be brought from Duisburg to Arnheim; shipments by water take 8 days for loading alone.

The objection may perhaps be made that level stretches can seldom be found. To which I reply that more power is needed in proportion to the gradient, or else the speed must decrease; but the return journey will be correspondingly faster and the average speed is therefore unchanged.

The steepest descent on the way to Killingworth was 1 foot in 840: and the steepest climb 1 in 327.

The railways will produce many a revolution in the commercial world. If Elberfeld, Cologne and Duisburg are linked with Bremen or with Emden, the duties charged by Holland will no longer exist for us!

The Rhenish–West India Company should regard Elberfeld as a port as soon as 1 cwt. can be loaded on ship at Bremen within 2 days at a cost of 10 silver pennies.

At this price it is impossible for the Dutch to undertake freight even by steam-boats.

How brilliantly would the trade of Rhineland and Westphalia develop, with such a link with the sea!

May the time soon come in our fatherland when the triumphal chariot of industry will be harnessed to smoking colossi, and will open the way for the welfare of all!

Wetter, March, 1825.

Friedrich Harkort.[151]

One year after his first essay on *Railways*, Harkort in 1826 wrote a memoir on *Railroads*. He sent copies to a number of influential personages, among whom was the Freiherr Heinrich Friedrich von Stein.

While we direct public attention to railways, we do not feel called upon to renew the old dispute as to whether Mechanics, by its advanced development of machinery, has been harmful or useful to society.

Enough for us is the warning: he who does not progress goes backward; we therefore consider that everyone is under an obligation to make an important improvement available to all, according to the best of his ability.

Industry no longer acts according to the niggardly methods of the guilds; no, it adopts every new invention, and in all forms of business a scientific life is developing of which olden days never dreamed!

No-one should be frightened of a well-thought-out scheme on account of its size, for public opinion is the giant who has brought into being the wonders of the world; without it there would have been no Trolhätta canal in Sweden, nor Brunel's subterranean road under the Thames.

The fundamental principle is established: that easy transport at home and abroad substantially increases a country's welfare.

For this reason, Holland and England formed a network of canals, and France made a way over the Alps; and for the same reason hundreds of steamboats traverse the seas and rivers of the young America.

Railways must today be reckoned as the chief means of speedy and cheap transport.

In the year 1680 the first railway line was laid at Newcastle-upon-Tyne. Until 1797 there was hardly a line to South Wales; but by 1811 there were 150 miles overland and 30 miles underground, so evident were the advantages of this discovery.

The Carron Company by the establishment of a railway reduced their monthly expenses from £1,200 sterling to £300.

The greater or lesser degree of usefulness depends mainly on appropriate construction.

From a summary of the English railways, it appears that one horse can move at worst 87 bushels and at best 170 bushels at a speed of 2½ English miles per hour.

Railways have hitherto had only limited application to mines.

In 1824 and 1825 it was decided in England to establish general railway connections over longer distances. More than 11 million pounds sterling was underwritten by a number of Companies for this purpose and from that moment the attention of the most distinguished mechanicians was more strongly directed to this object.

Every German mile costs in England about 150.000 dollars; there was therefore in the poorer country of Germany no means for similar expenditure—the railways of the German coal-mines are but weak imitations.

There were other obstacles to these German railways besides the heavy expense.

The foundations sink very easily as a result of frost and thaw—dust and snow are no less detrimental. Palmer states that light dust on the Cheltenham Railway set up a resistance of 19½ per cent.

In the year 1824 Palmer invented the floating Railway-line.

The importance of this new discovery stimulated the engineering works in Wetter to carry out a similar one, with certain changes, in order that the German public could be practically convinced of its utility.

Councillor for Mines, Heintzmann of Essen, and the Mine-superintendent, Honigmann of Bochum, who were present at the tests, as were later the Directors of the German-American Mining Union Co., were all completely convinced of the practical applicability of the inventions, as the output surpassed all previous results.

The advantages of this railroad are:

1. Cheapness. Apart from the cost of the land, a German mile would cost about 25,000 dollars. The number of waggons depends on the traffic. One is sufficient for 40 cwt. and would cost about 250 dollars.
2. The cost of the land is insignificant as, without any need for levelling the way, only a tow-path has to be laid down for the horse; streams and gullies are easily surmounted.
3. No existing communication is interrupted; and
4. Storm, snow and dust have no effect on it.

In this way we have obtained means to increase home and foreign trade immeasurably, since it is possible to lower freight-charges by 75 per cent and moreover to deliver goods three times as quickly.

Thus for example, a connection between Bremen–Hamburg and the Rhine would immensely increase trade with the North, and would furthermore force Holland to abolish her transit duties.

Austria has already attempted such connections. The establishment of a railroad between the River Moldau and the Danube would have fared better if Ritter von Gerstner had conducted it more prudently. Now the Government intends to connect Vienna and Trieste by a railroad.

In Germany such a plan could first be carried out in Elberfeld.

Industries there depend partly on cheap fuel; in spite, however, of the nearness of the coal-seams, prices have been forced very high owing to the bad transport connections.

If the daily consumption of Wuppertal coal is estimated at 4,000 bushels, about 600 dollars have to be added for freight. A railroad would reduce this at least by a half, and industry would save 90,000 dollars a year as against a capital expenditure of only 100,000 dollars.

If the railroad were connected with the Ruhr, goods from the Rhine and many building materials could come cheaply to Elberfeld.

Such facts are too obvious to warrant a further appeal for their consideration.

In order not to complicate the exposition, only the necessary figures are used here. Experts can obtain further particulars from the attached Tables.

It remains to be said that even canals lag behind railroads by 5 m.p.h. moreover the cost is as 5 to 1.

Whenever it is decided to proceed with an installation, the important point to bear in mind is that of gradient; once a suitable route is chosen, the actual execution is easy. According to English experience, the most appropriate gradient is 1 inch per 10 feet.

Everything new, however good, will be debated. Many interests will be injured in fulfilment of a higher aim. This is the usual fate of far reaching improvements.

So may this manuscript report set forth on its tour; if it merits attention, may it be well received and thus contribute its mite to furthering the general welfare.[152]

Harkort's proposals found at first but little support. By his side as pioneer of railroads for Germany must be classed the great national economist Friedrich List. List's efforts were directed toward economic and political unity for the shattered territories of Germany. As early as 1819 he emphasized that the power of Germany was ruined by thirty-eight separate toll and duty systems. List had to emigrate to North America in 1825. After his return from the New World in 1833, he combined his doctrine of a national economy with the idea of a national system of railways; for a unified German economic territory required also a unified comprehensive German transport and trade system which, besides leading to a rise in the economic standard, would also diffuse culture, overcome narrow particularism and strengthen all intellectual and political forces. In his work *On a Saxon Railway System as a foundation for a general German railway system*, and in other publications, List gave detailed expression to his thoughts. The Leipzig–Dresden Railroad, a stretch of some 71½ miles (115 km) which started operation in 1839, four years after the opening of the first German railroad—the line of 3¾ miles (6.1 km) between Nuremburg and Fürth—was due to the stimulus of List. In 1840 Germany had only 360 miles (580 km) of railroad; in 1850 there were already some 3,400 miles (5,470 km) and by 1904 the railway system of 1850 had increased tenfold to over 33,821 miles (54,430 km).

THE FACTORY SYSTEM

With the steam-engine as source of motive power, new production machinery also appeared on the scene. The meeting of these made possible the immensely rapid mechanization and industrialization which we call the Industrial Revolution. But machinery required for its manufacture further machinery. With the creation of increasingly efficient and complicated machinery both for power production and manufacture in ever increasing numbers, came the development of increasingly accurate machine-tools, especially for metal-working. Here also the English were pioneers. We have already mentioned that Watt's steam engine could hardly have been completed without John Wilkinson's cylinder boring machine. And the advance of the machine age in the first half of the nineteenth century depended, to a great extent, on the improvement of the lathe as the most important of all machine-tools. To Henry Maudslay is due the credit for having produced in the year 1800 the modern self-acting lathe with improved slide rest, which was constructed entirely of iron and enabled heavy components to be machined accurately. Maudslay's pupil the Scot James Nasmyth, who in 1839 invented the steam-hammer, has left us autobiographical records in which he speaks of the period when in 1829 as a young man he first entered Maudslay's workshop. Seldom has an engineer considered so thoroughly the potentialities of his material and its working as did Maudslay. Let us hear Nasmyth's enthusiastic account of him.

It was one of his favourite maxims, 'First, *get a clear notion* of what you desire to accomplish, and then in all probability you will succeed in doing it'. Another was, 'Keep a sharp look-out upon your materials; get rid of every pound of material you can *do without*; put to yourself the question, "What business has it to be there?" avoid complexities, and make everything as simple as possible'. . . .

He proceeded to dilate upon the importance of the uniformity of screws. Some may call it an improvement, but it might almost be called a revolution in mechanical engineering which Mr. Maudslay introduced. Before his time no system had been followed in proportioning the number of threads of screws to their diameter. Every bolt and nut was thus a speciality in itself, and neither possessed nor admitted of any community with its neighbours. To such an extent had this practice been carried that all bolts and their corresponding nuts had to be specially marked as belonging to each other. Any intermixture that occurred between them led to endless trouble and expense, as well as inefficiency and confusion,—especially when parts of complex machines had to be taken to pieces for repairs.

None but those who lived in the comparatively early days of machine manufacture can form an adequate idea of the annoyance, delay, and cost, of this utter want of system, or can appreciate the vast services rendered to mechanical engineering by Mr. Maudslay, who was the first to introduce the practical measures necessary for its remedy. In his system of screw-cutting machinery, and in his taps and dies, and screw-tackle generally, he set the example, and in fact laid the foundation, of all that has since been done in this most essential branch of machine construction. Those who have had the good fortune to work under him, and have experienced the benefits of his practice, have eagerly and ably followed him; and thus his admirable system has become established throughout the mechanical world.

Mr Maudslay . . . initiated me into his system. It was with the greatest delight that I listened to his wise instruction. The sight of his excellent tools, which he showed me one by one, filled me with an almost painful feeling of earnest hope that I might be able in any degree to practically express how thankful I was to be admitted to so invaluable a privilege as to be in close communication with this great master in all that was most perfect in practical mechanics. . . .

Mr Maudslay took pleasure in showing me the right system and method of treating all manner of materials employed in mechanical structures. He showed how they might be made to obey your will, by changing them into the desired forms with the least expenditure of time and labour. This in fact is the true philosophy of construction. When clear ideas have been acquired upon the subject, after careful observation and practice, the comparative ease and certainty with which complete mastery over the most obdurate materials is obtained, opens up the most direct road to the attainment of commercial as well as of professional success.

To be permitted to stand by and observe the systematic way in which Mr Maudslay would first mark or line out his work, and the masterly manner in which he would deal with his materials, and cause them to assume the desired forms, was a treat beyond all expression. Every stroke of the hammer, chisel, or file, told as an effective step towards the intended result. It was a never-to-be-forgotten practical lesson in workmanship, in the most exalted sense of the term. Illustrating his often repeated maxim, 'that there is a right way and a wrong way of doing everything,' he would take the shortest and most direct cuts to accomplish his objects. The grand result of thoughtful practice is what we call experience: it is the power or faculty of seeing clearly, before you begin, what to avoid and what to select.

High-class workmanship, or technical knowledge, was in his hands quite a science. Every piece of work was made subject to the soundest philosophical principles, as applied to the use and treatment of materials. It was this that gave such a charm of enjoyment to his dealing with tools and materials. He loved this sort of work for its own sake, far more than for its pecuniary results. At the same time he was not without regard for the substantial evidence of his supremacy in all that regarded first-class tools, admirable management, and thorough organisation of his factory.

The innate love of truth and accuracy which distinguished Mr Maudslay, led him to value highly that class of technical dexterity in engineering workmen which enabled them to produce those details of mechanical structures in which perfect flat or true plane surfaces were required. This was an essential condition for the effective and durable performance of their functions. Sometimes this was effected by the aid of the turning-lathe and slide-rest. But in most cases the object was attained by the dexterous use of the file, so that flat filing then was, as it still is, one of the highest qualities of the skilled workman. No one that I ever met with could go beyond Henry Maudslay himself in his dexterous use of the file. By a few masterly strokes he could produce plane surfaces so true that when their accuracy was tested by a standard plane surface of absolute truth, they were never found defective; neither convex, nor concave, nor 'cross winding',—that is, twisted.

The importance of having such Standard Planes caused him to have many of them placed on the benches beside his workmen, by means of which they might at once conveniently test their work. Three of each were made at a time, so that by the mutual rubbing of each on each the projecting surfaces were effaced. When the surfaces approached very near to the true plane, the still projecting minute points were carefully reduced by hard steel scrapers, until at last the standard plane surface was secured. When placed over each other they would float upon the thin stratum of air between them until dislodged by time and pressure. When they adhered closely to each other, they could only be separated by sliding each off each. This art of producing absolutely plane surfaces is, I believe, a very old mechanical 'dodge.' But, as employed by Maudslay's men, it greatly contributed to the improvement of the work turned out. It was used for the surfaces of slide valves, or wherever absolute true plane surfaces were essential to the attainment of the best results.

Maudslay's love of accuracy also led him to distrust the verdicts given by the employment of the ordinary callipers and compasses in determining the absolute or relative dimensions of the refined mechanism which he delighted to construct with his own hands. So much depended upon the manner in which the ordinary measuring instruments were handled and applied that they sometimes failed to give the required verdict as to accuracy. In order, therefore, to get rid of all difficulties in this respect, he designed and constructed a very compact and handy instrument which he always had on his bench beside his vice. He could thus, in a most accurate and rapid manner, obtain the most reliable evidence as to the relative dimensions, in length, width, or diameter, of any work which he had in hand. In consequence of the absolute truth of the verdicts of the instrument, he considered it as a Court of Final Appeal, and humorously called it 'The Lord Chancellor' (Fig. 56).

This trustworthy 'Companion of the Bench' consisted of a very substantial and inflexible bed or base of hard brass. At one end of it was a perfectly hardened steel surface plate, having an absolutely true flat or plane face, against which one end or one side of the object to be measured was placed; whilst a similar absolutely true plane surface of hardened steel was advanced by means of a suitably fine thread screw, until the object to be measured was just delicately in contact with it. . . . These two absolutely plane surfaces, between which the object lay, could easily have their distances apart read off from the scale engraved on the base of the instrument, in inches and tenth parts of an inch, while the disk-shaped head or handle of the screw was divided on its rim into hundredth or thousandth parts of an inch, as these bore an exact metrical relation to the pitch of the screw that moved the parallel steel faces of the measuring vice (as I may term it) nearer or farther apart.[153]

Examples of the great precision with which Maudslay worked were his 'Standards of absolutely true plane surfaces'. The above noted procedure for treble plane surfaces may perhaps have been followed in the Renaissance; but Maudslay now applied it to building large machines. Maudslay's revolutionary achievements were great accuracy in the manufacture of

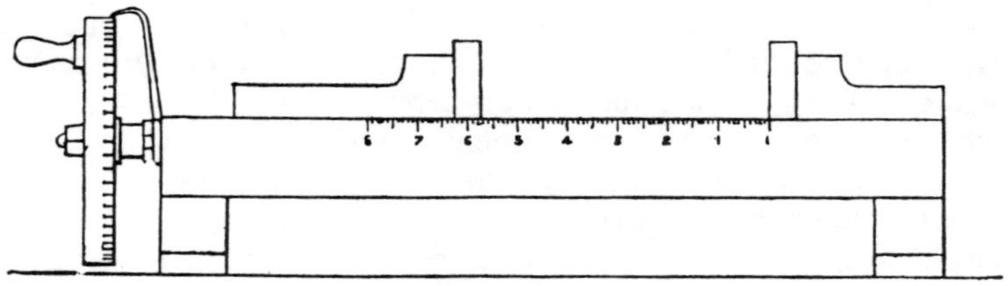

FIG. 56. *Maudslay's micrometer*, 1829.
From: Nasmyth, *Autobiography* (London, 1883)

the machine parts and the resulting possibility of mass production of identical parts—at first applied to screws—so that they could be exchanged one for another.

The most important raw material for the new machines running in the numerous factories that had arisen, was iron. Great progress in the foundries made it possible to produce this in ever greater quantity and ever better quality. The puddling process invented by Henry Cort in 1784 made it possible to produce serviceable wrought iron by the use of coal in the reverberatory oven. Coke was used more and more as a means of reduction in the blast furnace process. In the middle of the nineteenth century steel-casting was invented and Henry Bessemer in 1856 succeeded in producing steel merely by the introduction of air into molten pig-iron in the Bessemer converter named after him. Iron, the basis of industry, was available in sufficient abundance.

The mathematician and mechanical engineer Charles Babbage raised his voice in 1832 as promoter of the factory system with its increased division of labour and mechanization.

The object of the present volume is to point out the effects and the advantages which arise from the use of tools and machines;—to classify their various modes of operation,—and to trace both the causes and the consequences of applying machinery to supersede the skill and power of the human arm. . . .

. . . Perhaps the most important principle on which the economy of a manufacture depends, is the *division of labour* amongst the persons who perform the work. . . .

It will be readily admitted, that the length of time occupied in the learning of any task will depend on the difficulty of its execution; and that the greater the number of distinct processes, the longer will be the time which the apprentice must employ in acquiring it. Five or seven years have been adopted, in a great many trades, as the time required for an apprentice to acquire sufficient knowledge of the business, and to repay by his labour, during the latter portion of his time, the expense incurred by his master at the start. If, however, instead of learning all the different processes for making a needle, for instance, his attention be confined to one operation, a very small portion of his time will be consumed unprofitably at the commencement, and the whole of the rest of it will be beneficial to his master. . . .

A certain quantity of material will be consumed unprofitably or spoiled by every learner and, as he applies himself to each new process, he will waste a certain quantity of the raw material, or of the partly manufactured commodity. But the loss will be far greater if everyone has to learn all the processes in turn, than if each person learns only one; for this reason, therefore, division of labour contributes to a reduction in costs.

Another advantage resulting from the division of labour is, *that time is always lost in changing from one occupation to another*. When the human hand, or the human head, has been for some time occupied in any kind of work, it cannot instantly change its employment with full effect. . . . Another cause of the loss of time in changing from one operation to another, arises from the employment of different tools in the two processes. If these tools are simple in their nature, and the change is not frequently repeated, the loss of time is not considerable; but in many processes of the arts the tools are of great delicacy, requiring accurate adjustment whenever they are used. In many cases the time employed in adjusting bears a large proportion to that in using the tool. . . .

The constant repetition of the same process necessarily produces in the workman a degree of excellence and rapidity in his particular department, which is never possessed by one person who is obliged to execute many different processes. This rapidity is further increased from the circumstance that most of the operations in factories, where the division of labour is carried to a considerable extent, are paid for as piece-work. . . .

When each process, by which any article is produced, is the sole occupation of one individual, his whole attention being devoted to a very limited and simple operation, any improvement in the form of his tools, or in the mode of using them, is much more likely to occur to his mind, than if it were distracted by a greater variety of circumstances. Such an improvement in the tool is generally the first step to a new machine. . . .

When each process has been reduced to the use of some simple tool, the union of all these tools, actuated by one moving power, constitutes a machine. In contriving tools and simplifying processes, the operative workmen are, perhaps, most successful; but it requires a higher grade of intelligence to combine into one machine these separate processes.

Such are the principles usually assigned as the causes of the advantage resulting from the division of labour. . . . Now, although these are important causes, and each has its influence on the result; yet it appears to me, that any explanation of the cheapness of manufactured articles, as consequent upon the division of labour, would be incomplete if the following principle were not stated.

That the master manufacturer, by dividing the work to be executed into different processes, each requiring different degrees of skill and strength, can purchase exactly that precise quantity of both which is necessary for each process; whereas, if the whole work were executed by one workman, that person must possess sufficient skill to perform the most difficult, and sufficient strength to execute the most laborious, of the various operations. . . .

When (from the peculiar nature of the produce of each manufactory) the number of processes into which it is most advantageous to divide it is ascertained, as well as the number of individuals to be employed, then all other manufactories which do not employ a direct multiple of this number, will produce the article at a greater cost. This principle ought always to be kept in view in great establishments, although it is quite impossible, even with the best system of the *division of labour*, to carry it rigidly into execution. . . .

We have seen that the application of the *division of labour* tends to produce cheaper articles; it thus increases the demand, and gradually, by the effect of competition, or the hope of increased gain, causes large capital to be embarked in extensive factories. Let us now examine the influence of such accumulation of capital directed to one object. In the first place, it enables the most important principle on which the *division of labour* rests, to be carried to its extreme limits: not merely the precise amount of skill is purchased which is necessary for the execution of each process, but throughout every stage from that in which the raw material is procured, to that in which the finished product is conveyed into the hands of the consumer, the same economy of skill prevails. The quantity of work produced by a given number of people is greatly augmented by such an extended arrangement; and the result is necessarily a great reduction in the cost of the article which is brought to market.[154]

The marriage of steam and cotton, as it used sometimes to be called, led to those giant textile factories which produced the characteristic appearance especially of the city of Manchester. Edward Baines in 1835 gave a lively description of the conditions past and present of the English cotton industry. He hotly opposed those who deplored the expansion of mechanization.

Manufacture offered limitless scope to machine spinning, for henceforward yarn was available in as great quantity and as fine quality as could be desired. It was no longer necessary to use linen warps, since cotton warps were cheaper. Calicoes, muslins and all sorts of East Indian textiles could be attempted, for the machine provided the appropriate yarn for all. The high speed shuttles could work with ceaseless activity, since there was a superfluity of yarn.

And yet more far-reaching in its results was perhaps the system of manufacture or so-called factory system called into existence by Arkwright. A mechanical spinning mill, owing to the multiplicity and diversity of the machines employed, their productivity and the motive power that they required, became not only a great factory establishment, but also, as no other works, a veritable organic whole. The tendency of the cotton spinners was not so much to complete separation of the processes and a corresponding division of labour among the workers, but much more an analysis of the work of spinning into its elements so that more and more each individual operation was performed automatically by machines, that brought into action a new but common source of power; finally therefore there remained for the human being only the superintendence of the work. And this characteristic organic system must have offered so many advantages that it was soon adopted as far as possible in other branches of production also.[155]

A cotton-spinning establishment offers a remarkable example of how, by the use of very great power, an enormous quantity of the easiest work can be accomplished. Often we may see in a single building a 100 horse-power steam engine which has the strength of 880 men, set in motion 50,000 spindles besides all the auxiliary machines. The whole requires the service of but 750 workers. But these machines, with the assistance of that mighty power, can produce as much yarn as formerly could hardly have been spun by

FIG. 57. *Factories in Manchester*, 1826.
Drawing by K. F. Schinkel.—From: Schinkel's *Nachlass. Reisetagebücher, Briefe und Aphorismen.* (Berlin 1863)

200,000 men, so that one man can now produce as much as formerly required 266! Each spindle produces daily from 2½ to 3 hanks of yarn, and thus the 50,000 together will furnish in 12 hours a thread 62,000 English miles in length—that is to say which would encircle the whole earth 2½ times. . . .[156]

At the accession of George III, (1760), the manufacture of cotton supported hardly more than 40,000 persons; but since machines have been invented by means of which one worker can produce as much yarn as 200 or 300 persons could at that time, and one person can print as much material as could 100 persons at that time, 1,500,000 or 37 times as many as formerly can now earn their bread from this work. And yet there are still many, even scholars and members of Parliament, who are so ignorant or so blinded by prejudice as to raise a pathetic lament over the increase and spread of the manufacturing system. One would think that the history of cotton manufacture would have made an end to all these Jeremiads long ago, or that they would be heard only occasionally from a few classes of workers to whom, undoubtedly, certain changes may at first, and at least temporarily, bring disadvantage. But there are persons who regard it as a great disaster when they hear that 150,000 persons in our spinning works now produce as much yarn as could hardly be spun with the little hand-wheel by 40,000,000. These people appear to cherish the absurd opinion that if there were no machines, manufacture would really give employment to as many millions as now; nor do they reflect that the whole of Europe would be inadequate for all this work; and that in that case a fifth of the whole population would need to be occupied with cotton-spinning alone! Both experience and reflection teach us just the contrary; and we should certainly maintain that, if we still had to spin with the hand-wheel today, cotton manufacture would employ only a fifth of the present number. That one spinner can now produce as much yarn in a single day as formerly in a year, that fabrics can be bleached in two days to a pure white that would formerly have required six or eight months, is the reason why this industry can provide work and bread to incomparably more persons than formerly; and of such results we should not complain but rather we should greatly rejoice. . . .[157]

Thus the steam engine has clearly been a veritable servant and helper to the workers; and far from their having been forced by it to greater effort, rather it has, by means of the machines that it sets in motion, taken from them all the functions that demand the greatest accuracy, unceasing labour and great strength, and the workers are only required to supply the material to the various machines, superintend and control them, and correct their occasional faults. . . . [158]

The rise of our cotton industry occurred at a critical moment, for England had recently lost her American colonies. This new sphere of industry gave us generous compensation for this loss, and thus the inventive spirit of our mechanics repaired the damage brought on us by unwise statesmanship. And when, shortly after, the French Revolution involved us in a long and dangerous war which we could only maintain by the power we derived from trade, it was again this youngest branch of industry which produced by far the greatest activity for our trade. It may indeed be asserted that Watt and Arkwright made a greater contribution than Nelson and Wellington to England's victories. For we owe it to their inventions that we were not prostrated by such long and exhausting efforts.[159]

The chemist and technologist Andrew Ure was full of fulsome praise for the new factory system. He regarded the introduction of the machine as a positively philanthropic deed which freed the worker from heavy and soul-killing labour. It may be indeed that many of the factories offered conditions of work to some extent better than those of the old manufactories. But he overshot the mark in his praise of mechanization and the factory system; for the working folk everywhere complained about the long working day, the poor pay and the exploitation of children and of women. Of this we shall hear more later. Ure justly emphasized that unlike the old manufacturing, the factory system gave less attention to the suitability of the different parts of the work to each man's skill, striving instead to assign each piece of work requiring special skill to a separate mechanism. In the great machine composed of many mechanisms, the separate activities are then brought together again. We give here a few passages from Ure's *Philosophy of Manufactures* (1835).

The blessings which physico-mechanical science has bestowed on society, and the means it has still in store for ameliorating the lot of mankind, have been too little dwelt upon; while, on the other hand, it has been accused of lending itself to the rich capitalists as an instrument for harassing the poor, and of exacting from the operative an accelerated rate of work. It has been said, for example, that the steam-engine now drives the power-looms with such velocity as to urge on their attendant weavers at the same rapid pace; but that the hand-weaver, not being subjected to this restless agent, can throw his shuttle and move his treadles at his convenience. There is, however, this difference in the two cases, that in the factory, every member of the loom is so adjusted, that the driving force leaves the attendant nearly nothing at all to do, certainly no muscular fatigue to sustain, while it procures for him good, unfailing wages, besides a healthy workshop *gratis*: whereas the non-factory weaver, having everything to execute by muscular exertion, finds the labour irksome, makes in consequence innumerable short pauses, separately of little account, but great when added together; earns therefore proportionally low wages, while he loses his health by poor diet and the dampness of his hovel.[160]

The constant aim and effect of scientific improvement in manufactures are philanthropic, as they tend to relieve the workmen either from niceties of adjustment which exhaust his mind and fatigue his eyes, or from painful repetition of efforts which distort or wear out his frame. . . . In those spacious halls the benignant power of steam summons around him his myriads of willing menials, and assigns to each the regulated task, substituting for painful muscular effort on their part, the energies of his own gigantic arm, and demanding in return only attention and dexterity to correct such little aberrations as casually occur in his workmanship. The gentle docility of this moving force qualifies it for impelling the tiny bobbins of the lace-machine with a precision and speed inimitable by the most dexterous hands, directed by the sharpest eyes. . . . Such is the factory system, replete with prodigies in mechanics and political economy, which promises in its future growth to become the great minister of civilization to the terraqueous globe, enabling this country, as its heart, to diffuse along with its commerce the life-blood of science and religion to myriads of people still lying 'in the region and shadow of death.' When Adam Smith wrote his immortal elements of economics, automatic machinery being hardly known, he was properly led to regard the division of labour as the grand principle of manufacturing improvement; and he showed, in the example of pin-making, how each handicraftsman, being thereby enabled to perfect himself by practice in one point, became a quicker and cheaper workman. . . . But what was in Dr Smith's time a topic of useful illustration, cannot now be used without risk of misleading the public mind as to the right principle of manufacturing industry.

In fact the division, or rather adaptation of labour to the different talents of men, is little thought of in factory employment. On the contrary, wherever a process requires peculiar dexterity and steadiness of hand, it is withdrawn as soon as possible from the *cunning* workman, who is prone to irregularities of many kinds, and it is placed in charge of a peculiar mechanism, so self-regulating, that a child may superintend it.[161]

The grand object therefore of the manufacturer is, through the union of capital and science, to reduce the task of his work-people to the exercise of vigilance and dexterity,—faculties, when concentred to one process, speedily brought to perfection in the young. In the infancy of mechanical engineering, a machine-factory displayed the division of labour in manifold gradations—the file, the drill, the lathe, having each its different workmen in the order of skill: but the dextrous hands of the filer and driller are now superseded by the planing, the key-groove cutting, and the drilling-machines; and those of the iron and brass turners, by the self-acting slide-lathe. . . . [162]

The steam-engine is, in fact, the controller-general and main-spring of British industry, which urges it onwards at a steady rate, and never suffers it to lag or loiter, till its appointed task be done.

We have already stated that the labour is not incessant in a power-driven factory, just because it is performed in partnership with the workman's never-failing friend, the steam-engine. Those factory employments have been shown to be by far the most irksome and exhausting which dispense with power; so that the way to put the workman comparatively at his ease, is to enlist a steam-engine in his service. Compare the labour of an iron-turner at one of the self-acting lathes so common now in Manchester, and another at one driven by a power-strap as in London, where, however, the cutting-tools are held in the hands and regulated by the power of the arms and dexterity of the fingers. In the former case, the mechanism being once adjusted leaves the workman absolutely nothing to do but look on and study the principles of his trade, as the machine will finish its job in a masterly manner, and immediately thereafter come to repose by throwing itself out of gear. From the preceding details, the world may judge of the untruth and even absurdity of much of the pretended evidence scraped together and bespattered on the factories.[163]

THE CULTURAL PROBLEM IN TECHNOLOGY

The industrial revolution, which accomplished one of the most complete transformations known to history, had far reaching political, social and psychological results. Many of the leading men at the beginning of industrialization had a clear view of the cultural problem raised by technological development. They strove for a union of practical technology not only with the scientific spirit and general culture but also with an open mind towards social conditions.

The Swiss industrialist Johann Conrad Fischer, who visited England many times in the first half of the nineteenth century, observed the great technical advance there, but recognized as well the social problem raised by rapid industrialization.

Fischer, the tireless, creative, successful technologist expressed in his diary in London in 1825 his earnest conviction that Physics in the widest sense had greatly advanced, but that Metaphysics had made no corresponding progress. We give here extracts from Fischer's diary from the years 1825, 1845 and 1851.

My way led me to Piccadilly to the Exhibition of ancient and modern Mexico and then later to Leicester Square to the Exhibition of Egyptian monumental tombs discovered by Belzoni. I lingered long in these pseudo-mansions of death. . . .

I was alone, I had no guide with me to give me more information. Winged time forced me to depart from this place which attracted me indescribably when I reflected that we today know and understand many things no better after our experience of three thousand years—aeons in relation to our short life. We have progressed in Physics but not in metaphysics; the eternal laws to which created souls too are subject rule immutably. . . .

I was tired with gazing, and went away almost ill-humoured. Must we, I said to myself, always be mere imitators, and can we not ourselves produce something first rate and splendid in the realm of industry and particularly in iron founding in which, in my opinion and experience, so great and beneficial a revolution is yet to come. Will strength never be gained by union, and will wealth never associate with experience and with knowledge? What are water and dust? Two unstable bodies easily separated. But does not porcelain, combining them in bulk survive for thousands of years, and is it not in its highest perfection worth its weight in gold: We have now pretty well the key to the dissolution and formation of corporate bodies. But shall we forever carry them in our laps with regard to the commonest and yet the noblest of all metals? There is 24 carat iron as well as 24 carat gold. . . .[164]

Diaries 1794–1851

When I visited England for the first time 52 years ago (in 1794), I obtained work from a London mechanician, Mr Rhé of Shoe Lane, Fleet Street, and thereby to some extent had an opportunity to make various comparisons, and to judge how far and how completely this branch of industrial activity had developed, as against that which the Continent could show; and I must confess that the verdict fell entirely in favour of England; and almost nothing in this or other departments of industry need fear rivalry from without; the system of Patents so early introduced there may well have contributed considerably to this condition by its encouragement and protection, and have been responsible for manufactured goods possessing so high a degree of perfection combined invariably with complete solidity and suitability.

Yet nothing very new could be observed there at that period, so far as I was aware; the same things were to be found elsewhere, though not so good; but the exception to this was the steam engine, still almost unknown on the Continent.

Twenty years later, when I visited this remarkable island immediately after the destruction of the Continental System, I found great new developments in the above-mentioned field. Spinning mills, foundries, potteries among which Wedgwood was *primus inter pares* or rather *princeps*; steel and file factories, the plating works of Birmingham and Sheffield, the spinning and weaving mills of Manchester, and the cloth manufacture of Leeds, had acquired a size and perfection of which there can be no conception without actually seeing them.

Twelve or thirteen years later, when my inclination and my interests again urged me to visit this country, the scale of everything and especially the expansion of London had increased yet more. . . .

The already extensive steam navigation, the general installation of gas lighting, Perkins's steam-driven shuttles, Brunel's giant tunnel, apparently impossible and yet made possible by him, besides much else of the greatest interest . . . remain in my mind . . . as an ever fascinating picture.[165]

It is her Trade Unions that are already afflicting England grievously, though in part still secretly. It is necessary to have been in England's factory towns and to have visited this class of society in its low, dark dwellings, and to observe in comparison the vast gap separating them from the ostentatious luxury of a relatively small section of the nation, to realize that there is nothing in the world which has not its underlying cause; and moreover that the above-mentioned signs of the times, which may presently stand out even more harshly or may in any case last for a considerable time, are but the budding effects of deep-seated and long standing causes.[166]

The architect Karl Friedrich Schinkel undertook a journey through England in 1826 together with Christian Peter Wilhelm Beuth who rendered such service for the advancement of technical education and the promotion of industry in Prussia. Schinkel was deeply impressed by the enormous factory buildings of the industrial towns, which offered him an imposing yet sad spectacle. He also witnessed the revolts of the textile workers of Manchester in protest against their low rates of pay. Schinkel at that time wrote in his Diary as follows:

Sunday, 18*th June*, *Birmingham*

How sad a spectacle is presented by such an English factory town! Nothing was before us that could rejoice the eye, and there was something almost sinister in the Sunday stillness of a town of more than 100,000 inhabitants, usually so industrious and active.

Tuesday, 20*th June*, *Dudley*

The neighbourhood is pleasant; in the distance could be seen smoking the famous iron works, which stretch for miles. . . . The thousands of smoking chimney stacks presented a grandiose spectacle. For the most part these are pit head gears to raise coal, iron and lime from the mines. Only the cylinders of the steam engines are under cover; while the beam with crank and flywheel as well as the steam boilers of which there are two to each engine are in the open air.

Monday, 17*th July*, *Manchester*

The buildings are 7 or 8 floors high, and have the frontage and depth of the Berlin castle. They have completely fire-proof vaulting, and one canal with water runs beside them and another within. The streets of the town lead right through these massed buildings, and above the streets are bridges to unite the buildings. It is the same throughout Manchester. These are the spinning mills for the very finest cotton. No less magnificent are the bleaching works. . . .

The whole factory system in the town was just undergoing a severe crisis. Owing to the lack of work 600 Irish workers from the Manchester factories had just been repatriated at the expense of the city, and 12,000 workers assembled in a meeting of revolt, because many of them, though they worked 16 hours a day could earn but 2s. a week. Some of the works that had cost £500,000 sterling are now worth only £5,000 sterling. This is a terrible situation. 400 new factories have been established in Lancashire since the war with France. Buildings may be seen where three years ago there were still meadows, but these buildings are already so blackened that they might from their appearance have been in use for a century. Enormous buildings, constructed by a master builder of nothing but red brick, with a view to the barest utilitarian need, and with no attempt at architecture, make a most dismal impression. Here is a view of such a district of Manchester (Fig. 57).[167]

Beuth had already visited England once, in 1823. He had then described to his friend Schinkel in a vivid letter written from Manchester the 'marvel' of the English industrial establishments with their enormous spinning mills and the forest of boiler house chimneys.

The modern miracles, my friend, are to me the machines here and the buildings that house them, called *factories*. Such a block is eight or nine stories high, sometimes has 40 windows along its frontage and is often four windows deep. Each floor is twelve feet high, and vaulted along its whole length with arches each having a span of nine feet. The pillars are of iron, as is the girder which they support; the side-walls and the enclosing walls are as thin as cards,—attaining on the second floor a thickness of less that 2 feet 6 inches. It is said that a storm wrecked one such building in that neighbourhood before it had been completed; that may be true, but a hundred of them are now standing unshaken and exactly as they were erected thirty and forty years ago. A number of such blocks stand in very elevated positions which dominate the neighbourhood; and in addition a forest of even taller boiler house chimneys like needles, so that it is hard to imagine how they remain upright; the whole presents from a distance a wonderful spectacle especially at night, when thousands of windows are brilliantly illuminated by gas-light. You can imagine that bright light is necessary, where one worker must watch 840 threads so fine that 260 hanks weigh but one pound, and two threads twisted together form the selvedge strand of English lace.[168]

The introduction of machines with their greater efficiency was bound at first to bring unemployment to numerous workers. The machines and their inventors were therefore often regarded with suspicion, sometimes mounting to a demand for the destruction of new machines. The struggle against machinery had occurred as early as the seventeenth, but especially in the eighteenth century. John Kay, who in 1733 invented the flying shuttle which considerably accelerated the process of weaving, brought on himself the fury of the weavers, who in 1753 demolished his house. And Richard Arkwright, who in 1775 had provided his improved spinning machine with water power, had constantly to ward off attacks against his invention. Nor were matters much better for James Hargreaves, the inventor of the spinning-jenny, for Samuel Crompton, who developed the spinning-mule, or for Joseph Marie Jacquard, the builder of the standard loom.

Work up to 16 hours daily, low pay, bad housing, the ever-present threat of dismissal if the factory-owner considered that this was required by reason of the market conditions, led in the early days of industrialism to many risings of workers. In Nottingham, in the Midlands, unemployed stocking makers in 1811 attacked the machines in the factories. In 1812 Lord Byron intervened in the House of Lords on behalf of the despairing workers. But Parliament determined on the death-sentence for machine wreckers. Discussions between weavers and factory-owners at Rochdale near Manchester led in 1829 to violent tumults in which the machines were destroyed. A contemporary reports the affair:

The Factory Riots of 1829

. . . At Rochdale, now a model town, where co-operation and every other form of peaceful association flourish, the weavers broke into factories, destroyed looms and other machinery. Fifteen of the rioters were captured, and on an attempt at rescue being made, the military fired and killed six persons. The writer of these lines was present during the destruction by fire of one of the Manchester factories. The burning building was surrounded by thousands of excited people, who, faces reddened by the ascending flames, expressed a fierce and savage joy. As the fire forced its way from floor to floor, darting through the long rows of windows, cries of exultation were shouted by the crowds; and when, finally bursting through the roof, it went into the heavens, the maddened multitude danced with delight, shouting and clapping their hands as in uncontrollable thankfulness for a great triumph.[169]

Fierce social strife was especially kindled by the employment of children which had degenerated, particularly in the mines to limitless exploitation of young workers. A gradual improvement in social conditions did not begin here until the second third of the nineteenth century. The English Mines Law of 1842 at least forbade the labour underground of children under ten years old and of women.[38] Unbearable working conditions, long hours, low pay, exploitation of children, had no doubt existed in eighteenth century factories, but the rapid development of the cotton industry, of mines and of foundries during the Industrial Revolution extended these evils to a far greater number of workers. From the middle of the nineteenth century English industry at least was increasingly seeking high output not primarily through low pay and long hours of work, but by improvement in organization and technical methods.[39]

As we have repeatedly remarked, industrialization in England was far in advance of the Continent, which followed her falteringly. Critical voices were raised against the new technology to a far greater extent on the Continent than in England. The aged Goethe, who followed the great technical advances with sympathy, yet also recognized clearly that many high values of old would be swept away by increasing industrialization. In *Wilhelm Meister's Travels* he gave a vivid picture of the Swiss cotton home industry, with which he had become familiar through his friend Johann Heinrich Meyer. Here duty still bore a happy relation to ability and strength. The working population had preserved their solidarity. The contract system which was predominant here still bore decidedly patriarchal features. But the advancing machine industry would also invade this settled world.

The old people, on the other hand, were ready with quite a number of questions. Everyone wanted to know about the war which, fortunately, was carried on at a considerable distance, and even when

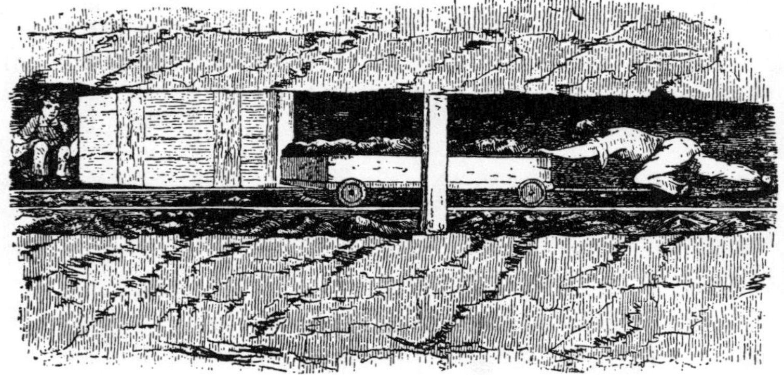

FIG. 58. *Boy labour in an English coal mine*, 1842.
From: Reports from Commissioners of children employment (London 1842)

nearer would scarcely have been dangerous to such regions. They were, however, delighted with the peace, though feeling anxious about another threatening danger; for it was not to be denied that working by machinery was always on the increase and threatening the hands of the workmen by degrees with inactivity. Yet they allowed themselves to be inspired with all kinds of reasons for hope and consolation. . . .

The spinner sits in front of the wheel, not too high up; several hold it in a firm position with their feet laid over one another, others only with the right foot, putting back the left. With the right hand she turns the sleeking board and stretches it out as far and as high as she can reach; by that arise beautiful movements and a slender form is distinguished advantageously by delicate turns of the body and fullness of the arms. The direction, particularly of the last way of spinning, preserves a very picturesque contrast, so that our most beautiful ladies need not fear to lose charm and grace, if they wished one day to take in hand the spinning-wheel instead of the guitar.

In such a surrounding new peculiar feelings pressed in upon me; the whirring wheels have a certain eloquence; the girls sing psalms, also, though less often other songs; *canaries* and goldfinches hung up in cages twitter between whiles, and there could not easily be found a picture of more stirring life than in a room where many female spinners are at work. . . . But I may not then deny that the female inhabitants of the mountains, aroused by the unusual guests, showed themselves to be friendly and agreeable. In particular, they were delighted that I informed myself so exactly about everything which they told me, observed their household stuff, and drew simple machinery work and depicted pretty limbs fleetingly with delicacy as were to be seen close by. Also, when the evening arrived, the completed work was directed farther on, the full spindles laid aside in the little boxes destined for them and the whole day's work carefully stored away. We were now already better acquainted; the work, however, went on its course; now they were busied with the hooks and already showed much more freely partly the machines, partly the treatment of them, and I carefully took note of it.[170]

I arrived just at the beginning of such a piece of work, the transition from spinning to weaving; and as I found no occasion for further distraction, I got them to dictate to me, straight on to my writing tablet as it were, the whole process exactly as it was proceeding.

The first work, gluing the yarn, was carried out yesterday. The yarn is soaked in a thin liquid glue consisting of starch and carpenter's glue, which makes the threads firmer. The threads of the yarn were soon dry, and preparation was then made for spooling, that is to say winding the yarn of spools on the wheel. The old grandfather, sitting by the stove, performed this light task; a grandchild stood near him and seemed anxious himself to handle the spooling wheel. Meanwhile the father ranges the spools on a frame divided by transverse rods for heddling, so that they move freely around strong perpendicular wires and allow the thread to run off. They are fixed with the coarser and the finer yarn ranged in the order required by the pattern or rather the lines of the fabric. An instrument (the Brittli), shaped rather like a drum, has holes at both sides through which the threads are drawn. This is in the loom setters right-hand; with his left hand he holds the threads together, and lays them, walking to and fro, on the heddle-frame. Once from above downward and from below upward again is called a shed, the number of these depending on the closeness and width of the piece. The length is either 62 or only 34 yards. At the beginning of each shed, one or two threads are always laid with the fingers of the left hand on top and the same number below, and this is called the *Rispe*; then the crossed threads are laid over the two pins fixed to the top of the heddle frame. This enables the weaver to keep the threads in their proper order. When the heddling is finished, the Rispe is tied underneath, and thus each shed is separated off, so that nothing can get out of place, then marks are made with dissolved verdigris on the last shed in order that the weaver may repeat the correct measurement. Finally it is removed and the whole is wound in the shape of a great ball, which is called the weft.[171]

I carefully watched the winding on process. For this the sheds of the warp are led in their order through a large comb of exactly the width of the loom on which they are to be wound; this has a notch in which lies a little round peg which is passed through the end of the warp and is firmly fixed in the notch. A little boy or girl sits under the loom, and holds tightly to the strand of the warp, while the woman weaving turns the beam violently with a lever, taking care at the same time that everything falls into its proper place. When the whole is wound, one round and two flat pegs called splints are thrust through the Rispe to hold it and now the winding in process begins. . . .

The smoothing and fanning are usually left to young people who are attracted to the weaving craft, or this service is rendered in the leisure of winter evenings by a brother or a lover of the beautiful weaver, or these will at least make the little reels of yarn for the weft. . . .

I was struck with the industry, indescribable animation, domesticity and peace of the whole atmosphere of such a weaving room; several looms were working, spinning and spool wheels were still in motion, and by the stove sat the old people, chatting intimately with visiting neighbours or acquaintances. From time to time songs might also be heard, usually Ambrosius Lobwasser's Psalms arranged for four voices—more rarely secular songs; then there bursts out the girl's happy ringing laughter if Cousin Jacob has an amusing idea. . . .

Domestic conditions based on piety, stimulated and maintained by hard work and good order, neither too restricted nor too lax, with the happiest relationship of duties to ability and strength. Around them moves a cycle of hand workers in the purest original sense; and far removed from narrowness and its effects, here is prudence and frugality, innocence and activity.[172]

We, as manufacturers ourselves, bring our wares which accrued during the week, on Thursday evening, to the market ship and so, in the company of others who are doing a like business, early on Friday morning we arrive in the town. Here everyone brings his wares to the buyer, who deals wholesale and endeavours to sell his goods as well as possible, takes also the necessary supply of raw cotton in place of payment.[173]

It is usual and arranged that the weaving be ready towards the end of the week, and on Saturday afternoon is carried to the *master of the stock*, who looks through it, measures and weighs it, to discover whether the work is in order and free from defect, also whether on weight and measure the right amount has been delivered, and if everything is found to be right he then pays the weavers' wages agreed upon. On his side he is now busied with cleansing the woven piece from all threads and knots hanging to it; to lay it out in the most delicate manner, bringing the side which is most beautiful and free from faults up in front of the eye and thus making the goods highly acceptable.[174]

[The woman weaver speaks to Wilhelm Meister] 'But what oppresses me is, however, a trade anxiety, unfortunately not for the moment, no, for all the future! The prevailing business of machines torments and causes me anxiety; it revolves like a storm, slowly, slowly; but it has taken its direction, it will come and reach me.

Already was my husband penetrated by this melancholy feeling. One thinks of it, one speaks of it, and neither thinking nor speaking brings help. And who would like to realize such alarms? Imagine that many valleys wind through the mountain like that by which you came down. There still floats before you the sweet, joyous life which you have seen there these days, of which the crowd in its decoration pressing in on every side bore most delightful witness to you yesterday. Imagine how that will gradually sink into ruin, die away; the desert, made alive and inhabited for centuries, will fall back again into its ancient solitude.

'Here remains only a double way, one as sad as the other: either oneself to seize upon what is new and to hasten the ruin, or to break off, carry away the best and worthiest things with oneself and look for a more favourable destiny on the other side of the sea. One like the other causes consideration; but who helps us to weigh the reasons which are to determine us? I know very well that in the neighbourhood they are thinking themselves of erecting machines, and drawing to themselves the nurture of the multitude. I cannot take it amiss of anyone that he regards himself as his own nearest; but it seemed to me contemptible that I should rob these good people and at last see them wandering poor and helpless, and wander they must early or late. They suspect it, they know it, they say it and no one decides on taking a solitary step. And yet, where is the decision to come from? Is not everyone just as much in difficulty as I am?'[175]

One of the leading spirits in building up the German industrial State, Friedrich Harkort whom we have mentioned as the founder of the 'mechanical workshop' at Wetter in the Ruhr, recognized with clear vision the social needs of the age, which one dared not ignore if one wished to avoid the State being caught unprepared by the 'hour of judgement'. Harkort, of whose efforts for German railways we have already spoken, wrote as follows in 1844 in his work *On the obstacles in the civilization and emancipation of the lower classes.*

320 years after Gutenberg (1763) there appeared James Watt, a simple Glasgow instrument-maker, with his improved steam-engine; Arkwright, a barber, added the spinning machine, and the industrial energy of the whole world was thrown headlong into new and unimaginable ways of life. Medieval conditions and limitations of industry were lost as in a sudden great flood, and statesmen were amazed at the grand spectacle, which they could neither fathom nor

21. American wood-working machines at the International Exhibition in Philadelphia, 1876. Woodcut from *The Scientific American*, Vol. 35, 1876, p. 335

22. Mass production at the end of the 19th century. The Westinghouse foundry in Pittsburgh, USA. Here, small castings for railway brakes were prepared on a travelling band. Woodcut from *The Scientific American*, Vol. 62, 1890, p. 369

follow. The machine obediently served men's will; for as it outstripped their strength, Capital triumphed over Labour, and thereby established a new subjection.

The great accumulations of capital have accrued mostly as a result of the sins of the administrations, from monopolies, government loans, and the evil speculation in paper money. They are the support of the gigantic installations, lead to extravagant projects which exceed the demand, and they displace the small workshops that formerly enabled a man without means by industry and judgement to rise to prosperity.

By the introduction of machines and the division of labour which, as Babbage has shown us, was carried to an incredible extent, (so that for example there are in watch-making 102 different branches, for each of which special apprentices must be engaged) both less strength and less general intelligence are required from the masses; and wages are forced down by competition to a bare subsistence level.

If those crises of oversupply which recur at ever shorter intervals occur; immediately wages fall below the subsistence level. Often work ceases entirely for a time, and since as a result of bad education the earnings of better times are dissipated, a mass of suffering persons are exposed to hunger and to all the agony of want.

The workers impelled by the urge to self-preservation, have often tried, especially in England where the evil is greatest, to defy the capitalist, by agreeing not to work below a certain rate of pay.

Usually wasted effort! Capital finds it easier to turn elsewhere and can hold out longer, while the worker is forced to yield at any price in order to live. His limited training and habits do not permit him to transfer to a new trade with new conditions. Large cities are usually the home of such industries which make the State richer and the populace poorer. They cause a race to grow up, which, estranged from church and school, renounces restraint and decency; dissipates the earnings of yesterday in the tavern to-day with no thought of the future; contracts frivolous unions or lives in sin, and without forethought, rapidly sinks into misery. This is the origin of the appalling fact that according to Villeneuve-Bargemont's table, the most civilized peoples of Europe are sunk deepest in pauperism! . . .

A generation already stunted cannot be reformed; nay, reform must begin at the root with the young. We demand of the State that it shall not only govern but shall intervene with help and advancement.

First, the government must establish and administer a strict law that absolutely no children of school age may be employed in factories.

The right to sell their children as slaves to industry must be absolutely removed from parents. . . . As things are at present, children are used to beat down the wages of adults; let infants be removed from the sphere of industry and let their elders gain a better reward for their handwork. I myself am a leader of industry, but I despise with all my heart every creation of values and wealth made at the expense of human dignity and based on the degradation of the working classes. The purpose of machinery is to raise mankind from brutish servitude, not to create worse drudgery. . . .

Just as the law appoints Sunday for rest, so it should determine the hour for ceasing work.

No hay is dried by gas-light, nor is sowing or harvesting carried on by lamp-light. Only in the workshops of industry do egoism and competition constantly exceed the limits imposed by Nature. . . .

By the introduction of machinery, man's animal strength has sunk in value; to restore the balance his intellectual powers should be increased, and we refer here to what has already been said in this connection concerning agriculture. . . .

It might be a practical suggestion if working-class districts were linked to the great cities by their own horse-drawn rail transport which would convey not only passengers but also building material and goods. Such rails can be laid for 30,000 dollars per mile, and far more cheaply if existing high-roads are used. Such suburbs would be no danger to the main city, their inhabitants would be healthy in body and mind, in contrast to the idle, starving dregs of the people in great cities today. . . .

Industry has already formed a powerful aristocracy, on a par with landed property and daily growing in influence; and like the English peerage it is constantly augmented from the ranks of the fortunate and the talented.

Its threads embrace the whole world, and even in those distant parts to which the power of kings does not extend, hand in hand with trade their security is accepted everywhere.

This prodigious industrial activity is a child of the modern age, and is the mighty heir of the Guild-industry of the eleventh to the eighteenth century.

The Guilds, once so powerful in the community life of the towns, have now disappeared; the many independent Masters, many of whom have gained a place in history are no more, gone are the secrets and Privileges, and the ranks of the workers, previously so well organized, are now dispersed.

From this crowd of traders arise the great industrialists; like the medieval *Condottieri*, they gather under their banners old and young wage earners from many nations, venturing the success or failure of the undertaking at their own risk.

Whilst waging the competitive struggle among themselves, they conduct a bloodless international war abroad.

The capital and ability of the leaders of the enterprise hold this loose crowd together; he dismisses or engages workers as may be required by fortune or circumstances; wages are the only attraction: if the leader falls, the follower is ground to dust.

This industrial army, often without a settled home, without hope and without future, squandering today to starve tomorrow, begins, by the considerable increase in its numbers to become a danger to the welfare of bourgeois society.

We have already remarked that it does not appear feasible to make the owner of the factory responsible for the support of his people. But they could be urged to assume the obligation to introduce the system of mutual aid (as we will show later) both in cases of illness and of invalidism, and to support it with proportionate contributions. If the State protects the owners by imposing customs duties, then something should also be done for their servants. . . .

Under present conditions, the worker performs certain services for a certain wage, and the strictest supervision has to be exercised over him; beyond this he cares nothing for the welfare either of the factory or of its owner. The labour force still appears too rough and uncultivated for any closer union with capital to be possible. Let us imagine, however, a decently educated body of individuals; there might then be a happy relationship. If, over and above his assured wage, the worker were conceded a share in profits, his industry and activity would work miracles. We can already see the result, in a small way, where the manufacturer grants a bonus for the lowest consumption of material. Working in their own interest, the people would be more ingenious and would keep a closer check on themselves than can be done by the best supervision. Overseers could be fewer, which would be a further economy. The same amount of capital would go further and the goods would become cheaper. . . .

The time will come when many industries will accept our proposal, for the harsh contrast between great superfluity and want becomes daily more serious. The population, however, which makes such an experiment must have had a more humane upbringing than the proletariat of our great industrial establishments today. . . .

Our desires for the industrial classes may be summarized as follows:

Better school and bodily training, exclusion of children from the factories, and maximum hours of work for adults; better housing, transferred to the country as far as possible; cheaper and healthier food; the formation of Unions for mutual support; a share in the profits of capital; societies for the dissemination of generally useful knowledge.

Most of these points are in more or less clear outline already for public consideration. They can be postponed, as has unfortunately happened in England, but the hour of judgment will strike one day, and it will be well for the State that has wisely prepared for it.[176]

Harkort, who wrote these words was an industrialist and a member first of the Diet of Westphalia and later of the Reichstag of the North German Confederation. He recognized the full extent of technology's social problem. He was especially concerned with better education for the workers and agreement between employers and workers. But his demands found acceptance but slowly, not least because many workers could not be induced even by higher pay to learn what was required for machine operation.

In the middle of the nineteenth century, Karl Marx raised a revolutionary voice. We need not discuss here Marx's doctrine of dialectical materialism and the proletarian revolution. But it must be pointed out that Marx was one of the first to draw urgent attention to the distress to which ruthless industrialization was leading the working class who, often amid inhuman conditions of work, and especially in times of economic crisis, were in increasing and continuous jeopardy of their very means of existence. As a student of social politics he recognized that the industrial worker was in danger of losing the characteristics of a human being. Marx emphasized especially that the machine presented not only a technical but also a sociological problem. He analysed thoroughly the influence of the steam-engine on human society. We give a few characteristic passages from Marx's work *Capital* of 1867.

A revolution in the method of production in one sphere of industry involves a similar revolutionary change in every sphere. This applies, first of all, to the branches of industry which, though they are isolated by the social division of labour (so that each of them produces an independent commodity), nevertheless are interconnected as phases of one integral process. Thus machine spinning made machine weaving necessary; and both together necessitated a mechanical and chemical revolution in bleaching, printing, and dyeing. In like manner, on the other hand, the revolution in cotton spinning made essential the discovery of the cotton gin for the separation of the seeds from the cotton fibre, for only then could the production of cotton reach the proportions which were now indispensable. The revolution in the method of production in industry and agriculture, likewise necessitated a revolution in the general conditions of the social process of production, that is to say in the means of communication and transport. In a society whose pivots (to use Fourier's expression) were, first, small-scale agriculture, with its subsidiary home industries, and, secondly, urban handicraft, the means of communication and transport were utterly inadequate to the requirements of the manufacturing period, with its extended division of social labour, its concentration of the means of labour and of the workers, and its colonial markets; communication and transport, therefore, had to be revolutionized, and were in fact revolutionized. In like manner, the means of transport and communication handed down from the manufacturing period into the period of large scale industry soon showed themselves to be an intolerable fetter upon this new type of industry with its febrile speed of production, its vast gradations, its continual transference of capital and labour from one sphere of production to another, and its newly created ties in the world market. Thus, over and above extensive changes in the construction of sailing ships, the means of communication and transport were gradually adapted, by a system of river steamships, railroads, ocean steamships, and telegraphs, to the methods of production of large-scale industry. But now, vast quantities of iron had to be forged, welded, cut, bored, and shaped. For this, in turn, huge machines were requisite, machines which the manufacturing system of machine production could not possibly provide.

Large-scale industry, therefore, had to gain control of its own most characteristic means of production, the machine itself had to produce machines by machines. Thus was it first able to provide itself with an adequate technical foundation, and to stand upon its own feet. With the growth of machine production in the early decades of the nineteenth century, machinery did in fact gain control, by degrees, of the fabrication of machine tools. But it was not until recent decades (i.e. before 1867) that the construction of railways and the building of ocean steamers on a hugh scale called into existence the gigantic machines now at work in the construction of prime movers.

The most essential condition for the fabrication of machines by machines was that there should be a machine competent to supply power to any extent, and under perfect control. This already existed in the steam-engine. But it was still necessary to gain the power of producing by machinery the perfectly accurate geometrical forms required for the separate parts of machines; straight lines, planes, circles, cylinders, cones, and spheres. This problem was solved by Henry Maudslay in the opening years of the nineteenth century by the invention of the slide rest, a tool that was speedily made automatic, and, having been first designed for the lathe, was soon applied in a modified form to other constructive machines. This mechanical appliance does not replace another tool, but the human hand itself, the hand which produces a particular form by holding, applying, and guiding the edge of cutting instruments against or over the material operated upon—iron or another. Thus it became possible to produce the geometrical forms requisite for the individual parts of machinery 'with the degree of ease, accuracy, and speed, that no accumulated experience in the hand of the most skilled workman could give'.

If we now turn to examine the part of the machine-building machinery which makes the actual mechanized tool, we get back to the handicraft instrument, but on a cyclopean scale. For example, the operating part of a boring machine is a huge drill, driven by a steam-engine, and it is an instrument without which the cylinders of large steam-engines and hydraulic presses could not be produced. The mechanical lathe is a titanic reproduction of the ordinary foot-lathe; the planing machine is an iron carpenter who works upon iron with the same tool used by the living carpenter when he planes wood; the implement which cuts veneers in the London shipbuilding yards is a gigantic razor; the tool of the shearing machine, which cuts iron as easily as a tailor cuts cloth with his shears, is an enormous pair of scissors; and the steam-hammer works with a head just like that of an ordinary hammer, but such a heavy one that Thor himself could not wield it. One of these steam-hammers weighs more than six tons, and strikes with a vertical fall of 7 feet on an anvil weighing 36 tons. It is child's play to such an instrument to crush a block of granite into powder, but the Nasmyth hammer is equally capable of delivering a succession of light taps which will drive a nail into a piece of soft wood.

The instruments of labour, when they assume the form of machinery, acquire a kind of material existence which involves the replacement of human force by the forces of Nature, and of rule-of-thumb methods by the purposeful application of natural science. In manufacture, the organization of the social labour process is purely subjective, is a combination of detail workers; in machine production, large scale industry has a purely objective productive organism, in which the worker is nothing more than an appendage to the extant material conditions of production. In simple co-operation, and even in the co-operation founded upon the division of labour, the substitution of the collective worker for the isolated worker still seems more or less a matter of chance. But machinery, with a few exceptions to be mentioned in due course, can only be operated by means of associated labour or joint labour. In the machine system, the co-operative character of the labour process has become a technical necessity dictated by the very nature of the means of labour.[177]

. . . Inasmuch as, the real wages of the worker sometimes fall below the value of his labour power and sometimes rise above that value, the difference between the price of the machinery and the price of the labour power it replaces may vary to a considerable extent, although the difference between the quantity of labour requisite to produce the machine and the total quantity of labour replaced by it remain constant. It is, however, only the former difference which decides the cost of production of the commodity for the capitalist himself, and influences his actions through the pressure of competition. That is why today machines are sometimes invented in England which can only be put into use in North America; just as, during the sixteenth and seventeenth centuries, machines were invented in Germany which were only put into use in Holland; and just as many French inventions of the eighteenth century were only utilized in England. In the older countries, machinery, when employed in some branches of industry, creates such a superfluity of labour ('redundancy of labour' is Ricardo's phrase) in other branches, that in these latter the fall of wages below the value of labour power hinders the use of machinery, and, from the standpoint of capital (whose profit comes, not from a diminution of the labour employed, but from a diminution of the labour paid for), renders that use superfluous and often impossible. In some branches of the English woollen industry the use of the labour of children has of late years been greatly diminished, and here and there almost completely suppressed. Why is this! Because the Factory Acts necessitate the use of two relays of children, one set working for six hours and the other for four hours, or each set working for five hours. But the parents refuse to sell the 'half-timers' cheaper than the 'full-timers.' That is why the 'half-timers' have been replaced by machinery.[178]

Great as was the advance of British industry during the eight years from 1848 to 1856, when the ten-hour working day was first in operation, that advance was greatly surpassed during the six years between 1856 and 1862. In silk factories, for instance, there were in 1856, 1,093,799 spindles; in 1862, there were 1,388,544; in 1856, there were 9,260 looms; in 1862, there were 10,709. But whereas in 1856 the number of operatives was 56,131, in 1862 the number had fallen to 52,429. Thus there was a 26·9 per cent increase in the spindles, and a 15·6 per cent increase in the looms, at the very time when the number of operatives fell by 7 per cent. In the year 1850, there were at work in worsted mills, 875,830 spindles; in 1856, there were 1,324,549 (an increase of 51·2 per cent); and in 1862, there were 1,289,172 (a decrease of 2·7 per cent). If, however, we deduct the doubling spindles that are included in the numbers for 1856, but not in the numbers for 1862, we find that, after 1856, the number of spindles remained almost stationary. On the other hand, after 1850, the speed of the spindles and the looms was in many instances doubled. In 1850, the number of power looms in worsted mills was 32,617; in 1856, it was 38,956; and in 1862, it was 43,048. In 1850, the number of operatives was 79,737; in 1856, it was 87,794; in 1862, it was 86,063. Children under fourteen were included in the above figures: in 1850, these children numbered 9,956; in 1856, they numbered 11,228; in 1862, they numbered 13,178. In spite, therefore, of the great increase in the number of looms in the year 1862 as compared with the year 1856, the total number of workpeople employed decreased, but that of the children who were being exploited increased.

On April 27, 1863, Mr Ferrand said in the House of Commons: 'I have been informed by delegates from 16 districts of Lancashire and Cheshire, on whose behalf I speak, that the work in the factories is, in consequence of the improvements in machinery, constantly on the increase. Instead of as formerly, one person with help tenting two looms, one person now tents three looms without help, and it is no uncommon thing for one person to tent four. Twelve hours' work, as is evident from the facts adduced, is now compressed into less than ten hours. It is therefore self-evident, to what an enormous extent the toil of the factory operative has increased during the last ten years.'[179]

In manufacture and in handicrafts, the worker uses a tool; in the factory, he serves a machine. In the former case, the movements of the instrument of labour proceed from the worker; but in the latter, the movements of the worker are subordinate to those of the machine. In manufacture, the workers are parts of a living mechanism. In the factory, there exists a lifeless mechanism independent of them, and they are incorporated into that mechanism as its living appendages. 'The dull routine of ceaseless drudgery and toil, in which the same mechanical process is incessantly repeated, resembles the torment of Sisyphus; the toil, like the rock, recoils perpetually upon the wearied operative.' While labour at the machine has a most depressing effect upon the nervous system, it at the same time hinders the multiform activity of the muscles, and prohibits free bodily and mental activity. Even the lightening of the labour becomes a means of torture, for the machine does not free the worker from his work, but merely deprives his work of interest. All kinds of capitalist production, in so far as they are not merely labour processes, but also processes for promoting the self-expansion of capital, have this in common, that in them the worker does not use the instruments of labour, but the instruments of labour use the worker. However, it is only in machine production that his inversion acquires a technical and palpable reality. Through its conversion into an automaton, the instrument of labour comes to confront the worker during the labour process as capital, as dead labour, which controls the living labour power and sucks it dry.

The divorce of the intellectual powers of the process of production from the manual labour, and the transformation of these powers into powers of capital over labour, are completed (as previously indicated) in large-scale industry based upon machine production. The special skill of each individual machine worker who is thus sucked dry, dwindles into an insignificant item as contrasted with the science, with the gigantic forces of Nature, and with the mass of social labour, which are incorporated into the machine system, and out of which the power of the 'master' is made. This master, in whose brain the machinery and his monopoly of it are inseparably intertwined, tells his 'hands' contemptuously whenever he is at odds with them: 'The factory operatives should keep in wholesome remembrance the fact that theirs is really a low species of skilled labour; and that there is none which is more easily acquired, or of its quality more amply remunerated, or which by a short training of the least expert can be more quickly, as well as more abundantly acquired. . . . The master's machinery really plays a far more important part in the business of production than the labour and the skill of the operative, which six months' education can teach, and a common labourer can learn.'[180]

The fight between the capitalist and the wage worker dates back to the very origin of capital. It continues to rage throughout the manufacturing period. But only since the introduction of machinery has the worker been at war with the instrument of labour itself, with the material embodiment of capital. He revolts against this particular form of the means of production, as being the material basis of the capitalist method of production.[181]

Let us suppose, for instance, that 100,000,000 persons would be needed in England in order to spin, with the old-fashioned spinning-wheels, the cotton that is now spun by machinery with the aid of half a million persons, this does not mean that machinery has taken the place of all these millions who never existed. It only means that millions upon millions of workers would be required to replace the spinning machinery. If, on the other hand, we say that in England the power-loom threw 800,000 weavers into the streets, we are not referring to extant machinery which could only be replaced by a definite number of workers, but to a definite number of workers who have actually been replaced or driven into the streets by machinery. In the manufacturing period, handicraft industry, although disintegrated, was still the basis. The new colonial markets could not be satisfied by the products of the comparatively small number of urban operatives who were a legacy from the Middle Ages; and manufactures properly so called opened new fields of production for the countryfolk who had been driven off the land when the feudal system broke up. At that time, therefore, the division of labour and co-operation in the workshop were regarded mainly from the positive outlook, that they made the workers actually in employment, more productive. It is true that co-operation and the concentration of the means of production into the hands of fewer persons, when applied to agriculture, gave rise to extensive, sudden, and forcible revolutions in the method of production, and therewith in the living conditions and the means of occupation of the rural population-in many countries, long before the period of large-scale industry began. But originally this contest was much more one between large landowners and small landowners, than between capital and wage labour. On the other hand, in so far as workers were driven off the land by instruments of labour, by sheep, by horses, etc., force was here directly resorted to in the first instance as a prelude to the industrial revolution. First of all the workers were driven off the land, and then the sheep came. Land grabbing on a great scale, such as was practised in England, is the first step for the establishment of agriculture upon a large scale. At the outset, therefore, this revolution in agriculture has rather the aspect of a political revolution.

In the form of machinery, the instrument of labour immediately enters into competition with the worker. The self-expansion of capital by means of machinery is directly proportional to the number of the workers whose means of livelihood have been destroyed by this machinery. The whole system of capitalist production is based upon the fact that the worker sells his labour power as a commodity. Thanks to the division of labour, this labour power becomes specialized, is reduced to skill in handling a particular tool. As soon as the guiding of the tool becomes the work of the machine, the use-value and the exchange-value of the worker's labour power disappear. The worker becomes unsaleable, like paper money which is no longer legal tender. That portion of the working class which machinery has thus transformed into superfluous population (this meaning a population which is no longer immediately required to promote the self-expansion of capital), either goes to the wall in the unequal struggle of the old handicraft and manufacturing industry against machine industry, or else floods all the more easily accessible branches of industry, glutting the labour market, and consequently reducing the price of labour power below its value. It is supposed to be a great consolation to these pauperized workers that their sufferings are to some extent no more than a temporary inconvenience; and also that only by degrees does machinery come to dominate a whole field of production, so that the scope and intensity of its devastating effects are mitigated. One of these consolations is neutralized by the other. When machinery invades a field of production by slow degrees, it produces chronic poverty among the workers that have to compete with the machines. When the transition is a rapid one, the effect of the machinery is massive and acute. History has no more pitiful spectacle to offer than that of the gradual decay of the English handloom weavers, a process which took several decades, and was finally complete by the year 1838. . . .

But even within the domain of large-scale industry, the perpetual improvements in machinery and the development of the automatic system exercise a similar influence. 'The object of improved machinery is to diminish manual labour, to provide for the performance of a process or the completion of a link in a manufacture by the aid of an iron instead of the human apparatus.' Again: 'The adaptation of power to machinery heretofore moved by hand, is almost of daily occurrence. . . . The minor improvements in machinery having for their object economy of power, the production of better work, the turning off more work in the same time, or in supplying the place of the child, a female, or a man, are constant, and although sometimes apparently of no great moment, have somewhat important results.'[182]

The enormous extensibility of the factory system, the way in which it increases production by leaps and bounds, and its dependence upon the world market, necessarily give a febrile impetus to production, with a glutting of the markets, a subsequent relative inadequacy of demand, and therefore a paralysis of industry. The life of industry becomes one characterized by a succession of periods of moderate activity, prosperity, over production, crisis, and stagnation. Thanks to this periodicity of the industrial cycle, the uncertainty and instability which machine production imposes upon the occupation of the worker, and therefore upon the general conditions of his life, now become habitual features. Except in the periods of prosperity, the capitalists are always fiercely competing one with another for a place in the market. The size of each one's share is directly proportional to the cheapness of his product. This need for cheapness causes rivalry among the capitalists in the use of improved machinery able to replace labour power, and in the application of new methods of production. Furthermore, there always arrives a moment when the attempt is made to cheapen commodities still more by forcing wages down below the value of labour power.[183]

An increase in the number of factory workers is also brought about by a proportionally much more rapid increase in the total amount of capital invested in factories. But this process goes on only within the ebb and flow periods of the industrial cycle. It is, moreover, continually interrupted by technical advances which sometimes replace the workers virtually, and sometimes actually drive them out of the factory. These qualitative changes in machine production are continually expelling workers from the factory, or shutting the factory door against the stream of new recruits; whereas a purely quantitative extension of the factories leads to the enrolment of fresh contingents, this process going on side by side with the expulsion of others. Thus the workers are insistently being repelled and attracted, hustled from pillar to post; and at the same time there are unceasing changes in the sex, age, and skill of the levies.[184]

It is in part due to Marx and the working class movement connected with him that the State since the last quarter of the nineteenth century has been concerned to improve the conditions of life and work of the productive workers in industry with ever advancing social and political legislation.

GERMAN SCIENTIFIC TECHNOLOGY

The advance of technology and the development of industry were founded in England entirely on private initiative. The English technologist was educated by experience. At the end of the eighteenth century, France created in the *École Polytechnique* with its associated specialist colleges the first great seeding ground of a technology based strictly on science. In and around these Institutes in the first half of the nineteenth century scientific architectural engineering and machine technology, associated with the names of S. D. Poisson, L. M. H. Navier, G. G. Coriolis and J. V. Poncelet, were established. We have already emphasized that the development of German industry was in part an educational task (p. 269). In Prussia after 1817 Christian Peter Wilhelm Beuth devoted his energies to Trade Schools. The formation of German Polytechnic Schools dates from 1825, the first that of Karlsruhe. Its formation betrays descent from the Paris *École Polytechnique*. But there was added the new spirit of a liberal and patriotic middle class devoted to technical progress and imbued with a sense of responsibility; which also left its mark on the new Institute. Further Polytechnic Schools followed rapidly in the other capital cities of the German States, in Munich in 1827, in Dresden in 1828, in Stuttgart in 1829 and in Hanover in 1831. They were, especially from the middle of the nineteenth century, the nurseries of a scientifically cultivated technology which vigorously advanced the industrial movement supported by middle-class culture.

In Austria, Polytechnics arose even earlier, for example in Prague (1806) and in Vienna (1815), but these were at first directed rather to practical technology.

At the Karlsruhe Polytechnic Ferdinand Redtenbacher[40] acted as Professor of Mechanical Engineering from 1841 and then as Director of the Polytechnic from 1857. In the middle of the century he founded in Karlsruhe scientific machine construction, which became the pattern in the form which he gave it for other Polytechnics. While the Paris *École Polytechnique* tended to pursue the theory of technology as a discipline in mathematics and physics, Redtenbacher stood for Machine Construction as an independent subject developing as purely technological thought from a study of machinery itself and leading to its scientific mastery. In order to attain the great objective of training men for the formation of a German industrial state he wished to link science more closely with practice than was done in Paris. Parallel with the scientific theory of machine construction practical construction was by no means to be neglected. And over and above technical specialization, he was concerned with the problem of general culture in an industrial community.

Besides specialized works, Redtenbacher wrote three books on the general technology of machines which form a unique trilogy. To take them in their logical rather than their chronological order, we mention first the *Principles of Mechanics* of 1852 which comprised the theoretical science of machinery. In 1862–5 Redtenbacher added to the *Principles* his great work *Machine Construction* which is devoted to practical construction. He regarded as the final work of the trilogy his *Conclusions Concerning the Construction of Machines* which he had published as early as 1848. As a help for exercises and for practical production, this volume gives only the results without the deductive processes. In the Introduction to his *Conclusions* Redtenbacher wrote in 1840 aptly as follows:

Conclusions Concerning the Construction of Machines

Whoever, equipped only with general principles, enters the practical arena, is like a ship provided indeed with a rudder, but with neither sails nor motor engine. The outcome of the voyage is beyond doubt. No machine can be invented merely by the principles of mechanics; for there is also needed, besides inventive talent, an exact knowledge of the mechanical process which the machine is to serve. No design for a machine can be made by aid only of principles, for an understanding of assembly, arrangement and proportion is also needed. Nor can a machine be produced in practice by the principles of mechanics; for practical knowledge of the materials to be used, dexterity in handling the tools and in the manipulation of the auxiliary machines are also needed. No industrial business can be conducted on the principles of mechanics, for this requires also a suitable personality, and knowledge of commercial business. It will be seen that for the manifold technical activities the principles of mechanics are not entirely adequate; nevertheless, if used with intelligence, they render excellent service, for they always indicate what should happen, often determine the most important dimensions and lead to a correct judgement. But invention, assembly, arrangement and proportion, and practical work with the file and turning tool is another matter.

A School, which wishes to give suitable training for a mechanical-technical career should therefore by no means follow a one-sided scientific course, but must strive to arouse and to exercise all the powers which are important for the calling of draughtsman, constructor, engineer and manufacturer.[185]

Redtenbacher's foresight may again be recognized in the following passages:

Apart from the dictates of sentiment, religion or philosophy, pure egoism, and the ordinary desire for gain oblige us not always to employ men as motive power, but whenever possible to use instead motive power derived from inorganic nature.

In antiquity there was no physics, no chemistry, no science of mechanics and man had no inkling that a multitude of forces are available in the world of Nature which can be applied to perform work for human purposes; hence the ubiquity of slavery which indeed is not entirely abolished in modern times; it still exists in fact though only partially; in principle it is no longer tolerated and there is a struggle for its abolition as far as practicable. Natural science, technology and Christianity have led towards making work a free activity, so that man-power is never the sole motive power but is now only applied to a task which requires not merely physical force but also intelligence, which cannot be replaced by a machine. Horticulture is performed by men, and agriculture by draught animals supervised by men. The colossal labours involved in building railways are carried out partly by draught animals but chiefly by human acitivity, as man has no means as yet nor perhaps ever will have whereby these tasks can be performed by the insensible motive forces of inorganic Nature. In industrial and factory work man is no longer used solely for motive power; he performs only such work as cannot be carried out by machines, or he assists, supervises and directs a machine that is driven by water or steam. Man as motive power is weak and expensive, but is equipped instead with intelligence, albeit this also is sometimes of a weak order. Thanks to his bodily form and thanks to his mind, he is a universal machine, capable of an infinite diversity of movement. He is a self-propelling motive power, that can betake himself to the places of work, can at his pleasure start, interrupt or complete the work; and can deal with the many accidental disturbances and obstacles, so that there is a great variety of work in which he can be replaced by no other form of power. But human labour is and remains disadvantageous in those cases (1) when great power has to be developed and (2) when a high degree of uniformity is required in the work. Spinning, weaving, turning, planing, filing etc. are activities where uniformity yields the best product; in such a case therefore machine work is preferable to manual work, for even the greatest virtuosity of the manual worker can never attain such uniformity as is easily produced by a well-equipped machine.[186]

Innumerable things, are not serviceable for human purposes in the condition in which they are freely produced. In order to be used for nourishment, clothing or other purposes, they must first be transformed and altered and for that purpose the exercise of greater power than man himself can develop is often required. But here again his intelligence finds the remedy; first of all he recognizes the existence of mighty forces hidden in flowing water and in steam, and then he knows how to discover the means whereby these forces can be constrained to effect exactly the desired changes in the natural products without violating their innate principles. The water of a stream can be harnessed to convert grain into meal, if it is conducted to a water-wheel that is connected with a mill. Thus arises mechanical technology.

This guidance, mastery and control of the forces of Nature by which they can be caused to act and work in accord with our purposes has become really effective only in our own day. Within a short time it has been developed to a high degree, and history will one day not fail to recognize the achievement in this respect of the first half of the eighteenth century. Unfortunately much that should be fostered and developed for the attainment of a happy existence has been neglected or completely omitted in the course of all this technological activity, and in consequence the fruits of these strenuous activities are to some extent open to adverse criticism.

But here again bitter experience, will lead to an intelligent and wise application of this mastery of technology, and surely it may then be confidently hoped that technology having developed its full potentialities, will be a blessing and not a curse to mankind.[187]

Redtenbacher describes below the special technical mentality needed to produce mechanical contrivances.

The manifold mechanical movements needed for the arrangement of machinery need not always be invented anew. This was, however, necessary when spinning machines and steam engines were invented, for at that time very few of the mechanisms for effecting conversion of motion were known. But today a great number and diversity of these are known, and as a rule one or another is found adequate to meet any given need, so that really new inventions are required only for most unusual movements. A very exact and complete knowledge of mechanisms already invented is therefore most important in the arrangement of machines. Scientific knowledge is actually of little help, for complex mechanisms are evolved not through general

23. The first Diesel Engine, on the test bed at the Maschi-nen-fabrik Augsburg, 1893. *Centre*, Rudolph Diesel. *Right*, Director Heinrich Buz

24. The dynamo room at the first electric power station in New York (Pearl Street) installed by Th. A. Edison. Woodcut from *The Scientific American*, Vol. 47, 1882, p. 127

powers of thought but by quite special powers of understanding of form, of disposition and of assembly of parts. Whoever is gifted with these powers and has developed them by varied practice will therefore be able to produce many and very ingenious inventions even though almost totally lacking in previous intellectual education; while he who lacks these powers, even though he have other most remarkable and diverse gifts, will yet not be in a position to devise even the most insignificant mechanism.[188]

Notable means for scientific technical production were provided by the integral calculus and solid geometry. The first systematic scientific treatise was that of G. Monge (1799), the organizer of the *École Polytechnique.* In 1811, J. N. P. Hachette teacher at this advanced School of Technology was the first to employ the methods of solid geometry comprehensively in the construction of machinery. Redtenbacher in his Principles of 1852 strongly emphasised the need for technical draughtsmanship for the designer of machines.

Drawing is a means by which the mechanician can represent his thoughts and ideas with a clarity and distinctness that leave nothing to be desired. A machine that has been drawn is like an ideal realisation of it, but in a material that costs little and is easier to handle than iron and steel.

To draw a machine requires an expenditure of time and effort incomparably less than that needed to build the machine in iron and steel, especially if one considers the usefulness of the drawing both for the plan and its execution.

If everything is first well thought out, and the essential dimensions determined by calculation or experience, the plan of a machine or of an installation of machines can be quickly put on paper and the whole thing as well as the detail can then most conveniently be submitted to the severest criticism. If the whole is found to be unsatisfactory, the drawing is discarded and a second, better one is soon ready. If only certain changes of detail are desirable or necessary, the parts to be changed can be readily removed and replaced by others that are better. If at first there is doubt as to which of various possible arrangements is the most desirable, then they are all sketched, compared one with another, and the most suitable can easily be chosen.

But drawings are of the utmost importance not only for planning but also for execution, since by means of them the measurements and proportions of all the parts can be so sharply and definitely determined from the beginning that when it comes to manufacture it is only necessary to imitate in the materials used for construction exactly what is shown in the drawing. Every part of the machine can in general be manufactured independently of every other part; it is therefore possible to distribute the entire work among a great number of workers, and to organize the whole business of manufacture in such a manner that each job can be completed at the right time, in the most suitable position, with the least expenditure of time, money and material, and with an accuracy and reliability which leave almost nothing to be desired. No substantial errors can arise in work organized in this manner, and if it does happen that on a rare occasion a mistake has been made, it is immediately known with whom the blame lies.[189]

In the mind of Redtenbacher the German technological Institutes were to develop into places of research and teaching that would also always remain aware of the requirements of practical technological production and indeed the rise of German technology from the second half of the nineteenth century was due in no small measure to the work of the German Technical Institutes.

PART VII

TECHNOLOGY BECOMES A WORLD POWER

PART VII

THE RISE OF AMERICAN TECHNOLOGY

FROM the second half of the nineteenth century, technology spread with giant strides over wide areas of the world. Modern technology, a creation of the West, was adopted and developed not only by the American continent, which was colonized by people of European descent, but also by those who lacked the basic experience needed to build up gradually a technology such as that of the West. Spengler sees in the surrender of western technology to other peoples a 'Betrayal of technology'. Technology in the hands of others would be turned against the West itself.[41]

In 1868, with the accession to the throne of the Emperor Mutsuhito (Meiji), Japan by imperial decree opened her door wide to European science and technology. The State energetically urged forward the building up of industry on the European model. Especially after the First World War, the industrial potential of Japan increased enormously. If in 1863 the Island Kingdom had a spinning industry with but 6,000 spindles, yet by 1914 there were 2.7 million, and by 1934 8.8 million spindles in Japan. Japan's share of world trade amounted in 1900 to about 1 per cent; by 1930 it had risen to 3.5 per cent.[42]

In Russia, where western thought on enlightenment, liberalism, socialism and technological progress had long been adopted and developed, with the rise of Soviet power after the First World War, a process of intensive industrialization began, favoured by rich natural resources, the object of which was to convert this neglected agrarian land into a highly developed industrial realm. From 1920 Lenin energetically pursued the electrification of Russia, vigorously emphasizing its 'immense usefulness' and 'indispensability'.[43] Between 1913 and 1938, the industrial production of Soviet Russia had increased ninefold.

But nowhere outside Europe did western technology succeed in rooting so strongly or developing so independently as in young America. After the American colonies had won their independence from England in 1783, began the process of technical development and industrialisation. The Puritanical activity of the English immigrants encouraged this materially. Technical achievements especially the steam-ship, the railway and the telegraph, played a prominent part in the expansion to westward. Technology was therefore closely connected with the life of the young state. Especially after the American States had united, following on the Civil War (1861–5), a great technological development began. The American was obliged at an early stage to use machines extensively in agriculture, since but small man-power was available for the vast stretches of farm land. This development began as early as 1834 with the invention of the reaping-machine by Cyrus McCormick. European capital took a great part in this introduction of technology. But as a result of the First World War, America was transformed from a debtor state to the greatest creditor state in the world. The extremely rapid industrial advance of North America was reflected in the growing value of American industrial production, which multiplied thirty-three fold between 1859 and 1919.[44] The population trebled in this period, the number of workers and officials employed in technology and industry rose from 1.3 millions to 9.1 millions, that is to say sevenfold. The year 1919 as compared to 1859 showed a thirty-three fold increase of production of goods as against a seven fold increase in man-power. These figures reveal the increasing replacement of man-power by the machine.

In 1876, in Philadelphia, for the first time on the American continent one of the great International Exhibitions was held—typical expression of industrial enterprise and of middle-class faith in progress. Old Europe pricked up her ears on reading the reports of the exhibition; for it was generally realized that young America was outstripping Europe in technology. The interchangeability of parts in the manufacture of steam-engines was admired; which had only been known previously in the manufacture of arms (John Hall, E. Whitney) and for about fifty years in the production of sewing-machines; astonishment was aroused by the extensive automation in industrial production processes; amazement was excited by the gigantic Corliss steam-engines (Plates 20, 21). Franz Reuleaux wrote at the time in his *Letters from Philadelphia* as follows:

It was here fully demonstrated that, as could already be observed and understood in Paris in 1867, and as soon afterwards became fully manifest in Vienna, North America had begun to take her place and indeed was in some respects unrivalled in the very first rank in machine construction. First, certain details of the steam-engine have been further developed, and then they have been able to give it an external finish and appearance which is really admirable. This is a significant sign. For when beauty of form has been developed as the object of especial even critical care, the difficulties of the purely utilitarian design must already have been overcome. Confidence and peace of mind concerning them must at least be established. The method of manufacture of the machine has also been brought to a high state of perfection. Thus several firms exhibited steam-engines in various sizes, the parts of which are mass-produced automatically on machines, so that, like the parts of American sewing machines and those of several German firms, they can be interchanged one with another. In machine-tool construction, American engineering industry is brilliantly represented. Here it merits the palm not only at the exhibition, but apparently in every respect. American production in this department is characterized by a wealth of practical new ideas, astonishingly clever adaptation to special requirements, accuracy of the component parts, and an increasing elegance in the outward appearance of the machine. Perhaps Germany is best able to compete with the machine tool construction here. The construction of machine tools calls for a gift and an interest to follow technological processes which are inherent in the German character, and have often made themselves felt among us. Nevertheless only the most strenuous industry and the exertion of all our power will put us in a position to overtake the lead that America has won. Not long ago we had made the English type of tool our own, and moreover had further developed it in accordance with its special nature, and had thus begun to stand on our own feet in this matter. A German type of machine-tool was gradually developing. But now the American with quite new ideas has unsaddled the Englishman, and we must unhesitatingly follow the new system if we do not wish to fall behind. Excellent beginnings have been made as is well known. But repose is far, very far from beckoning to us yet; only ceaseless labour in order just to follow up and keep pace.[190]

German industry must relinquish the principle of competition merely in price and must decide whether to turn instead to competition in quality or value. Nevertheless in order to produce cheaper and more saleable goods, she must adopt machines, or to speak generally the scientific-technological apparatus, in all those cases in which the human hand can be replaced with advantage to the product, that is to say when bodily efforts can thereby be abolished or lightened, and when extensive repetition work forms the basis of production; on the other hand, she must use the intellectual power and the skill of the worker to finish off the product, and this to a greater degree the more it approaches to art.[191]

The Swiss A. Göldy gives an enthusiastic report of the American machine Division of the International Exhibition at Philadelphia.

One of the most striking aspects of the American Machine-Section (of the Exhibition) is the variety of special tools for every conceivable purpose. Here also William Sellers & Co. are the leading firm. An example is a metal lathe for the manufacture of water injectors. This is a combination resulting from the application of fine intelligence for a positive, concrete purpose. It enables an intelligent worker to produce far more in a given time than was possible with an ordinary slide-lathe and at the same time is far less dependent on the worker for strict accuracy of work. This type of machine, since it reduces the cost of production, will gradually be introduced in all countries where there is keen competition. Sellers' gear cutting machine is known pretty well universally throughout Europe; it is perfect and amazingly automatic in its work. It receives the job and carries it through from one process to another to its conclusion, without any worker being required. One man can supervise four or five of these machines, and attend to other things at the same time.

The solution of this enigma will gradually be found in all machine-tools: the mind of the worker directs, and the iron slave carries out its duty.[192]

If the human mind will quietly contemplate again the ideas and achievements of the wonderful American exhibition of machines, it will not fail to recognize the approach of an immense change. The productive power of automatic machines will rapidly develop and become magnificent, and will overwhelm all mankind with a quantity of products which, we will hope, will bring them blessing.[193]

Another reporter on the Exhibition, the German F. Goldschmidt, showed that American technology was especially stimulated by four factors, namely the effort to replace human power by the machine, practical training, extensive division of labour, and a sound Patent law.

Let us turn to the heart of the exhibtion, to the Machine Hall, that vast airy building of glass and iron where the pulse-beat of ceaseless industrial activity was most audible. Here the diligence, energy and inventive gift of the North Americans celebrates its triumph over all that had ever been achieved by other nations in the invention and construction of machines. Here among hundreds of thousands of whirring wheels and rotating shafts, all moving and transmitting motion, which they had received from a gigantic flywheel, powered by two mighty steam engines, there could be seen those winding engines which extracted ore from the depths, pumping installations, motors of every sort, machine-tools and apparatus, machines of every size and shape which produce those things necessary for our livelihood; indeed these could be clearly seen arising before our eyes. The essential element in the life of North Americans—the MACHINE—which has enabled them to replace laborious hand-work, to mass-produce everything, and to acquire immense wealth in such a short time, is overwhelmingly presented here. Certainly it is unlimited mastery of material which speaks from the endless machinery here; certainly it is a picture of a wild chase, in which all energies are concentrated; a chase after material gain. But who can deny that there is herein greatness and power?

Involuntarily one is seized with the desire to look round at a people that has been able to accomplish anything so mighty, and that has a commercial life and an industry such as is possessed by no other people on earth. Let us glance back for a moment to the period preceding the Declaration of Independence, the hundredth anniversary of which was celebrated by the exhibition. We see a little band of colonists with axe and with cocked rifle laboriously opening a way through the forest. Inland they make war upon the Indians, northward against the French and later against the English, southward against the Spanish. But in the midst of these struggles and countless obstacles, that manliness and self-reliance that became the main heritage of the next generation was strengthened. Also to the new immigrants who streamed in hordes to the land which began to disclose ever more of its wealth, the strife with the crude forces of Nature, the harsh struggle for existence became the basis of their spiritual power and development. Towns shot almost straight out of the ground. Canals, railways, were established, factories were founded. There was a shortage of man-power, that which is available is very expensive, and unless one wished to be left standing half-way, or to abandon the course started, it became absolutely necessary to replace arms of flesh and blood by those of iron and steel which never weary and which produce in an instant that which the others can make but slowly and laboriously.

The characteristic education of the North Americans, to prepare the boy early for a calling and to concentrate all his thoughts on a single aim (wherefore the American lack of general education should cause no astonishment) has proved very advantageous for inventive talent, as has also the wise division of labour, which has been too little valued in German industry. It must not be understood by this that it is a question of limiting the scope either of the manufacturer or of the worker. Nothing is further from the American idea. But when either the manufacturer or the worker oversteps the amount of his capital, of his mental power or of his technical skill, naturally he must disrupt and instead of carrying out perfectly a more or less limited line of production, he will achieve only medium quality work. A patent law, if indeed not quite faultless, but nevertheless sound and effective, protects the inventor from unauthorized imitation and assures him the prospect of enjoying the fruits of his research and diligence. These four factors, the widest possible development of the replacement of man power by machines, education directed towards practical ends, division of labour, and a sound patent law have combined to bring machinery to the perfection at which we now see it, and to stimulate those great and small inventions which engender ever fresh undertakings and creations.

The most important branch of the machine industry is the manufacture of the steam engine, that is to say of those prime movers or motors which are above all designed to supply motive power to other machines. There were innumerable examples of these, of every size, shape and description. An imposing sight, standing in the middle of the hall on a raised platform was afforded by the two coupled units of the Corliss Steam-engine (Plate 20) named after the builder and inventor of the principle adopted, a manufacturer of Providence in Rhode Island. This twin giant delivers 1,500 horsepower. . . .

Next to the steam-engines were numerous other prime movers or motors, namely those driven by gas or petroleum. Here must be reported at once the pleasing fact that the gas-engine in the German Division manufactured by a Firm in Deutz near Cologne, was adjudged by experts to be the best of this sort and the one that consumed least material. . . .

Wherever the eye turned, the constant effort to remove purely mechanical work from the human hand and to apply it only where artistic shapes and forms were required was obvious. Think of our nail-smiths! How soul destroying and how laborious is the work, and how little, perhaps 100 nails a day, can be manufactured by one worker. In America there are no more nail-smiths; the nail there is manufactured by the machine. The American auxiliary machines and machine-tools which furnish all the needs of industry,—produce tools, make mortices, tenons, pegs,—are incomparable.[194]

The tour through the main building that we have reported above has shown us that in many other branches American industry is also superior to ours. In a short time it has risen to a height of which those in Europe had no real idea and which filled with astonishment most of us who had failed to recognize the conditions prevailing on that side of the Ocean. American industry, both as regards quality and output is continually advancing.[195]

The increased need of machines resulting from the scarcity of man-power led to mass-production earlier in North America than in Europe. Already at the end of the Middle Ages indeed we find the beginnings of mass-production of certain simple objects. The casting of type by Gutenberg, or the mass-production of useful objects cast in brass in Milan during the first half of the sixteenth century as reported by Biringucchio in 1540 (cf. p. 139),[45] may be mentioned in this connection. But not until the second half of the nineteenth century, after the improvement of machine-tools, did the method of mass-production, of assembled units, especially of machines, make headway in America. The extreme precision with which parts could now be made on the new machine-tools made it possible to introduce to mass-production the important principle of interchangeable parts. Arms, clocks, sewing-machines, since the end of the century bicycles also, were mass-produced in factories. For the precise and rapid production of single parts strictly accurate to specification, the introduction, after the seventies, of steel alloys for machine tools was of the greatest importance, especially in the manufacture of arms.

At the end of the nineteenth century, the Americans began to plan mechanised production by accurate time and motion study. Especially Frederick W. Taylor from the turn of the century led the way with his scientific and rational system of works management.[46] The beginnings also of mechanized handling of materials and parts by the moving belt system were in America by the end of the nineteenth century. We will learn by word and illustration from a report of the year 1890 concerning an *American Works with flow production of small castings for railway brakes.*

. . . The plant has been erected for the making of small castings for the Westinghouse Company. The air brakes for trains and a complicated interlocking switch and signal mechanisms that these companies are concerned with require a multiplicity of small parts. The foundry plant to which we refer is peculiarly adapted for the production of such pieces (Plate 22).

Its distinctive feature is a series of tables carried on wheels and linked together so as to constitute an endless chain. These move upon an endless track up and down the main foundry and through a smaller room adjoining the casting and moulding room, and carry the flasks or moulds from moulder to founder, and to the room where the moulds are broken up. In this compartment the castings are removed from the moulds. The arrangement of wheels and tracks is peculiar. The front ends of the tables are supported on two wheels, each wheel being journaled in a pedestal, the whole being underneath the table, so as to sustain it at a convenient height from the floor. A single wheel supports the inner end of the table. This wheel is also journaled in a pedestal, but is placed on the top surface of the table so as to rise well above it. An elevated rail is provided for these wheels, which rail, on both the straight sides of the run, rests on short columns. On these rails the inner wheels run, so that the tables are kept horizontal. The elevated rail bends in the arc of a semicircle at both extremities of the straight portion, and thus completes the circuit. At the end nearest the observer, . . . a shaft extends from the floor up toward the ceiling, connected above with a countershaft and friction clutch. Below the level of the table the shaft is provided with a sprocket wheel. This shaft occupies the centre of the circle described by the bent rail. The sprocket wheel gears into the endless chain of tables beneath their working surfaces. If rotated, it causes them to travel around the circuit. . . .

. . . On one side of the foundry is the moulding table, to be provided with mechanical moulding machinery. A conveyor trough runs from the room in the rear toward this table. . . . As fast as the trough is supplied with moulding sand it is carried to the vicinity of the moulding bench and is charged ready for use. A chain and bucket elevator is arranged in the rear room, which supplies the conveyor with the sand in question. Adjoining the revolving chain of tables on the side opposite the moulding bench are the cupolas, where the metal is melted.

The operation of the apparatus is as follows: The moulders turn out quite rapidly the moulds by the aid of the machinery. As soon as it is finished, the operator places the mould on one of the travelling tables, which are constantly moving behind him toward the cupolas. The mould is carried around the curve until it reaches the founders. They pour the metal into the mould as it passes. If necessary, the whole series of tables can be stopped for this process.

The mould, now filled with metal, goes on its way and enters the rear room. There it is thrown off the table and the casting is extricated from the sand. The sand is sifted in a revolving sifter, and is elevated to the conveyor, which returns it to the moulders. . . .[196]

The moving belt system was widely used by Henry Ford in the flow production of automobiles from 1913. It gained ground rapidly from the period of the First World War onward. Ford reported in his autobiography in 1922 on the moving belt system:

Along about April 1, 1913, we first tried the experiment of an assembly line. We tried it on assembling the fly-wheel magneto. We try everything in a little way first—we will rip out anything once we discover a better way, but we have to know absolutely that the new way is going to be better than the old before we do anything drastic.

I believe that this was the first moving line ever installed. The idea came in a general way from the overhead trolley that the Chicago packers use in dressing beef. We had previously assembled the fly-wheel magneto in the usual method. With one workman doing a complete job he could turn out from thirty-five to forty pieces in a nine-hour day, or about twenty minutes to an assembly. What he did alone was then spread into twenty-nine operations; that cut down the assembly time to thirteen minutes, ten seconds. Then we raised the height of the line eight inches—this was in 1914—and cut the time to seven minutes. Further experimenting with the speed that the work should move at cut the time down to five minutes. In short, the result is this: by the aid of scientific study one man is now able to do somewhat more than four did only a comparatively few years ago. The assembling of the motor, formerly done by one man, is now divided into eighty-four operations—those men do the work that three times their number formerly did. In a short time we tried out the plan on the chassis.

About the best we had done in stationary chassis assembling was an average of twelve hours and twenty-eight minutes per chassis. We tried the experiment of drawing the chassis with a rope and windlass down a line two hundred fifty feet long. Six assemblers travelled with the chassis and picked up the parts from piles placed along the line. This rough experiment reduced the time to five hours fifty minutes per chassis. In the early part of 1914 we elevated the assembly line. We had adopted the policy of 'man-high' work; we had one line twenty-six and three quarter inches and another twenty-four and one half inches from the floor—to suit squads of different heights. The waist-high arrangement and a further subdivision of work so that each man had fewer movements cut down the labour time per chassis to one hour thirty-three minutes. Only the chassis was then assembled in the line. The body was placed on in 'John R. Street'—

the famous street that runs through our Highland Park factories. Now the line assembles the whole car.

It must not be imagined, however, that all this worked out as quickly as it sounds. The speed of the moving work had to be carefully tried out; in the fly-wheel magneto we first had a speed of sixty inches per minute. That was too fast. Then we tried eighteen inches per minute. That was too slow. Finally we settled on forty-four inches per minute. The idea is that a man must not be hurried in his work—he must have every second necessary but not a single unnecessary second. We have worked out speeds for each assembly, for the success of the chassis assembly caused us gradually to overhaul our entire method of manufacturing and to put all assembling in mechanically driven lines. The chassis assembling line, for instance, goes at a pace of six feet per minute; the front axle assembly line goes at one hundred eighty-nine inches per minute. In the chassis assembling are forty-five separate operations or stations. The first men fasten four mud-guard brackets to the chassis frame; the motor arrives on the tenth operation and so on in detail. Some men do only one or two small operations, others do more. The man who places a part does not fasten it—the part may not be fully in place until after several operations later. The man who puts in a bolt does not put on the nut; the man who puts on the nut does not tighten it. On operation number thirty-four the budding motor gets its gasoline; it has previously received lubrication; on operation number forty-four the radiator is filled with water, and on operation number forty-five the car drives out onto John R. Street.

Essentially the same ideas have been applied to the assembling of the motor. In October, 1913, it required nine hours and fifty-four minutes of labour time to assemble one motor; six months later, by the moving assembly method, this time had been reduced to five hours and fifty-six minutes. Every piece of work in the shops moves; it may move on hooks on overhead chains going to assembly in the exact order in which the parts are required; it may travel on a moving platform, or it may go by gravity, but the point is that there is no lifting or trucking of anything other than materials. Materials are brought in on small trucks or trailers operated by cut-down Ford chassis, which are sufficiently mobile and quick to get in and out of any aisle where they may be required to go. No workman has anything to do with moving or lifting anything. That is all in a separate department—the department of transportation.

We started assembling a motor car in a single factory. Then as we began to make parts, we began to departmentalize so that each department would do only one thing. As the factory is now organized each department makes only a single part or assembles a part. A department is a little factory in itself. The part comes into it as raw material or as a casting, goes through the sequence of machines and heat treatments, or whatever may be required, and leaves that department finished. It was only because of transport ease that the departments were grouped together when we started to manufacture. I did not know that such minute divisions would be possible; but as our production grew and departments multiplied, we actually changed from making automobiles to making parts. Then we found that we had made another new discovery, which was that by no means all of the parts had to be made in one factory. It was not really a discovery—it was something in the nature of going around in a circle to my first manufacturing when I bought the motors and probably ninety per cent. of the parts. When we began to make our own parts we practically took for granted that they all had to be made in the one factory—that there was some special virtue in having a single roof over the manufacture of the entire car. We have now developed away from this. If we build any more large factories, it will be only because the making of a single part must be in such tremendous volume as to require a large unit. I hope that in the course of time the big Highland Park plant will be doing only one or two things. The casting has already been taken away from it and has gone to the River Rouge plant. So now we are on our way back to where we started from—excepting that, instead of buying our parts on the outside, we are beginning to make them in our own factories on the outside.[197]

The machine was in fact after the American Civil War, as was expressed by a reporter of the 1876 International Exhibition (p. 328) the sole element in North American life. All America was filled with a living sense of practical technology.

THE POSITION OF TECHNOLOGY IN GERMANY AT THE CLOSE OF THE NINETEENTH CENTURY

In Germany at the close of the nineteenth century, quite contrary to England and America, technique and the technician were regarded by the upper classes with considerable prejudice. In the general culture of the eighteenth century, the natural sciences and technology had had their place. In the nineteenth century, the cultural ideal of the upper classes narrowed to the philological, literary and aesthetic sphere. The engineer and poet, Max Maria von Weber, son of the composer Karl Maria von Weber, characterized the situation excellently in a *Dialogue* in 1882.

COUNT C.: Dear Baron, I congratulate you on your son. He is a quite charming young man. I was astonished at the good knowledge of literature, art and science that he betrayed in conversation with fine tact and without pretension. Beautifully brought up. What do you propose to make of him?

BARON E.: He will be a technologist, Count C., and will soon enter the Commercial Academy at B.

COUNT C.: You are joking! With your ancient name, your connections in the best circles! This elegant young man, cut out for a diplomatic or military career to be a sort of superior workman!—Forgive me, if I laugh.

BARON E.: Your laughter would astonish me, if I were not forced to believe that we associate very different ideas with the word *Technologist*.

COUNT C.: You are mistaken if you think that I am quite unversed as regards substances. I regarded as too terrible all these affairs of commerce, industry and trade and such modern catch-words still in use, and I held myself as aloof from them as possible. Clamouring machines, dirty hands, sweating fellows, boring figures—such is technology!

Then I married as you know the cousin of Princess Albina. There was then the most stringent obligation in high places to make me in some fashion. There was some hesitation as to whether I should be made the stage-manager of the Court Theatre at D. or whether I should take over the presidency of the management of the railway here.

I had not the least knowledge either of the theatre or of railways, but precisely this enchanted his lordship the Minister, who with the sound, fine wisdom of the statesman of the good old school, held fast to the view that administration could only be really impartial by those who knew nothing of the work. In June he made me Director of Railways.

What would you? I learnt to rule, and I ruled well, for when the Altdorf-Bergen Railway was opened, which I had never seen but which came under my control, the Duke decorated me. I then made the acquaintance of your technicians.

BARON E.: And this acquaintance has given you so unfavourable an impression of the technologists that you find fault with my desire to educate my son to be such a person? . . . Undoubtedly they are persons greatly cultured, with both knowledge and experience.

COUNT C.: Oh, just listen to this! The locomotive superintendent was a Frisian by birth, stout, broad shouldered; he looked as though cut out to grapple with machines, and his hands looked as though they had just laid aside the file. He rose so early and went so betimes to work that he never had time for proper washing and combing—in short, a splendid practitioner!—He had a pair of saucy eyes and a half-chewed cigar always stuck in his cheek. So he sat craftily smiling until everyone had had his say. Then like an oracle: 'That is all nonsense; it goes like this. I have calculated it'. And naturally, since he knew it so certainly, therefore it must be correct:

BARON E.: Wisdom tests and measures;
Foolishness yells: Thus goes it.

COUNT C.: I beg your pardon? Yes, he was a splendid practical fellow, the Minister said so, and we could shake hands over it.

It is true that the machines invented by him never went properly, but he declared that their efficiency, or whatever he called it, was splendid. He proved it by experiments and rows of figures—enough of them to give you vertigo—we were astonished, and naturally believed. He was most capable. He was regarded as the born builder of machines.

BARON E.: And the chief engineer?

COUNT C.: A man as dependable as gold in his work! He was a thick-headed phlegmatic countryman and thirty years earlier he had built a railway, whereby he had gained such excellent experience that he rightly adhered conservatively to it. He laughed in a superior fashion at all new constructions and daring innovations as 'theoretical swindle'. In that I thoroughly agreed with him for Goethe says: 'Only the golden tree of life is green', and why should not that which was best thirty years ago be of value also today?

He never risked his career as a distinguished practitioner nor his country's money by study of the questions under discussion, or by clever building, by the use of new building methods and tools, or by experiments whereby the technologists expected to advance their so-called art, but nearly always only ministered to their own vanity; but he followed, looking neither to right nor to left, the orders given him from above—he was a proper official, his sphere of control in exemplary order—My God, was it his fault that in spite of exact writing and obedience and firm adherence to the old tested methods, sound progress of the work on the roads and buildings was absolutely impossible, and often—I will confess—much money had to be lost?

If only from the way he laughed over innovations one realized that he was a splendid practical man—had not the Minister himself said so?

Baron E.: You met some peculiar specimens on whom to base your idea of technologists. Which department did the consulting engineer represent?

Count C.: Oh, there is not much to say of him. He was too like others of the party to be able to be technologically differentiated from them. Oh, I execrate these men. He was the most insupportable of all the technologists. The fellow dared to look like us, to have manners like one of ourselves and to speak to us as though he were one of ourselves. He could really have been taken for a civil servant or government officer. In order to become a technologist, the idiotic lad had wasted the years of his youth with Caesar, Horace and Xenophon instead of working at the bench; he was the child of good people, he had had a French nurse, he had attended the University lectures of Dove and Hegel—good God, what was a technologist doing there? That he had passed with high distinction through the famous Polytechnic at Z., that he had done practical work and construction in large factories, that he had built railways under the so-called great masters, that he had worked through almost all of the departments of a railway from the bottom up, that could not possibly turn him, after such antecedents, into a practical technologist!—So also thought the Minister.

Baron E.: Marvellous. It would have seemed to me that after such an education, this man was cut out to be a practical Director.

Count C.: Baron, I cannot tell you how clumsy and conceited this fellow was.

For example, if any great technological question—God knows what—came under discussion; then the chief engineer just had to be asked how it went, and as a really practical man he told me without more ado.—Thus the Minister as an intelligent man always behaved. —No, our consulting engineer insisted on a fundamental study of the question, on correspondence with neighbouring establishments, on experiments of all sorts, on despatching and journeying of experts hither and thither, and at last brings from the mass of information and results an elaborate report for better or worse so dull, wordy and unpractical as to be laughable,—I carefully avoided reading it, I had it filed.

I ask you was that not an execrable fellow, a thorn in the flesh of the management? . . .

Baron E.: You start apparently from the assumption that technologists have seldom or never enjoyed a good education, meaning by well educated, not to have learnt something, but to be a *gentleman.* Your consulting engineer proves exactly the contrary. Should I consider him an unpractical technologist because he was a gentleman?

COUNT C.: Rare exceptions prove nothing. Say what you will, a fellow who has to touch so much dirt during twelve hours of the day that he fears to soil his clothes with his hands, holds his hands, even when they are washed, as though they were dirty. The manner in which a technician stands at a vice, gazing downward, stands with his fellows round a piece of machinery, his manner of moving his arm as of a person accustomed to handle a file or a lathe, this can never be effaced and infallibly betrays a lowly occupation.

While the officials and therefore in Germany naturally also private persons can imagine a reliable, practical technician only as a fellow of fairly ordinary appearance, wearing waterproof boots, with a ruler under his arm, and a roll of paper in his hand, they in a sense force the technicians, for the sake of peace and quiet, to bear themselves in outward appearance in the manner corresponding to this picture. . . .

BARON E.: Educate real men, who in general culture and manner of life are at the peak of our national life and of civilized society, and then make technicians of these men—that is the whole secret and the sole solution of the problem.[198]

The social prejudice against the engineer began, however, around the turn of the century slowly to weaken even in the land of poets and thinkers. An important step in this development was the victory gained in 1899 for the right to graduate through the Technical Colleges, a condition corresponding to that of the Universities. German industry increased rapidly from about 1900 not least as a result of technology scientifically conducted in the Colleges, and it became an important rival of England and America. If Reuleaux in 1876 had called the German industrial products shown at the International Exhibition at Philadelphia 'cheap and nasty', yet already at the International Exhibition at Paris in 1900, Germany was able to win special recognition for the high quality of her technological products.

NEW PRIME MOVERS

The large and expensive steam-engines were in general, as was shown by G. Reichenbach in 1816 of use 'to rich individuals and to large manufacturers'.[47] Reichenbach laboured at that time, certainly without great success, to produce a cheap high-pressure steam-engine occupying little room and easily movable, for the use of the smaller concerns. It was thought at first that the small engine for industry had been found in the hot-air machine on which many inventors, in conjunction with S. Carnot worked in the first half of the nineteenth century and of which a few were distributed in the fifties owing to the work of John Ericsson; but it was not able to survive. In 1860 Étienne Lenoir constructed a gas engine with electric ignition, which attracted attention. It was already being produced in great numbers. Stimulated by the reports of Lenoir's engine, Nikolaus August Otto worked on building a gas engine. In 1867, Otto—who had meanwhile in 1864 founded a gas-engine factory in partnership with Eugen Langen—came forward with an atmospheric gas-engine that needed only a third as much gas as Lenoir's motor. In ten years, 5000 atmospheric engines of from a quarter to 3 horse-power were sold. The atmospheric engine was used chiefly for driving pumps and in printing works. But it was very noisy. Moreover this machine which was jerky and shaky stood 5 feet 6 inches high. It could not develop more than 3 horse-power.

Franz Reuleaux gave eloquent expression to the general need at the time for a small engine for the artisan. He wrote as follows in 1875:

> Only capital is able to build and operate the powerful steam-engines around which is grouped the remainder of the establishment, which indeed also requires capital but from which it is not inseparable. . . .
>
> How to make power independent of capital. The little weaver . . . would be relieved from the oppression of capital if we could give him the necessary motive power to drive his loom. A similar attempt could be successful in the spinning trade. . . . Other trades . . . are cabinet making, locksmithing, saddlery, tinsmithing, brushmaking, pump building, etc. What is lacking in these trades is partly power and partly machines. But the individual artisan could even now procure the latter, as they can be obtained at a really low cost; but he is always without motive power. The cabinet maker to whom motive power was supplied cheaply for a circular saw, a band-saw, a planing machine or a moulding spindle, could work on these machines in his own home quite as well as he now does in the furniture factory which has absorbed him. He would then, while having utilized his group of machines for many purposes, have retained or re-acquired the skill which would be lost to him as a factory worker. The process would be similar in the other trades mentioned. The small masterman, in spite of certain advantages of large scale industry, would be able to compete with it because in home industry the mutual help of members of the family, and above all the moral

element is an effective factor. The little masterman, with his surrounding helpers and pupils would form a closed working organization with its leader and members, and its higher and lower grades of ability; this would be similar to the handicraft of earlier days but would differ from it by the introduction of the machine. Once small scale industry becomes competitive it would quickly improve his standing because at the same time the demand for workers for large-scale industrialists, that is to say for capital, would show an increase.

To combat most of the evil, engineers must provide cheap, small engines, or in other words, small engines with low running costs. If we give a power supply to the little masterman as cheaply as the great powerful steam-engines can be obtained by capital, and we thus support this important class of society, we shall strengthen it where it happily still exists and we shall re-create it where it has disappeared. It will do us nothing but good, that the pressing call shall echo among us to help the little masterman once more on to his feet also in other callings, for example in artistic work. . . .

The feeling that the subdivision of basic power is desirable, is current in various places and in many forms. One is the leasing of power which has been successfully tried in large towns. It has resulted indeed in the crowding together of workers in a single building, in the voluntary crowding together of families and of work-people in unhealthy rooms, and therefore brings the old evil in a new form. This, however, is far behind the principle of separate engines in small installations. Already there are many excellent examples of this nature. Above all the gas-engines, then the hot air engines, the small ram pump and in a very promising experimental stage the oil engines. . . .

They must therefore be reckoned among the most important of all new machines; in them is the germ of a complete revolution in one section of industry. . . .

Air and gas engines can be . . . used almost everywhere; are approaching final perfection. These little engines are the true power units of the people.[199]

Otto and Langen worked indefatigably for a gas-engine without the drawbacks of the atmospheric engine. In 1877 Otto could at last offer a workable four-stroke engine which was distinguished from the atmospheric engine by its quiet action and remarkable economy in space and weight. The invention pointed the way for modern motor construction. Reuleaux was enthusiastic over the new engine, which he considered to be the greatest invention in the power machine industry since Watt. Otto and Langen, when in 1889 they celebrated the twenty-fifth year of the foundation of their firm, wished that their machine should place a dam 'to hold back the overbearing and arbitrary power of capital, to strengthen small industry for the severe struggle of economic life and to direct the method of production back into paths which would guarantee the further development of culture'.[48]

The gas engine of Otto and Langen at Deutz was at first dependent on gas supply, although as early as 1875 experiments were made there with benzine as fuel, and the question of driving vehicles by gas engines was already considered. But it was reserved for Gottlieb Daimler and Wilhelm Maybach, following up the work of Otto and Langen, to manufacture in 1883 a light, high speed benzine motor which was capable of development. Otto's four-stroke motor made 150–80 revolutions per minute. But Daimler's new motor ran at 900 revolutions per minute. This introduced the possibility both of motorizing vehicles and aerial transport. A benzine powered vehicle built as early as 1875 by Siegfried Marcus in Vienna, was quickly forgotten. Only with the motor vehicles with which Gottlieb Daimler and Carl Benz had come forward independently of one another since 1885, ensued a development which led to the modern motor vehicle which—as we have already heard—was in mass production as early as 1913 in America, and created a new revolution in transport.

Besides the internal combustion engine there appeared at the end of the nineteenth century the Diesel oil engine, working with higher compression pressure, with self-ignition and combustion at constant high pressure. This developed in the following decades thanks to its great economic efficiency as the most promising engine of the future. Diesel wrote in 1913 shortly before his death a clear account of the creation of his engine:

An invention consists of two parts: the idea and its execution.

How does the idea originate?

It may be that it sometimes emerges like a flash of lightning; but usually after laborious searching it will hatch itself out of innumerable errors; and by comparative study will gradually separate the essential from the non-essential, and will slowly permeate the senses with ever greater clarity, until at last it becomes a clear mental picture. The idea itself originates neither from theory, nor deduction, but by intuition. Science is merely the means of investigating and proving, but is not the creator of the idea.

But even when the idea has been scientifically established, the invention is not yet complete. Only when Nature herself has given an affirmative reply to the question, which the test has put to her, is the invention completed. Even then, it is only a compromise between the imagined ideal and the attainable reality.

When my revered teacher, Professor Linde, explained to his audience at the Munich Polytechnic in 1878, during his lecture on thermo-dynamics, that the steam-engine transforms into effective work only 6–10 per cent of the heat value of its fuel, when he explained Carnot's theorem, and pointed out that with isothermic changes of condition all heat conducted to a gas would be converted into work, I wrote in the margin of my college note-book: 'Study whether it is not possible to realize in practice the isothermal curve'? Then and there I set myself the task! This was no invention, not even the conception of one. Thereafter the wish to materialize the ideal Carnot cycle dominated my being. I left College, went into practice, was obliged to win myself a position in the world. The thought pursued me incessantly.

At that time, all hope for improvement in the use of the heat of the steam-engine was placed on super-heated steam. Since I, as the man in charge of the refrigerating machinery, was familiar with ammonia vapour, I thought over the possibility of using, instead of aqueous steam, super-heated ammonia vapour which, because in normal working conditions far from its point of condensation, is much less sensitive to the cooling influence of the cylinder-walls. In Linde's ice factory in Paris, I installed a laboratory for fundamental study of super-heated ammonia vapour and ammoniac solutions, as well as for the construction of small ammoniac-motors which would absorb the exhaust vapour. The theoretical researches conducted at the same time showed that, in order to obtain full use of the super-heat, it was necessary to use very high pressures.

Such high pressure and greatly super-heated vapour already exist almost in the very nature of a gas. But I cannot say whence came the basic idea of replacing ammonia by a real gas, namely super-heated air under high pressure, gradually to introduce into such air a finely divided fuel, and to let the air expand simultaneously with the consumption of the fuel particles, so that the utmost possible amount of heat engendered should be converted to work. But from the incessant chase after the desired result, from researches into the relationships of countless possibilities, the correct idea was at length developed and I was filled with unutterable joy. After I had, on my way to the super-heating of steam, come upon a process of combustion of a special sort, I tested the idea by means of thermodynamics and published these views, at first purely theoretical, in a little work which appeared in 1893, fourteen years after that marginal observation in my College note-book; in this work, after investigation of every sort of combustion curve, I declared isothermal combustion to be the most rational. The German patent No. 67,207 was registered a short time previously.

It would lead us too far to reproduce here the contents of this work; those who are specially interested in the gradual transition from theory to practical machines must be referred to the book itself; it must be added that by going more deeply into these studies, especially on the practical side and with particular consideration of the loss of mechanical power, I recognized that Carnot's cycle only theoretically deserves its fame as 'completely perfect', and that for the practical machine, not the maximum temperature but the maximum compression is the deciding factor. Therefore in practice not only in the compression, as I had assumed in my theoretical work, but also in the combustion, the isotherm must be relinquished in order to attain greater specific outputs and improved mechanical efficiency, though undoubtedly against considerable sacrifice of the heat efficiency originally calculated. I therefore registered in the same year, 1893 a second German patent, No. 82,168 in which in addition to the isotherm, every other form of combustion curve in the diagram was protected, By this means complete freedom was gained for the development of the original and actual inventive ideas, which were as follows:

1. Heating of pure air in the cylinder of the engine, by mechanical compression by means of the piston, to a point far above the ignition temperature of the fuel to be used.

2. Gradual introduction of finely divided fuel, and its combustion in this super-heated and compressed air, and simultaneous performance of work from it on the outward moving piston.

Since fuel can only burn if it has first been vaporized, for all nongaseous fuels the immediate outcome from this second fundamental thought was:

3. The gradual vaporization of the fuel in the working cylinder itself, though only very little at a time for every stroke of the piston, especially with the removal of the heat of vaporization for inducing combustion through the heat of compression. The third fundamental idea would abolish the complicated and wasteful gas generator.

It is often asserted without hesitation by the laity even in scientific circles that the important point of the Diesel process is the self-ignition of the fuel; and that the object of the high compression is that the fuel, injected at the dead centre, shall ignite itself, the high degree of compression being demanded by this self ignition.

Nothing is more incorrect than this superficial view, which is directly contrary to the facts and especially to the historical development (of the idea).

Motors with self-ignition of the fuel had existed before. Neither in my patents nor in my writings did I mention self-ignition as a goal to be sought. I was seeking a process with the highest heat efficiency, and as it turned out, self-ignition was naturally involved in this. If the air is heated by compression to a point far above the ignition point of the fuel, the ignition of the fuel in contact with this air follows automatically; but it is not the object of this high compression. Self-ignition of every liquid or vaporized fuel within an engine already heated by work takes place at quite low pressure, of 5–10 or at most 15 Atmospheres. It would thus be much simpler to construct lighter and cheaper engines for these degrees of compression, and overcome the difficulty of ignition in a cold engine by employing artificial ignition temporarily. It would be absurd merely for the sake of ignition when the machine is cold to construct such heavy and unhandy engines to withstand a pressure of 30–40 Atmospheres, since, once the engines are warmed up by work, they run quite as satisfactorily under lower pressure, as has often been shown by experiment.

The object of the system, sought through long years and realized with such difficulty, is however quite different, namely the attainment of the highest possible fuel efficiency; this object necessitates highly compressed air. Since, however, this produces self-ignition much too soon in the fuel mixed with the air, therefore self-ignition by compression as it was known in the gas engines of those days was an obstacle to the process and had to be avoided in that form. It was necessary that the air alone should be so highly compressed by mechanical means as to give the desired heat efficiency.

The degree of this compression is determined not by the ignition of the fuel but, corresponding to the original object, by the maximum economic utilization of the fuel.[200]

In masterly fashion, guided by his own experience, Diesel embraced the whole essence of inventive creation. He showed also that the German patent law—for at length in 1877 Germany had enacted valid protection for the inventor throughout the whole kingdom—much as it may encourage inventiveness generally could yet have a restrictive influence, if a patent for a real invention to the industrial realization of which the inventor has devoted much exhausting work, can yet not be granted because it can be proved that 'the idea was already mouldering in some forgotten publication'.

No idea can ever be designated solely as an invention; if a random selection is made from a list of inventions, the telescope or the Magdeburg hemispheres, the spinning-machine, the sewing machine or the steam engine, it is only the application of the idea that ranks as an invention. An invention is never a purely mental product, but is the result of a struggle between thought and the material world. Therefore it can be demonstrated regarding every complete invention that similar ideas had often been in the minds of others more or less definitely and consciously, long before.

Between the conception and the completed invention there lies always the time spent by the inventor in work and suffering.

It is always only a small part of the exalted ideas which can be established in the material world; and the completed invention always appears quite different from the original imagined ideal which will never be attained. That is why every inventor works amidst an enormous number of rejected ideas, projects and experiments. Much must be attempted for anything to be achieved. Very little of it is left standing at the end.

Our Patent Law in general only protects an idea but not an invention, and it can therefore destroy the most valuable and practical inventions if only it can be proved that the idea is somewhere mouldering in a forgotten publication.

The origin of an idea is the happy period of creative mental work; when everything seems possible because so far it has had nothing to do with reality.

The carrying out is the time for preparing all the means that will assist in the realization of the idea; still creative, still happy, the time when natural obstacles are overcome; from which one emerges steeled and exalted even when beaten.

The introduction is a time of struggle against stupidity and envy, apathy and evil, secret opposition and open conflict of interests, the horrible period of struggle with man, a martyrdom even if success ensues.

Invention then involves carrying to practical success through innumerable failures and compromises a sound basic idea born of a great number of errors.

Therefore every inventor must be an optimist. The force of the idea only exercises its influence in the single soul of the originator; only he has within him the holy fire to carry it through.[201]

When in 1893 Diesel published his work on the *Theory and Construction of a rational heat-engine* and sent a copy of it to Franz Reuleaux, he received from him a letter of thanks which makes it clear that Reuleaux immediately and justly grasped the great importance of the invention.

To Herrn Rudolf Diesel, Engineer

I have read with great interest your work on your new motor which you so kindly sent to me. The theoretical work appears to me faultless and most convincing. A few practical points will at first present difficulty. These will certainly give you trouble for a time—e.g. the distribution of the injected pulverized fuel, the arrangement of components, the tight closure of vessels subject to large variations of temperature. But nowhere as it seems to me is there an alarming obstacle. One may therefore wish you luck; also that endurance demanded by every new task. Again, your machine inflicts a blow on the powerful steam-engine, since it exceeds it in heat efficiency. Technology must ultimately eliminate the fault so long recognized in the old steam-engines. The increase of power that we can gain thereby is so great that it is worth every effort.

I express once more my best thanks to you for having sent your work to me. Yours most sincerely

F. Reuleaux[202]

The Diesel engine, developed after 1893 in the engineering works at Augsburg (Plate 23), soon won a wide field of operation both as a stationary engine for various purposes or for ship propulsion, for road and rail vehicles, and since 1928 even for aeroplanes. The speed with which the Diesel engine gained ground in marine work, thanks to its high efficiency, is revealed by the fact that by 1939 something like a quarter of the world's mercantile marine was equipped with Diesel engines. With the increased use of benzine and oil engines, mineral oil became a raw material of the highest political importance. In the short century after boring began in 1859 in Pennsylvania, production rose steeply.

The ever-increasing need of power since the last third of the nineteenth century directed the attention of technologists also to the water turbine which was greatly improved, and soon used for the generation of electric current. By the side of the steam-engine there stood at the end of the nineteenth century the steam turbine, whose efficiency clearly surpassed that of the reciprocating engine. Especially the high-pressure steam turbine of Charles A. Parsons, who built his first machine as early as 1884, was from 1900 extensively used in electricity works which had been run on coal. In this case the turbine could be coupled directly to the shaft of the dynamo, thus achieving the highest economy of work. The beginnings of the new sort of turbine, the combustion or gas turbine, go back—if we ignore mere projects of an earlier period—to the researches of Holzwarth in 1906. The actual development of the gas turbine, which is distinguished by its high power-weight ratio, was completed in the second quarter of the present century. Reduction of the weight per horse-power is an essential feature in the development of all power generators. In 1915 the weight of an aeroplane engine was about $3\frac{3}{4}$ lb. per horse-power; but by 1950 it was only about 1·1023 lb. per horse-power.

ELECTRO-TECHNOLOGY

L. Galvani had discovered the electric current in 1786. A. Volta, J. W. Ritter, H. Christian Oersted, A. M. Ampère, G. S. Ohm and M. Faraday sought during the first third of the nineteenth century to discover the properties and behaviour of the electric current. Experiments in the technical application of the electric current began primarily in the laboratories of the physicists. The beginning of electro-technology was the discovery by Schilling von Canstadt in 1832 and by F. Gauss and W. Weber in 1833 of the electro-magnetic telegraph. Samuel F. B. Morse followed in 1843 with the first workable teleprinter which won practical adoption. In the second half of the nineteenth century there followed the telephone. Already in 1832 in the year following the discovery of induction by Faraday, H. Pixii, Ampère's mechanic, built the first rotary generator. And in 1834 M. H. von Jacobi constructed the first electric motor which could give a useful power output; this machine was fed by a battery. The Dutch physicist Vorsselmann de Heer reported in 1839 on electro-magnetism as a propulsive power. In this he drew on the writings published by Benjamin Silliman in the *American Journal of Science* in 1838, and also on a little electro-magnetic vehicle which the chemist, Professor S. Stratingh had built in Groningen in 1835.

. . . The possibility of using electro-magnetism as a new propulsive power has aroused the attention of several scientists in the United States. We only hear what Silliman says on the subject in his much respected scientific journal which appears in New Haven. . . . Science, he says, has most unexpectedly put into our hands a new force of great, but unknown power. It neither lures the winds from their caves, nor moves water by the propulsive power of heat, nor directs for production muscular animal power, nor works with complicated mechanisms, nor does it accumulate the forces of water by taming wild torrents, nor does it invoke any other form of gravity, but in mysterious fashion, by the simplest means, the mere contact of small metallic surfaces with weak chemical agents, there is mysteriously developed a power dispersed throughout Nature though in general hidden from our minds. And by means of circulation in insulated wires, this power is in even more mysterious fashion increased many thousand fold, until it produces an unbelievable force. No significant period of time has elapsed between the first development and the full accomplishment, and the child rises as it were like a giant.

Such is the present position as regards electro-magnetism. Will this new power, at least in certain cases, really be able advantageously to supplant steam? I hear that skilled mechanics in our native land reject such a hyphothesis as complete nonsense. . . . Such an attitude is unbecoming in a man of scientific culture, and is blameworthy in so far as the desire of others to support a good thing is belittled by this view. Electro-magnetism is a power 'of great but unknown force'. That which is not known should not be rejected; or if mechanics have in fact such solid reasons that they can support their adverse judgement, why do they not publish these reasons in order that others may be freed from error, and may apply their time to something better? In the realm of natural science, only factual considerations are admissible; and since, so far as I know nothing has yet been decided in this fashion, I therefore venture to describe an apparatus in which electro-magnetism is applied in a remarkable fashion as propulsive power. Perhaps this device if correctly applied will bestow certain advantage; and if it is never applied, it is perhaps worthy to hold a place among the toys of physicists.[203]

. . . Will these electro-magnetic capstans one day serve to set in motion the shafts of our factories? Shall we see Herr Stratingh's carriage only on the tables of our audience chambers, or perhaps travelling along the road from Amsterdam to Harlem? In short, it is possible, even if only in certain cases, to replace steam by electro-magnetism; and if that is the case, will the replacement be advantageous?

I think that for a definite and fundamental answer to this question, science is entirely without the necessary data. Meanwhile much has been done which may serve towards a complete solution of the problem and what is lacking can easily be supplied by further experiments. But the experiments must be to some extent on a large scale, and thus exceed the expenditure which most of our physicists have at their disposal. But the nature of the experiment does not appear to me impossible, and I think that for a few thousand gulden the matter could be accomplished. Certainly the object is sufficiently important to merit some support from the Government. The Emperor of Russia has already set an example worthy of imitation; and if the knowledge and enthusiasm of Russian scholars corresponds to the abounding protection afforded to them by their monarch, we may soon expect something admirable from there. Let us look a little more closely meanwhile at the nature of the task.

For the application either of steam or of electro-magnetism, water is necessary; in the first case it must be evaporated, in the second disintegrated. For evaporation heat is needed, and therefore coal, wood, fuel. For disintegration chemical action by zinc is required. The question now is by which method the greatest power can be evolved with least expense. How many pounds of zinc are required to drive a machine of one, ten, or a hundred horse-power?

That is the vital question for electro-magnetism, and so long as it is not answered, all hopes and conclusions, whether favourable or unfavourable, are merely idle and deceptive. But is there any hope of attaining a solution to this problem? I think there is. The weight of zinc required to confer a definite degree of magnetic power on a definite quantity of iron must be as easily determined as the calculation of the weight of coal necessary to heat a definite quantity of water to a definite temperature. . . .[204]

Finally, an important point remains to be examined. It is a question of using as a motive force the magnetic power conferred. It is necessary to know how many pounds can by this means be raised to a given height in a given time. Clearly this is not a question of the magnet's lifting power but of the attraction and repulsion which will be exercised across a space. If two magnets of known size and power, that is to say, which under the influence of terrestrial magnetism, perform a definite number of oscillations and approach one another either at right angles or in parallel motion, what power will they exercise on one another, what weight is needed to cancel the attraction or the repulsion? How far does the weight depend on the respective positions of the magnets? This point too can be clarified by experiment, and thereby can be determined the motive power which can be produced by a given magnetic force. . . .

I have only shown superficially the course of the research. May more skilful physicists and those commanding greater resources fill these gaps in our knowledge, so that both favourable and unfavourable prejudices may be removed by clearer and more exact ideas. And should it follow that this new power can set no ship nor any vehicle in motion if its efficiency be calculated not in dynes but in millidynes, even then it can be usefully applied for countless purposes. It is only a question of knowing whether it will prove a cheap motive power. Steam power can be used with advantage only when heavy loads have to be moved. A steam-engine, for example, which is to perform the work of a man would be useless if it were to be used not continuously but only intermittently with interruptions. Just in these circumstances electro-magnetism could be used in many cases on the smaller machines. Such a machine is rapidly set in motion; as soon as the zinc is immersed in the fluid, the effect begins; it ceases as soon as the zinc is removed. Waste takes place neither beforehand nor subsequently; work is obtained from every grain of zinc which undergoes voltaic oxydisation; the machine costs nothing unless it is at work. How many cases can be imagined even in the ordinary needs of a house in which such a power could be usefully applied. The only question is: can work be achieved more cheaply by men or by magnets? May these considerations contribute to direct general attention to this subject. With this wish, I bid farewell to my readers. . . .[205]

After M. H. von Jacobi had invented galvano-plastics in 1839, there arose the need for large electro-magnetic machines, as work with batteries was too expensive. To meet the needs of electro-technology, a large electro-magnetic machine, driven by steam, was constructed in 1844 by John S. Woolrich, professor of chemistry at Birmingham. The cousins George R. and Henry Elkington used Woolrich's current generator in their electro-technical factory. The Russian State Councillor J. Hamel, who undertook an educational journey to England by order of the Russian government reported on the machine in 1847.

Among the interesting and splendid technological applications made in England, during the last decade, of the results of scientific research is the liquid process of gilding and silver-plating metal objects, and indeed there are three methods:

(1) Gilding by mere immersion;

(2) silver-plating and gilding by means of the galvanic battery; and

(3) by electro-magnetic machines. . . .

What speaks most highly for the magnetic machines for silver-plating and for gilding introduced by Mr Woolrich is perhaps the fact that Messrs Elkington, undeterred by their special patent for silver-plating and gilding by means of the battery, have bought the patent right from Woolrich and at the present moment are getting him to install in their works a really colossal magnetic machine. It has eight horse-shoe magnets each of which consists of twelve plates which are 2½ feet long from the ends of the poles to the outer edge of the curve; they are 2½ inches wide and form together a thickness of 4 inches. Between the poles is a space of 6 inches. These eight magnets are held by brass devices between two circular cast iron discs in such a fashion that all the poles are directed towards one centre at the axis of the wheel which is 2½ feet in diameter and has on its periphery no less than sixteen armatures, wound on iron cylinders nearly 6 inches long and 2½ inches thick, which fly around between the poles of the magnets at a speed of seven hundred or more revolutions per minute. Mr Woolrich thinks that one H.P. would suffice to turn the wheel bearing the armatures.

The machine described here will soon be installed at Messrs Elkingtons. Even if Mr Woolrich carries his enthusiasm too far in the expectation that from sixteen to twenty ounces of silver may be deposited in an hour, that is to say up to 27 lb. per day, yet this enormous apparatus will accomplish more than all previous magnetic machines used for electrolytic work.

In factories where great quantities of metal, whether silver or copper, are continuously deposited, especially where a steam-engine is at hand for other purposes, magnets should certainly be preferred to batteries. . . .

I have already spoken of the great quantity of gold which is used in gilding work in Birmingham. A comparatively notable quantity of quicksilver which in the old method was used to carry the gold in the form of an amalgam, must have been driven off again by the heat as vapour. These fumes, drawn into the lungs of the workers busied at the furnace, caused the well known terrible trembling and twitching of all the limbs, profuse salivation and other diseases. Thanks to the gilding methods introduced by the Elkingtons and J. S. Woolrich, not a single case could be shown me on my last visits to Birmingham of such pitiable sacrifices to the quicksilver gilding method of which I had previously seen so many and always with a heavy heart.[206]

Besides galvano-technology it was the arc-light which necessitated serviceable electro-magnetic machines such as were constructed in the fifties of the century by the 'Société d'Alliance'. The technique of real high tension current began in 1866 with the construction by Werner

Siemens of the dynamo which was the first generator working on the electro-dynamic principle. Thomas A. Edison erected in 1882 in New York the first electrical power station in the world, and soon it was supplying current to some thousands of electric lights (Plate 24). A few years earlier in 1879, Edison had introduced electric carbon filament lamps with screw sockets. And in 1884 the first electric motor was added to Edison's power station. The first German central power station was opened in Berlin in 1885. W. Siemens, who had put the first electric locomotive in commission in 1879 and the first electric lift in 1880, undertook for the first time in 1881 the direct connection of a steam-engine to an electric generator. We reproduce the following three letters written by Werner Siemens during the 1880's; they reveal his far-seeing views.

To Karl Siemens in St. Petersburg

Charlottenburg, 6th December 1887

Now is the time to build electric power stations throughout the world. So far this has been done only by the Americans and ourselves, that is to say, by Siemens and Halske and by the General Electric Co (formerly Edison), In England there is as yet not a single power station for electric light, and we are building the first in France at Lyons, through the agency of the Mülhausen-Belfort Engineering. In London also they decline the establishment of power stations in foreign lands, that is to say, they renounce it in favour of the Americans.[207]

To Karl Siemens in St. Petersburg

Charlottenburg, 25th December 1887

. . . Certainly I have striven for profit and wealth, but not mainly in order to enjoy them; rather to gain the means for the execution of other projects and undertakings, and by my success to win recognition of the correctness of my procedure and the usefulness of my work. Therefore from my youth upward I have yearned to establish a world-wide firm such as that of Fugger, which would assure not only to myself but also to my successors power and esteem in the world, and the means also of raising my sisters and other near relatives to higher standards of life. . . .

I regard our business as only secondarily a source of wealth; for me it is rather a kingdom that I have founded and that I hope to leave intact to my successors for further creative work. . . .[208]

The following letter to Count de Bylandt reveals to us how Siemens, with a mind open to all the problems of the age, felt strongly on the social significance of the small engine of which we have already heard something in connection with internal combustion engines. Attention, he said, should be directed not only to electric light but also to the 'socially far more important electric power transmission'. The electric motor in fact offered a possible solution of the problem of the provision of small power units for industry.

To Count Bylandt at the Royal Netherlands Embassy in Berlin

Electricity has hitherto not been suitable for the actual generation of power, such as is provided by steam and gas-engines; but only to conduct power generated in a different fashion to places where it will be used for work. The power of a waterfall can thus be conducted by electric current to neighbouring places where it can be used for electric lighting, for driving machines and for other electro-technical operations. But such great natural sources of water power occur almost exclusively in mountainous regions. Where they are available, as in Switzerland, they are already widely used for the transmission of electric power to neighbouring towns such as Zürich, Geneva, etc. This has happened to a much less extent in Germany, which lacks great water power available both in summer and in winter. In the plains we are therefore compelled to generate, in steam and gas-engines, power which can be transmitted electrically. The great expense involved in a cable network to distribute electric power from central points to towns has hitherto been an obstacle to the establishment of such power stations. Everywhere, main attention has been given to electric light, since the introduction of electric lighting has everywhere been regarded as a necessity. It is, however, well known to electrical engineers that electric light is merely the preparation for the change to the socially far more important transmission of electric power. By the transmission of electric power, a cheap source of power can be conveyed to city populations by the least laborious method. Thereby the small workshop and the individual working by himself in his own home will be in a position to employ his own capacity more effectively and to compete with the factories that generate their power cheaply by steam-engines or by gas-engines. This circumstance will in course of time produce a complete revolution in our conditions of work, in favour of small-scale industry. It follows that the ease with which power can be transmitted to those places needing it will result in innumerable installations in houses and streets which will add to the amenities and ease of life—by ventilators, lifts, street tramways, etc. Furthermore, electricity has yet other application to industrial life by the electro-chemical action of the electric current. By its means electricity can be stored in accumulators for whatever purpose is desired,—it can carry out gilding, silver-plating, make galvano-plastics etc. A certain time will be needed before the public becomes accustomed to this use of electricity, but it will undoubtedly come. Thereby an installation to distribute electric current will be useful also by day, whereas an installation providing only electric light is fully used solely during a few hours in the evening. This will greatly increase the profitableness of the installations. At present many towns hesitate whether to install direct or alternating current. Thus my Firm in the Hague is installing direct current, while in Amsterdam an installation of alternating current is being constructed. But the alternating current plants are not suited, or at least are very imperfectly suited for the transmission of power; they therefore constitute an obstacle to the future development of the social life of the town.[209]

In addition to the competing systems of direct and alternating current, of which we have heard in the conclusion of Siemens' letter quoted above, there soon appeared a new system superior to them both, the multi-phase current. In the very year when Siemens wrote the above letter, Monsieur von Dolivo-Dobrowolski developed the practically serviceable multi-phase current motor and the current transformer, after preparatory work had been accomplished also by others. With the completion, at the electro-technological exhibition at Frankfurt-am-Main in 1891, through the stimulus given by Oskar von Miller, of multi-phase transmission of electric power over nearly 109 miles from Lauffen to Frankfurt-am-Main, the age of overland supply had begun. The ease of transformation of the multi-phase current and the simple construction of the multi-phase motor contributed to the victory of the multi-phase current.

By the end of the nineteenth century, enthusiasm was aroused by the imposing achievements of this new electric technology. It was thought that the age of steam had completely passed. The young giantess, electricity, would bring ruin to the slave, steam. In 1893, A. Wilke wrote as follows:

Not long ago, our period was described as 'the century of steam,' and rightly so; for the steam-engine has been in the front rank in moulding modern conditions of industry and commerce. But hardly had the slave *Steam* grown to its full strength when there appeared for the service of mankind a young giantess who as it seems desires to work in harmony with Brother Steam for their masters, but in fact is proceeding completely to displace him. This giantess is *Electricity*. After remaining behind for decades in the process of growth, she suddenly about 25 years ago began to develop; and in this brief time she has made such growth that she begins to revolutionize technology. It is no longer observed that she spent long childhood years in laboratories and that her first steps were guided by quiet scholars. But just this former concealment and her present successes rivet the interest of all educated persons; for her unexampled speedy race to victory causes even the ignorant to expect yet greater things from her in the future.

And thus it has happened that poor *Steam* is already dethroned in the mind of the multitude and is deprived of the value that caused the century to be named after it. No longer 'the century of Steam,' no, the present period will be named 'the age of electricity'. That is ungrateful but it is comprehensible.

We shall see that electricity combines within itself a number of brilliant qualities that far outstrip those of steam which fundamentally is able only to push hard. In the first place electricity's far greater independence of the place where it is generated enables it, however distant from the latter, to carry out its work, while steam can only venture forth a short distance away from its boiler, and even in this restricted case requires the laborious construction of a route. But electricity flies with lightning speed along thin wires over heights, depths and round corners, and is easily distributed, performing its duties in many places simultaneously. But this is the lesser of its advantages. It is above all the universality of its performances which distinguishes electricity. Do we need light? Electricity distributes it to us, and does so better than any other agent. If we require motive power, there is the electric motor, a mere dwarf in comparison with the steam-engine of equal power. We freeze with cold: the same wire which has already afforded us help also yields us heat, and though this is a little too expensive in the circumstances of today, that too will change in course of time. And so we can continue and relate all that electricity may be able to do now or in the future, but we prefer to condense it all and to say: We possess in electricity every required form of energy, motive-power, light, warmth etc., in the most convenient and the concentrated form. We have thus defined its value, which will be instantly recognized on all sides as soon as the present difficulties in cheap generation of the current have been overcome.[210]

FLIGHT AND RADIO

Of immemorial age is the human longing to be able to rise in the air like a bird. In vain Leonardo da Vinci struggled at the transition period between the Middle Ages and the Renaissance by simple experiments and ingenious contrivances to solve the problem of using muscular power for human flight. And in 1673 Christian Huygens was convinced that one day the gunpowder-motor invented by him could, if properly developed further, be used owing to its relatively light weight as motive power for an aeroplane. His good technological sense enabled him to make this correct forecast. But in spite of all longing, effort and research, it was reserved for Otto Lilienthal in the last decade of the nineteenth century to provide the first definite information on flying from his gliders (Fig. 59). In 1895, Lilienthal wrote on the subject of flight in *Prometheus*:

It has certainly not been made easy for human beings such as ourselves to traverse the realm of the air freely like a bird. But the

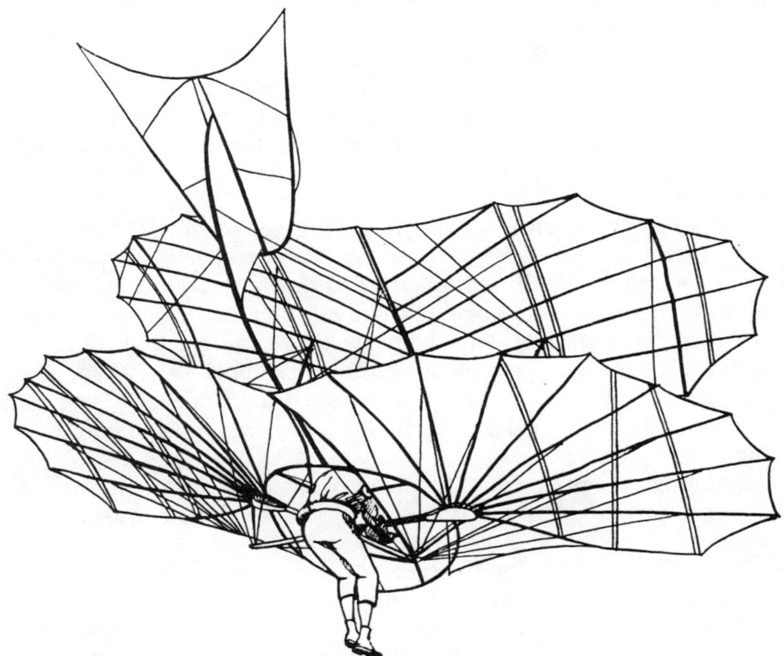

FIG. 59. *Otto Lilienthal in his glider.*
Original glider (Double decker) 1895 in German Museum, Munich

longing to achieve this leaves us no repose; a single great bird, circling above our head arouses in us the desire to soar through the firmament as he is doing.

The physical understanding of every ordinary person is sufficient for him to conjecture that only the correct key to this matter needs to be discovered in order that a completely new world of communication may be opened to us. How restfully and with what complete safety and amazingly simple means do we observe that bird gliding through the air. Can that not also be accomplished by man, with his intelligence and with the mechanical aids to power that have already enabled him to perform veritable miracles? And yet it is hard, extraordinarily hard, even to approach what Nature achieves so brilliantly. What immense efforts have been vainly exerted in the attempt to make the skill of the bird available to mankind. Science too has turned seriously to the question of flight. The phenomena of natural flight have been analysed both anatomically and mechanically, optically by means of instantaneous photography and graphically by electrically recorded graphs. Now at last we have the bird in such a position that theoretically he cannot cheat us any more but in practice he still gives us the wrong result. As soon as we wish to apply our knowledge to actual flight, our clumsiness is deplorably obvious, and the swallows fly over our head laughing at us. Perhaps there is no department of technology in which the pathway from theory to practice is so hard to find. . . .

Whether direct imitation of natural flight is one of many ways or is the only way to our goal is today still a matter of dispute. For example, to many technologists the flight of birds by means of wings appears too difficult to be practicable by machinery, and they would not dispense with the use of the propeller for air traction which has won such approval for work in water. But almost all of them agree that flight, if it is to be achieved, must be at high speed; and that brings us to a main difficulty in the discovery of (a method) for flight. . . .

Although the principle of the flight of birds has now been taken as a basis for most of these projects, in which forward moving wing surfaces provide the lift, nevertheless the methods employed to imitate this natural flight mechanically are as numerous both as regards the production of the apparatus and the nature of the experiments as are the aeronautical engineers who are working on the problem, for each goes his own way in the matter. But all these individual ways lead as a rule to one and the same rock on which usually the idea, if not the ingenious flying machine itself is shattered before it has had time to serve its purpose. Unfortunately progress is hardly ever made beyond the first attempt which usually ends either by failure to rise in the air or, if this is achieved, by inability to land with undamaged apparatus.

Everyone can easily picture to himself what it would mean, to fly through the air with the speed of an express train and then, to land safely without damaging the machine. But if such a feat be demanded from a large heavy and complicated machine, the prospect of a safe landing is all the less. It seems absolutely presumptuous to reckon on a happy ending to such a first attempt at flight, especially with complicated equipment.

If we could not convince ourselves every day of the ease and safety with which birds rush through the air and know how to move with the wind, we should positively despair of discovering the art of flight. But is there really a prospect that we can attain this skill? What then are actually the aims of the technology of flight? To what degree of perfection will it be possible to develop free human flight? Yes, *develop*! that is the right expression: *development* the right idea which we must take to heart in order to make progress in the technology of flight.

None can forsee today to what a degree of perfection man will succeed in developing free flight, because up to now far too little work has been done on this special development. If now and again any idea for a flying machine is carried out and it is shattered on the above mentioned rock, it will be of little significance for the development of dynamic flight. Moreover, work on flight is mostly pure theory, and that is of little help at the present stage of flight technology.

The theory of flight is today really no longer in too bad a state. Since we have elucidated the aerial resistance encountered by the bird's wing, and the economy of power effected by the curve of its profile, we can understand very well all the phenomena of natural flight. But what we must now develop is actual flight. We now have to eliminate purely practical difficulties, but these are greater than is apparent at the first glance. We must therefore devote specialist study to these practical difficulties. We must consider methods of gaining fundamental understanding of them, in order to be able to overcome them successfully. Only thus shall we be able to plant the seed of useful effort in this hitherto so unprofitable field.

For, however primitive they may be at first, the methods which are to lead us to free flight must be capable of development. And for that purpose the experiments which are made must provide opportunity to collect information which will lead to real even if at first only to limited flight through the air; whereby we can accumulate experience on stability in flight, on the influence of the wind and on safe landing, so that by continual improvement we may gradually approach sustained free flight.

Perfection cannot be achieved by violence. Just because the inventors of flying machines have usually demanded far too much at once from them, their positive success has been so small. To remain in the air without a balloon, and to achieve free flight through the atmosphere are tasks which present such a new field of work that it will only be possible to become orientated in it gradually. He who discounts sound development through ever increasing experience of free, stable and safe movement through the air will get nowhere in this matter. . . .

After I had established that glider flights with quite simple apparatus can be carried out from elevated points with stability and safety over long distances in moderate winds, the next need was on the one hand to extend this gliding practice in ever stronger winds in order as far as possible to achieve the long flights that we so much admire in birds; and on the other hand to attempt to assist simple glider-flight by dynamic means so as to lead gradually to continuous flight through stiller air. . . .

Also an apparatus for steering flight was soon attempted in practice. To produce the beating of the wings a machine driven by compressed CO^2 was used. . . . For the rest, the handling of the apparatus is similar to that for simple glider flight, though the first timid attempts convinced me that if I had merely hurled myself straightway into the air with beating wings, the apparatus would probably not have landed without damage. New and unaccustomed phenomena are constantly appearing, and a single unlucky landing is enough to ruin the whole appliance. Here again it is a question of *not demanding too much at once*. I had therefore to resign myself at first to making only the usual glider flights with this larger and heavier apparatus which weighs 88¼ lb. or twice as much as an ordinary glider, whereby I could practise certain and safe landing. Only now, when that stage has been successfully accomplished, can I cautiously attempt free flight with beating wings.

There may of course be other methods for the proper development of free human flight. But if so they will involve similar problems which must be dealt with in order to reach the solution of this difficult undertaking.

But whichever way you take, there is no hope of progress unless the experiments which are made permit instructive observation of a body moving in really free flight, for we are dealing here with quite new phenomena which we never encounter in other fields of technology. Stable, free flight in opposition to changes of wind, and safe landing from dynamic flight are factors for which at present very little practical experience is at our disposal, though they constitute the very essence of flight technology.

Nevertheless this circumstance only makes the solution of the problem of flight more difficult but by no means impossible. When it has been generally recognized in which direction flight research is needed, the forces now so scattered will be concentrated on the right objects for successful work on the steady development of free flight.[211]

Stimulated by Lilienthal, the brothers Orville and Wilbur Wright turned, in 1900, to the problem of flight. In those days, Count Zeppelin's first dirigible airship, equipped with two Daimler motors was already cruising through the air. This vessel, lighter than air, continued until 1936, to be built in ever larger dimensions, though its further evolution was arrested by the successful development of the aeroplane. An ancestor of the Zeppelin was the air balloon invented by the brothers Mongolfier in 1783; its first ascent into the air took place over France and greatly stirred the world. In 1903 the brothers Wright succeeded in making the first guided flight in a bi-plane weighing over 6 cwt driven by an air-screw powered by a light benzine-motor; 284 yards against the wind were traversed in 59 seconds. The spell was now broken. The motor aeroplane with air-screw pursued its course of conquest of the air. But only after the world war of 1914–18, began an incredibly vigorous development which was pursued with feverish enthusiasm during the Second World War. The aeroplane became a terrible weapon of destruction. Mankind suddenly realized the connection between technological progress and catastrophic destruction. Technology, the hope of the future, was recognized as also a danger. But technological achievement pressed forward. The period following the Second World War saw the real development of the jet planes of which the first, a machine built by Heinkel, had already in 1939 traversed the atmosphere in an experimental flight. In 1947 the speed of sound (764 m.p.h.) was for the first time exceeded by a rocket aeroplane. These high speeds introduced fundamental new problems in design and material. In our own days a jet plane has at length attained a height of nearly 66,000 feet. The earth has become small; its peoples pressed close together. Nevertheless, or perhaps for that very reason, they often confront one another in greater hostility than ever before.

Of immense importance for the transmission of news and for trade and the communication of ideas, though sometimes also for misleading the masses by one-sided propagandist views, was radio-technology which after the discovery by H. Hertz in 1888 of electric waves, began with the work of G. Marconi. In May, 1897, he succeeded for the first time in transmitting messages over a distance of over three miles by wireless-telegraph. In the same year he went further, over nine and thirteen miles. Marconi's original set up was greatly improved in 1898 by Ferdinand Braun who introduced the induction transmitter. The prospects of development of the new wireless telegraphy were at first approached cautiously by Braun, the careful man of science. He spoke as follows in a lecture in the Winter of 1900:

If one were asked for one's views on the practical value and probable development of wireless telegraphy, as far as one can see today the reply would probably be as follows:

Its value is already recognized as an improved service of signals, independent of weather, time of day, of fog, rain or snow. This will hold good in many cases, even if the hope to make the stations independent of one another is not fulfilled to the desired extent.

Stronger transmission, and synchronized reception must be the first aims. Even if they are only attained within certain limits, there is yet ample opportunity for practical application. There are enough coasts, islands adjacent to one another, sparsely populated regions where a cable connection is not worth while or when telegraphic communication by wires is threatened by storms, by wild animals or ignorant human beings. In all countries great interest is shown in the utilization of wireless for military purposes.

But it must probably be described as for ever an illusion to cherish the hope of superseding cable telegraphy by it. Just as a sealed letter is the safest form of written communication, so the wire, when available, is the most discreet means of communication between two points.

The problem of making it impossible for an unauthorized person to pick up a message, which, as is known, can also happen in cable telegraphy, is not hopeless although it has not yet been achieved.

The best wishes are laid in a child's cradle, and there is joy if he develops accordingly. But who can say with certainty even five years hence what sort of man he will become? The child will grow and will achieve something, even if he proves to be no Hercules![212]

In his Nobel Address at Stockholm in 1909, Braun reported on the difficulties of the preliminary period.

So far as I am aware, Marconi began his experiments in 1895 on his father's estates, and continued them in 1896 in England. In the year 1897 his experiments reached among other places the harbour of Spezia where he reached a distance of over 9 miles. In the autumn of the same year Slaby covered over 13 miles overland with substantially the same installations, but only by the use of air balloons carrying wires 328 yards long. Why, it must have been asked, were there so many difficulties in increasing the range? If, we say, the whole apparatus functions over 9 miles, why could not double and indeed many times the distance be covered by increasing the initial voltage—for which means were available? It appeared, however, that this would require ever larger aerials. I was under this impression when, in the autumn of 1898, I turned to the subject. I undertook the task of attaining more powerful transmission.[213]

I concluded; if it is possible to stimulate a wireless aerial from a closed electric circuit of great capacity to potential oscillations whose average value is equal to that of the initial load on the Marconi transmitter, then one would possess a serviceable transmitter. The only question was whether this would be possible. Moreover the experiment must be decided by long distance transmission to discover whether any disturbing element had been overlooked. By the use of excitation circuits of suitable size it was possible to fulfil the first requirement; and comparative experiments over long distance decided in favour of the new method. . . . With my arrangement the so-called induction system was everywhere introduced to wireless telegraphy.[214]

George, Count von Arco who had since 1897 developed, together with his teacher, A. Slaby, his own system of wireless telegraphy, wrote in 1904 as follows concerning the stage of development reached at that time:

Thanks to the enormous mental effort devoted by distinguished scholars and engineers during recent years to wireless telegraphy, it is now beyond the stage of preliminary development in which assertiveness and secretiveness were able to conceal imperfect knowledge. Today we can compare the processes and separate the possible from the impossible by calculation.

Hostile interruptions of wireless reception and conversely, tapping of our messages by the enemy can in certain circumstances be excluded, though only under certain conditions. . . .

There are . . . methods of connection between transmitter and receiver by means of which the area free from interruption can be increased. But the possibility of interruption is not excluded. The only certain means of diminishing the danger of interruption lies in a form of construction both of transmitting and receiving apparatus to allow a rapid alteration both in the transmitted and received electric oscillations.[215]

LARGE-SCALE SYNTHETIC CHEMISTRY, ATOMIC ENERGY, AUTOMATION

If we pass over the establishments for the production of chemical pharmaceutical preparations in the twenties of the nineteenth century which had descended from the apothecaries, there arose in Germany from about the middle of the century a chemical industry, at first on an inorganic basis. The work of J. von Liebig on the application of chemistry to agriculture had promoted the founding of factories, between 1840 and 1860, for soda, sulphuric acid, potassium salts, as well as for nitrogenous and phosphate fertilizers. Synthetic manures first made possible the provision of sufficient bread for rapidly increasing populations. England and France had, however, pioneered the development of an inorganic chemical industry, especially the manufacture of sulphuric acid and soda which were very necessary for the cotton industry in England. These countries had a start of nearly half a century. After 1860 a German industry based on organic chemistry was developed for the production of artificial dyes, which was also greatly stimulated by the work of Liebig but especially by A. W. von Hofman and later by A. von Baeyer. The production of coal-tar dyes in Germany was at first small, since the raw material was lacking as the manufacture of gas was but little developed. But by 1878 German factories produced 40 million marks' worth of coal-tar dyes, whereas in England but 9 million marks' worth was produced, and in France and Switzerland about 7 million marks' worth each. Fundamental scientific chemistry and the technical application of the results of research were from the outset closely linked in large-scale chemical industry. The development of large-scale pharmaceutical industry was directly linked with the establishment of the coal-tar dye industry.

The chemical industry of this century is characterized by large-scale chemical synthesis, so that very valuable substances are produced by catalysis from simple raw materials. We mention only such synthetics as ammonia, artificial mineral oils produced under high or low pressure, Buna rubber, and methyl alcohol, from highly polymerized artificial materials and chemical fibres. Thus the modern synthetic chemist is able today, on the basis of scientific knowledge of the relationship between the constitution and properties of macro-molecules, to achieve planned production of new organic materials, on which within certain limits he can bestow the desired technical properties.

At the beginning of large-scale synthesis stands C. Bosch's process of 1913 for the synthetic production of ammonia from the nitrogen of the atmosphere and hydrogen. The scientific foundation of the synthesis was the work of F. Haber in1908–9. His colleague R. Le Rossignol reported as follows on the first demonstration of the high-pressure synthesis on a small scale in Haber's laboratory at Karlsruhe.

It was an exciting day when in July 1909 two representatives of the aniline and soda factory at Baden, Dr Bosch and Dr Mittasch, came to Karlsruhe, invited by Haber, to see the first experiments with the small test apparatus. As so often in important demonstrations something went wrong! A bolt broke on the high pressure apparatus whilst being tightened, and the demonstration had to be postponed for several hours. Dr Bosch did not return, so only Dr Mittasch was present when synthetic ammonia produced for the first time in the small model apparatus, rose in the beaker. He enthusiastically shook Haber's hand, and from that moment he was entirely won over in favour of the new process.

The Baden Company under the excellent leadership of Dr Bosch and Dr Mittasch now embarked on the gigantic task of developing the process for large-scale industry. Their work resulted in the magnificent plants at Oppau and Merseburg.[216]

After prolonged preparatory work Bosch and Mittasch succeeded in 1913, as has already been related, in the aniline and soda factory at Baden, in the large-scale industrial production of synthetic ammonia, which was then developed further, chiefly for nitrogenous fertilizer, so that the import of saltpetre from Chile soon became unnecessary. In his Nobel address in 1932 Bosch gave a vivid description of the course of his process and of the difficulties that had to be overcome.

When I was entrusted by the directors of the business with the task of developing this high pressure synthesis from technical to economic application, it was clear that three problems were in the immediate foreground, and that their solution was absolutely necessary before we could embark on the construction of a factory installation. These problems were: to procure the raw materials, that is to say the gases hydrogen and nitrogen, at a cheaper rate than had been possible up to that time; then the manufacture of effective and stable catalysts and finally, the construction of the apparatus. The solution of these problems was undertaken simultaneously.

First sufficient pure hydrogen was available for the tests from chloride electrolysis from which, by combustion with air, could be obtained the required mixture of nitrogen and hydrogen. For large-scale manufacture, none of the processes then known for manufacture of hydrogen was suitable. They were all examined by us, but without exception they worked out as too expensive, or they provided gases that were not sufficiently pure. It must not be forgotten that the production of hydrogen is the biggest item in the cost of production at least today; for now, once the mixture is available, its conversion into ammonia has become, thanks to the efficiency of high-pressure synthesis, but an inconsiderable item in the costs. As we were directed to use carbon as a basis, we turned our attention immediately to the sole source which in our circumstances entered into the question, namely water gas; after a short transition period of the separation of hydrogen by low temperature liquefaction by Linde's method, hydrogen is now produced in great quantity by a catalytic process evolved by ourselves.

The solution of the second task was equally important, Osmium, a very good catalyst, was difficult to handle; for if in its active, that is to say in its finely divided form, it comes into contact with air (which can in practice never be completely avoided) it very easily volatilizes as osmiumtetroxide; an even greater difficulty is that the osmium supply of the whole world amounts to but a few kilograms; at best, therefore, we should have been able to build up by its means only a very modest output. Uranium, indeed was expensive but obtainable in considerable quantity; but it proved extraordinarily sensitive to oxygen or water and even today it has not been possible to convert it to any form serviceable for large-scale industry. But by dint of our widely ranging series of experiments based on our rapidly developed experimental technique, we did succeed fairly quickly in manufacturing catalysts technically faultless, easy to handle, stable and cheap; chiefly those derived from iron which today are used exclusively throughout the world for synthesis of ammonia. These catalysts are of a quite new type which may be recognized by the fact that they no longer consist of pure elements in more or less refined form, but represent mixtures. Only after many years of work and at first only on the basis of the theoretical discoveries of recent years have we been able to explain the special effect of these mixed catalysers: although at the very outset we made for ourselves a generally correct empirical picture which has been of no little help in our work.

The solution of the third problem of which alone I will treat in greater detail today, concerned the apparatus. There were no models for us in existing technology. The only process—and indeed it was of a practical physical nature—that was worked by high pressure, was that of Linde for liquefaction of air. The apparatus used for this was made of copper soldered with soft solder, and was therefore out of the question for temperatures up to 1,112 degrees Fahrenheit and occasionally even higher. . . .

We began by building an apparatus of which the essential and to us the most important and the most interesting part beside a circulating pump and the ammonia separator was a contact-tube, wall thickness 1 inch and length 8·2 feet. This tube was heated externally.

As careful people, we had installed it in a strong, armoured, concrete casemate, far from all the operations, for we had meanwhile become acquainted also with the dangers of fire and of tongues of flame which might occur, often by self-ignition on the release of hydrogen under high pressure.

The two pipes obtained from Mannesmann could each run for 80 hours, after which they burst. Had we filled them with osmium instead of with our new catalyst, the world's whole supply of this rare metal which we had already bought up, would have disappeared. . . .

The long-sought solution lay in the provision of a steel cover that took the pressure and was furnished with a thinner lining of soft iron, in such a manner that the hydrogen entering this thinner lining would, since it diffuses spontaneously, be able to escape without pressure before it could attack the outer steel cover at high temperature. This is easily arranged, since the lining tube on being turned outside acquires grooves, and the steel cover is furnished with many little drilled holes, through which the hydrogen can escape freely. Immediately from the beginning the thinner lining tube attaches itself, owing to the high pressure within, firmly against the cover; and later, after it has hardened, it cannot possibly yield, so that no cracks can occur. The loss by diffusion is negligible. . . .

The contact tube was heated externally by gas. With this apparatus, work was for the first time performed for a longer period. But the hope that we were now at the end of all our difficulties was not fulfilled. The steel covers of the heaters, though not destroyed by hydrogen, were not capable of withstanding for any length of time the high pressure at the high temperatures produced by the external heating. They bulged, cracked and the result was a series of violent explosions which indeed we rendered harmless by installing the plant in explosion-proof chambers; but the constant heavy expenses for repair threatened the economy of the process. Another solution had therefore to be found, and this consisted in the introduction of internal heating. Even then we had not faith in electric internal heating, as we calculated that we should encounter great unavoidable heat loss to prevent overheating of the cover. So we tried gas heating; in which hydrogen (synthetic gas) with additional air supply, was burnt inside the contact heater. Some water indeed was thereby produced which, however, did not appear to interfere with the contact. At any rate, this was the lesser evil. . . .

It was obvious, as we installed ever larger plants, that above a certain size it was possible with good heat exchange conditions, to replace heat losses by reaction heat. That was a great step forward; for we could now completely discard continuous heating throughout the process. It was merely necessary to bring the heater at the beginning to the correct reaction temperature. Gas heating had in practice the drawback that during the heating-up period the gas circuit had to be reversed from the normal direction, in order to keep the products of combustion at a distance from the contact. This often caused trouble from cracks in the lining of the heater. Moreover the heat had to be transferred to the gas in a comparatively small space with high temperature drop.

The high-pressure compressors gave us at first great anxiety. Large compressors had hitherto been known only for the purpose of compressing air. They had mostly been used for the production of compressed air for driving compressed air locomotives in mines and for liquid air installations on the Linde system. These high-pressure compressors were only made in moderately small sizes. And above all, little attention had been given to the question of packing glands. In the case of air, leaks did not matter and short stoppages were expected. It was a different matter with hydrogen and the sensitive contact process. Losses of hydrogen meant danger both as regards money and explosion, and the contact apparatus could not tolerate stoppages.

We tested all the types which had been made at that time, in our works and from the experience gained from those small compressors which never ran more than half a day without breaking down, after years of work we succeeded in building the powerful 3,000 horsepower plants which would run safely for 6 months without interruption, and would then be cleaned in regular rotation. I may mention here how important it is, especially for the production of synthetic ammonia, to have stable operating conditions. Every interruption in a single part affects the whole process, and it takes hours before all is again in order. I may say, without exaggeration that without completely uniform and uninterrupted work the plant cannot be operated profitably. It took us years to achieve this. We were specially helped in this by the *Control-instruments* to which, from the beginning, we gave special attention, as it soon became clear that only by continuous recording was it possible to follow the processes taking place in the heaters. . . .

Special study is of course required for the erection of an apparatus, so large and interdependent in its parts, such as stands today in our works at Oppau and at Leuna, which includes many thousands of yards of pipes and thousands of flanges and valves. Above all, everything must be gas-tight. In this we were so far successful that the losses of gas were exactly sufficient to prevent too great an increase in the content of argon absorbed from the air and in the methane collected in the circulation.

The apparatus must be absolutely reliable in operation, and it must be possible, in the event of interruptions, to disconnect and empty any part as quickly as possible. In the course of years we have been able to collect sufficient experience. . . .

I now reach my conclusion: The development of ammonia synthesis, which was accomplished in a relatively short time by harnessing all the available energy and resources, partly under the pressure of Germany's isolation by war, naturally, gave rise to the development of other reactions which work far better under high pressure than merely under atmospheric pressure. And as experience has abundantly shown, it is often far more economical to work with a high-pressure system if the technique of high-pressure synthesis has been mastered; for it must be borne in mind that the subsidiary pressure, whether of the reaction gases or of impurities is also correspondingly high.

Very soon after we began work we examined the reduction of carbon monoxide by hydrogen, and thereby we were able to establish that liquid reaction products could be obtained which appear as a combination of aliphatic alchohols, aldehydes and ketones and the acids derived from them. Only later, in the beginning of 1923, was it recognized that the first reaction-product, *Methylalcohol* could be generated in quantity only if gas is used which is free from iron carbonyl, and the contact masses must similarly be free from iron, (especially $ZnO—Cr_2O_3$). And therefore the effect of iron walls to the containers must be avoided since they would produce condensation of methylalcohol. The best conditions of work were soon found, which afforded high outputs of pure methylalcohol. Thus the synthesis could be developed very quickly, so that today methyl alcohol or *methanol* has become one of the cheapest products of the chemical industry. At least 40,000 tons per year are probably produced by high-pressure synthesis.

Another reaction which was very soon tackled was the preparation of carbamide (urea) from carbonic acid and ammonia, by extraction of the water.

Besides other pressure processes of less importance, carbo-hydration aroused great interest in recent years. Bergius had, as is known, and as he will report . . . discovered that by heating charcoal with hydrogen under pressure, great quantities of fluid reaction products are generated. It was then established in our laboratories that by the use of catalysts, these reactions can be better regulated and above all that they can lead to any desired end-product, from light hydrocarbons to lubricating oil. After acquiring the master patents from Bergius this process was widely used in Oppau and Leuna, being developed from our experience of high-pressure technique and catalyser construction. It now produces annually 120,000 tons of benzine.[217]

With the first use of atomic nuclear energy in the Second World War, a new technical epoch began. Until the late Middle Ages, almost all engineering work was performed by man-power: then animal-power, water and wind-power all came more to the fore. Since the eighteenth century, the chemical forces slumbering in coal and later in oil have been utilized in heat-engines of most varied nature. And in our own day attention has been turned both to chemical energy and to atomic nuclear energy. The energy to be obtained from water, wind, coal and oil, is derived ultimately from the sun, whose abounding wealth of energy is based on nuclear processes. But now man has been able to release nuclear energy from the elements constituting our earth, and thus to gain access to an unlimited treasury of energy. In 1938 O. Hahn and F. Strassmann achieved fission of the element uranium into elements in the middle range of the periodic system by bombarding it with neutrons. In 1939 Lise Meitner and O. R. Frisch recognized that an immense development of energy takes place with these fissions. A scientific basis was thus obtained for the production of nuclear energy. It was soon shown that fission takes place as a chain reaction which can either be directed to the uniform provision of heat in the sense of a static discharge, or can be applied unregulated to induce violent explosions. Inevitably the war weighted the scales in favour of explosive chain reaction. In America, as a result of intensive work and vast expenditure of material it led to the development of the atom-bomb, which was first experimentally tested in its vast power on July 16, 1945. The official American report dramatically portrays the experiment:

Mankind's successful transition to a new age, the Atomic Age, was ushered in, July 16, 1945, before the eyes of a tense group of renowned scientists and military men gathered in the desertlands of New Mexico to witness the first end results of their $2,000,000,000 effort. Here in a remote section of the Alamogordo Air Base 120 miles southeast of Albuquerque the first man-made atomic explosion, the outstanding achievement of nuclear science, took place at 5.30 a.m. Darkening heavens, pouring forth rain and lightning immediately up to the zero hour, heightened the drama.

Mounted on a steel tower, a revolutionary weapon destined to change war as we know it, or cause the end of all wars, was set off with an impact which signalized man's entrance into a new physical world. Success was greater than the most ambitious estimates. A small amount of matter, the product of a chain of huge, specially constructed industrial plants, was made to release the energy of the universe locked up within the atom from the beginning of time. A fabulous achievement was completed. Speculative theory, barely established in pre-war laboratories, had been projected into reality.

This phase of the Atomic Bomb Project, which is headed by Major General Leslie R. Groves, was under the direction of Dr J. R. Oppenheimer, theoretical physicist of the University of California. It is due to him that the apparatus for applying atomic energy to military purposes was completed.

Tension before the actual detonation was at a tremendous pitch. Failure was an ever-present possibility. Too great a success, envisaged by some of those present, might have meant an uncontrollable, unusable weapon.

Final assembly of the atomic bomb began on the night of July 12 in an old ranch house. As various component assemblies arrived from distant points, tension among the scientists rose to an increasing pitch. Coolest of all was the man charged with the actual assembly of the vital core, Dr R. F. Bacher, in normal times a professor at Cornell University.

The entire cost of the project, representing the erection of whole cities and radically new plants spread over many miles of countryside, plus unprecedented experimentation, was represented in the pilot bomb and its parts. Here was the focal point of the venture. No other country in the world was capable of such an outlay in mental and technical effort.

The full significance of these closing moments before the final test was *not* lost on these men of science. They fully knew their position as pioneers into another age. They also knew that one false move would blast them and their entire effort into eternity. Before the assembly started a receipt for the vital parts was signed by Brigadier General Thomas F. Farrell, General Groves' deputy. This signalized the formal transfer of the irreplaceable material from the scientists to the Army.

During the preparations for the final assembly, a bad few minutes developed when the assembly of an important section of the bomb jammed. The entire unit was machined to the finest limits. The insertion of the part was partially completed when it apparently wedged tightly and would go no farther. Dr Bacher, however, was undismayed and reassured the group that time would solve the problem. In three minutes' time, Dr Bacher's statement was verified and assembly was completed without further incident.

Speciality teams, comprising the top men of the appropriate branches of science, all of whom were bound up in the whole, took over their specialized parts of the assembly. In each group was centralized months and even years of concentrated endeavour.

On Saturday, July 14, the unit which was to determine the success or failure of the entire project was elevated to the top of the steel tower. All that day and the next, the job of preparation went on. In addition to the apparatus necessary to cause the detonation, complete instrumentation to determine the pulse beat and all the reactions of the bomb was rigged on the tower.

The ominous weather which had dogged the assembly of the bomb had a very sobering affect on the assembled experts whose work was accomplished amid lightning flashes and peals of thunder. The weather, unusual and upsetting, prevented aerial observation of the test. It even held up the actual explosion scheduled for 4.00 a.m., for an hour and a half. For many months the approximate date and time had been set and had been one of the most important of the best kept secrets of the entire war.

Nearest observation point was set up 10,000 yards south of the tower where in a timber and earth shelter the controls for the test were located. At a point 17,000 yards from the tower, at a point which would give the best observation, the chief figures in the atomic bomb project took their posts. These included General Groves, Dr Vannevar Bush, head of the Department of Scientific Research and Development and Dr James R. Conant, President of Harvard University.

Actual detonation was in charge of Dr K. T. Bainbridge of Massachusetts Institute of Technology. He and Lieutenant Bush representing the Military Police Detachment, were the last men to inspect the tower with its cosmic bomb.

At three o'clock in the morning the party moved to the control station. General Groves and Dr Oppenheimer consulted with the weathermen. The decision was made to go ahead with the test despite the changeable weather. The time was set for 5.30 a.m.

General Groves joined Dr Conant and Dr Bush, and just before the test time they joined the scientists gathered at the Base Camp. Here all present were ordered to lie on the ground, face downward, heads away from the blast direction.

Tension reached a tremendous pitch in the control room as the deadline approached. The several observation points in the area were connected to the control room by radio and with twenty minutes to go, Dr S. K. Allison of Chicago University, took over the radio net and made periodic time announcements.

The time signals, minus 20 minutes, minus 15 minutes and so on increased the tension to breaking point as the group in the control room which included Dr Oppenheimer and General Farrell held their breath, all praying with the intensity of the moment which would live forever in each man who was there. At 'minus 45 seconds', the robot mechanism started and from that point on the whole great complicated mass of intricate mechanism was in operation without human control. Stationed at a reserve switch, however, was a soldier scientist ready to attempt to stop the explosion should the order be issued. The order never came.

At the appointed time there was a blinding flash lighting up the whole area brighter than the brightest daylight. A mountain range three miles from the observation point stood out in bold relief. Then came a tremendous sustained roar and a heavy pressure wave which knocked down two men outside the control centre. Immediately thereafter, a huge multi-coloured surging cloud boiled up to an altitude of over 40,000 feet. Clouds in its path disappeared. Soon the shifting substratosphere winds dispersed the now grey mass.

The test was over, the project a success.

The steel tower had been entirely vaporized. Where it had stood, there was a huge precipitous crater. Dazed but relieved at the success of their tests, the scientists promptly marshalled their forces to estimate the force of America's new weapon. To examine the nature of the crater, specially equipped tanks trundled into the area, one of which carried Dr Enrico Fermi, the noted nuclear scientist. The answer to their findings is found in the destruction effected in Japan today in the first military use of the atomic bomb.[218]

As described above, the first application of the new weapon was the American attack on Hiroshima and on Nagasaki in Japan in August, 1945. The destructive effect was terrible. After the war, a beginning was made with the development of atomic energy for peaceful purposes. Great installations are being built. A small experimental atom-powered motor started work in the United States in 1953; it drives a turbo-generator with an output of 250 kilowatts. In 1956 the first atomic power station operating on a commercial basis was opened at Calder Hall in England. But strenuous as are the efforts to achieve an atomic engine, no less enthusiasm is applied to the further development of the atom bomb. The hydrogen bomb represents a further stage in the 'perfecting' of that most fearful weapon ever created by human technology. May humanity at last attain mutual trust in one another so that the terrible power of atomic energy that has been unleashed by science and technology may in future serve only for co-operation in peaceful tasks.

With the mighty technological achievement of the utilization of atomic energy, which opened a new epoch in technology, there must be ranged another technological development, no less profound in influence, namely, the construction of electronic installations for calculation and control —a development which will, among other results, lead to the automisation of factory organization. This process, American writers especially are convinced, will involve a second industrial revolution which will produce results, psychological, sociological and economic, no less important than those produced in the first industrial revolution by the introduction of steam power and new manufacturing machines.

Electronic calculating apparatus rendered important service to America during the war, especially for the solution of ballistic problems. J. W. Mauchly and J. P. Eckert built, in 1946, the large electronic calculating machine known as ENIAC (Electronic Numerical Integrator and Computer) with its 18,000 electronic valves. Complicated mathematical calculations are accomplished with extraordinary speed by such apparatus. The electronic calculating machine 'records' also intermediate results, and works further with them for a given period of time. It is even able to reach 'decisions' as to the desirable further course of the calculation, choosing by means of the intermediate results one out of several methods which are provided in its programme.

Moreover, by use of an apparatus such as the electronic calculating machine, the mechanized output of a factory can be regulated. The manufacturing machine relieves men of mechanical work, but the electronic apparatus replaces also routine human mental work. In the factory equipped with automization, the machine corrects its own mistakes. Once the automization apparatus has been installed in the factory, man need provide only for the maintenance and any necessary repairs to the many diverse machines; he needs only to be hygienist and physician to the machines. It will be readily understood that such automization of the factory organization must lead to changes both economic and sociological. The office work will also undergo automization as a result of electronic installations. The distinction between office and factory will thereby be removed.

People like nowadays to call the great electronic calculating installations 'thinking machines' or 'electronic brains'. But these machines are not characterized by the power of thought, nor are they human in other respects.[49] If indeed they reach decisions, this is not done by the exercise of free will: they merely complete a logical process for which they have previously been designed.

Excellent insight into the problems of the automized factory is furnished by the American mathematician Norbert Wiener who has himself taken an important part in the development of machines to effect communication and control. He writes as follows:

The computing machine represents the center of the automatic factory, but it will never be the whole factory. On the one hand, it receives its detailed instructions from elements of the nature of sense organs. I am thinking of sense organs such as photo-electric cells, condensers for the reading of the thickness of a web of paper, thermometers, hydrogen-ion-concentration meters, and the general run of apparatus now built by instrument companies for the manual control of industrial processes. These instruments are already built to report electrically between remote stations. All they need to enable them to introduce their information into an automatic high-speed computer is a reading apparatus which will translate position or scale into a pattern of consecutive digits. Such apparatus already exists, and offers no great difficulty, either of principle or of constructional detail. The sense-organ problem is not new, and it is already effectively solved.

Besides these sense organs, the control system must include operating, or adjusting members which act on the outer world. Some of these are of a type already familiar, such as valve operating motors, electric clutches, and the like. Some of them will have to be invented, to duplicate as nearly as possible the functions of the human hand as supplemented by the human eye. It is possible in the machining of automobile frames to leave on certain metal lugs, machined into smooth surfaces as points of reference. The tool, whether it be drill or rivetter or whatever else we want, may be fed to the approximate neighbourhood of these surfaces by a photo-electric mechanism, actuated for example by spots of colour. The final adjustment brings the tool on to the reference points, so as to establish a firm, but not too close contact. This is one way of doing the job. Any competent engineer can think of a dozen more.

Of course, we assume that the instruments, which act as sense organs to the system, record not only the original state of the work, but also the results of the work in all previous processes. Thus the machine can exercise control of the simple types, already frequently described, or those involving more complicated processes of discrimination, regulated by the central control as a logical or mathematical system. In other words, the whole system will correspond to a complete living being with sensory organs, operative powers and discrimination, and not, as in the ultrarapid computing machine, to an isolated brain, dependent for its experience and for its effectiveness on our intervention.

The speed with which these new devices are likely to come into industrial use will vary greatly with the different industries. Automatic devices, which may not be precisely like those described here, but which perform roughly the same functions, have already come into extensive use in flow production industries like tin container factories, steel-rolling mills, and especially wire and tin-plate factories. They are also operating in paper factories, which likewise produce a continuous output. Another place where they are indispensable is in those production processes which are too dangerous for any considerable number of workers to risk their lives in their control, and in which breakdown is likely to be so serious and costly that its possibilities should have been considered in advance, rather than left to panic judgment in case of emergency. If a policy can be thought out in advance, it can be transferred to a programme which will control the plant as previously planned in accordance with the readings of the instrument. In other words, such factories should be under a régime rather like that of the railway signal box with its interlocking signals and points. This régime is already followed in oil-refineries, in many other chemical works, and in the handling of the sort of dangerous materials found in the exploitation of atomic energy.

We have already mentioned that these methods are used on the assembly line. In the assembly line, as in the chemical factory or the continuous-process paper mill, it is necessary to exert a certain statistical control on the quality of the product. This control depends on a sampling process. These sampling processes have now been developed by Wald and others into a technique called *sequence analysis*, in which one does not take previously determined samples, but they are extracted continuously hand in hand with the production. As this testing is now so standardized that it can be put into the hands of a statistical computer who does not understand the basic theories behind it, it may also be executed by a computing machine. In others words, excluding special cases, the machine takes care of the routine statistical controls, as well as of the production process.

In general, factories have an accounting procedure which is independent of the production. All the data for cost-accounting could, so far as they come from the machine or assembly line, be fed directly into the computing machine. Other data may be fed in from time to time by human operators, but the bulk of necessary clerical work could be cut to the handling of special cases. For example, employees will be needed to take care of outside correspondence and the like. Even a large part of this may be received on punched cards by the correspondents or transferred to punched cards by extremely low-grade labor. From this point everything can be done by machine. This mechanization also may apply to a not inappreciable part of the book-keeping and filing facilities of an industrial plant.

In other words, it is all one to the machine whether it performs overall work or white collar work. Thus the possible fields into which the new industrial revolution is likely to penetrate are very extensive; where it will take over all the work which needs only simple decisions in its execution, in much the same way as the earlier industrial revolution displaced manpower everywhere. There will, of course, be trades into which the new industrial revolution of judgment will not penetrate: either because the new control machines are not economical in industries which cannot afford the considerable capital costs involved, or because their work is so varied that a new programme would be necessary for almost every job. I cannot see automatic machinery of the type described coming into use in the corner grocery, or in the backyard filling station, although I can very well see it employed by the wholesaler and the automobile agent. The farm laborer too, although he is beginning to feel its influence, is protected from its full effect because of the ground he has to work, partly because of the variety of the crops he must harvest, and partly because of the special weather conditions and the like that he must meet. Even here, the plantation farmer is becoming increasingly dependent on cotton-picking and weed-burning machinery, as the wheat farmer has long been dependent on the McCormick reaper. Where such machines may be used, some use of selective machinery is not inconceivable.

The introduction of the new devices and when they are to be expected are, of course, largely economic matters in which I am not an expert. Short of any violent political changes or another great war, I should give a rough estimate that it will take the new apparatus ten to twenty years to come into its own. A war would change all this overnight. If we should engage in a war with a major power like Russia, which would make serious demands on the infantry, and consequently on our manpower, we may be hard put to it to keep up our industrial production. Under these circumstances, the matter of replacing human productivity by other modes may well be a life-or-death matter to us. We have already reached the stage in the development of a unified system of automatic control machines as we were in the development of radar in 1939. Just as the emergency of the Battle of Britain made it necessary to attack the radar problem vigorously and thereby accelerate the natural development of a field of evolution by what may have been decades, so too, the need to replace labor is likely to act on us in a similar way in the case of another war. The circle of skilled radio amateurs, mathematicians, and physicists, who were so rapidly turned into competent electrical engineers for the purposes of radar design, is still available for the very similar task of automatic-machine design. There is a new and skilled generation coming up, which they have trained.

Under these circumstances, the period of about two years which it took for radar to get on to the battlefield with a high degree of effectiveness is scarely likely to be exceeded by the period of evolution of the automatic factory. At the end of such a war, the 'know-how' needed to construct such factories will be general. There will even be a considerable surplus of equipment manufactured for the government, which is likely to be on sale or available to the industrialists. Thus a new war will almost inevitably see the automatic age in full swing within less than five years.

I have spoken of the actuality and the imminence of this new possibilty. What can we expect of its economic and social consequences? In the first place, we can expect an abrupt and final cessation of the demand for the type of factory labor performing purely repetitive tasks. In the long run, the deadly uninteresting nature of the repetitive task may make this a good thing, and the source of the leisure which is necessary for the full cultural development of man on all sides. It may also produce cultural results as trival and wasteful as the greater part of those so far obtained from radio and the movies.

Be that as it may, the intermediate period in the introduction of the new methods, especially if it comes in the incisive form to be expected from a new war, will lead to an immediate period of disastrous confusion. We know well enough how the industrialists regard a new industrial potential. All their energies are turned towards ensuring that it must not be considered as the business of the government but must be left open to whatever producers wish to invest money in it. We also know that they have very few inhibitions when it comes to taking all the profit out of an industry that there is to be taken, and then letting the public pick up the crumbs. This is the history of the lumber and mining industries, and is part of what we have called in another chapter the traditional American philosophy of progress.

Under these circumstances, industry will be flooded with the new apparatus to the extent that they appear to yield immediate profits, irrespective of what long-term damage they can do. We shall see a process parallel to the way in which the use of atomic energy has been exploited. Its use for bombs has made it problematical whether we can use it as a substitute for our oil and coal supplies, which are within centuries, if not decades, of utter exhaustion. Note well that atomic bombs do not compete with power companies.

Let us remember that the automatic machine, whatever we think of any feelings it may have or may not have, is the precise economic equivalent of the slave. Any labor which competes with slave labor must accept the economic conditions of slave labor. It is perfectly clear that this will produce an unemployment situation, in comparison with which the present recession and even the depression of the thirties will seem a harmless joke. This depression will ruin many industries—possibly even the industries which have taken advantage of the new potentialities. However, there is nothing in the industrial tradition which forbids an industrialist to make a sure and quick profit, and to get out before the crash touches him personally.

Thus the new industrial revolution is a two-edged sword. It may be used for the benefit of humanity, assuming that humanity survives long enough to reach a period in which such a benefit is possible. If, however, we proceed along the clear and obvious lines of our traditional behavior, and follow our traditional worship of progress and the fifth freedom—the freedom to exploit—it is practically certain that we shall have to face a decade or more of ruin and despair.[219]

It is just the new technical developments of the last half century, during which technology has become an ever more influential power in our life that make the question of the sense and the limits of technology most pressing for us. Abundant as have been the earnest and apt statements in our day on this problem, let us turn our gaze once more toward Goethe, and close with those ever applicable words from the *Wanderings of Wilhelm Meister*:[50]

On and within the earth we find material to fulfil the highest earthly needs, a world of matter devoted to the cultivation of man's highest faculties. But on that road of the spirit will be found always sympathy, love and well-regulated free activity. To stir these two worlds to confront one another, to manifest their reciprocal qualities in the transient phenomena of life, that is the highest stature to which man must develop.

FOOTNOTES REFERRED TO BY SUPERIOR NUMBERS IN THE TEXT

1. LIPS, J. E.: *Vom Ursprung der Dinge*, Leipzig 1951. p. 92 ff.
2. DINGLER, HUGO: *Über die Geschichte und das Wesen des Experiments*, Munich 1952. p. 15.
3. MENNICKEN, P.: *Die Technik im Werden der Kultur*, Wolfenbüttel, 1947. p. 35 ff.
4. GUIRAUD, P.: *La main-d'œuvre industrielle dans l'ancienne Grèce*, Paris 1900. p. 37 ff.
5. BREUSING, A.: *Die Lösung des Trierenrätsels*, Bremen, 1889, p. 116.
6. BALDI, B.: *In Mechanica Aristotelis Problemata*, Moguntiae, 1612. Extract in: D. Mögling, *Mechanische Kuntstkammer*, Part 1, Frankfurt-am-Main, 1629, p. 127–184.
7. REHM, A.: *Zur Rolle der Technik in der griechisch-römischen Antike*, Arch. f. Kulturgesch, Vol. 38, 1938, p. 158.
8. ROSTOVTZEFF, M.: *Gesellschaft und Wirtschaft im römischen Kaiserreich*, Vol. 1, 2, Leipzig, 1930.
9. SCHMIDT, W.: *Heron von Alexandrien im* 17, *Jahrhundert*, Abh. z. Gesch. d. Mathem, Vol. 8, 1898, p. 195–214.
10. DRACHMANN, A. G.: *Ktesibios, Philon and Heron*, Copenhagen, 1948.
11. REHM, A.: *Zur Rolle der Technik in der griechisch-römischen Antike*, Arch. f. Kulturgesch, Vol. 38, 1938, p. 161.
12. MAIER, ANNELIESE: *Die Vorläufer Galileis im* 14, *Jahrhundert*, Rome, 1949, Ibid., *An der Grenze von Scholastik und Naturwissenschaft*, 2nd Imp. Rome, 1952.
13. W. RYFF, German Translation of Vitruvius, 1548.
14. HUIZINGA, J.: *Das Problem der Renaissance*, in Huizinga, *Wege der Kulturgeschichte*, Munich, 1930.
15. OLSCHKI, L.: *Geschichte der neusprachlichen wissenschaftlichen Literatur*, Vol. 1, Leipzig, 1919.
16. BABINGER, F. and HEYDENREICH, L. H.: *Vier Bauvorschläge Leonardo da Vincis an Sultan Bajezid II*, (1502/03) Nachr. Akad. Wiss. Göttingen. Phil.-hist. Kl., 1952, No. 1.
17. GALILEI, G.: *Le mecaniche*, written 1593; *Galilei*, *Le opere*, Vol. 12, Florence, 1891, p. 147 ff.
18. GOETHE, J. W. V.: *Meteore des literarischen Himmels. Erfinden und Entdecken.* Goethe, *Werke*, Sophien Edn. Pt. 2, Vol. 11, Weimar, 1893, p. 255.
19. GOETHE, J. W. V.: *Über Naturwissenschaft im Allgemeinen*, Goethe, *Werke*, Sophien-Edn. Pt. 2, Vol. 11, Weimar, 1893, p. 134.
20. MERTON, ROBERT K.: *Science, Technology and Society in Seventeenth Century England*, Osiris, Vol. 4, 1938, p. 473.
21. HUYGENS, CHR.: *Letter to ?* 24/5/1686, Huygens, *Œuvres complètes*, Vol. 9, The Hague, 1901, p. 78 f.
22. LEUPOLD, JAKOB: *Theatrum machinarum*, Vols. 1–9, Leipzig, 1724–39 and 1788.
23. REULEAUX, F.: *Lehrbuch der Kinematik*, Vol. 1, Brunswick, 1875, p. 11 ff.
24. CALVÖR, H.: *Beschreibung des Maschinenwesens ... auf dem Oberharze*, Part 1, Brunswick, 1763, p. 112.

25. *Encyclopédie ou Dictionnaire raisonné des sciences, des arts et des metiers*, D. Diderot and J. L. d'Alembert, 35 vols., Paris, 1751–80.
26. TORLAIS, J.: *Réaumur*, Paris, 1936, p. 249.
27. MÜLLER-ARMACK, A.: *Genealogie der Wirtschaftsstile*, Stuttgart, 1941, pp. 107, 173, 191.
28. BECKMANN, J.: *Anleitung zur Technologie*, Göttingen, 1777.
29. JASTROW, J.: *Die Stellung der Technologie an den deutschen Universitäten*, Zs. f. angew. Chemie, 1924, pp. 953 ff.
30. WOLFF, CHRISTIAN: Preface to the 1740 edition of the German translation of B. F. de Bélidors *Architecture hydraulique*.
31. BÉLIDOR, B. F. de: *Architecture Hydraulique*, German edn., Pt. 1, issue 2, p. 83.
32. MÜLLER-ARMACK, A.: *Genealogie der Wirtschaftsstile*, Stuttgart, 1941, p. 259 f.
33. MERTON, ROBERT K.: *Science, Technology and Society in Seventeenth Century England*, Osiris, Vol. 4, 1938,p. 482 ff.
34. SCHNABEL, FRANZ: *Deutsche Geschichte im* 19 *Jahrhundert*, Vol. 3, 2nd Imp. Freiburg, 1950, p. 292 ff.
35. SCHNABEL, FRANZ: as above, p. 297.
36. FREYER, HANS: *Weltgeschichte Europas*, Vol. 2. Wiesbaden, 1948, p. 910.
37. WAGEMANN, ERNST: *Menschenzahl und Volkerschicksal*, Hamburg, 1948, p. 276.
38. BRENTANO, L.: *Geschichte der wirtschaftlichen Entwicklung Englands*, Vol. 3, 1, Jena, 1928.
39. MICHEL, ERNST: *Sozialgeschichte der industriellen Arbeitswelt*, Frankfurt a.M., 1947, p. 73 ff.
40. SCHNABEL, FRANZ: *F. Redtenbacher*, Blätter f. Gesch. d. Techn., No. 4, 1938, p. 66–71.
41. SPENGLER, O.: *Der Mensch und die Technik*, Munich, 1931.
42. GLÜCK, K.: *Japans Vordringen auf dem Weltmarkt*, Würzburg, 1937.
43. Lenin's letter to the Chairman of the Electrification Commission, 14/3/1920. *Böttcherstrasse*, 1929, No. 9 (*Weltreich der Technik*) p. 48.
44. FAULKNER, H. U.: *Amerikanische Wirtschaftsgeschichte*, Vol. 2, Dresden, 1929, p. 268.
45. BIRINGUCCIO, V.: *Pirotechnia*, Venice, 1540.
46. TAYLOR, F. W.: *The Principles of Scientific Management*, New York, 1911.
47. MATSCHOSS, C.: *Entwicklung der Dampfmaschine*, Vol. 1, Berlin, 1908, p. 174.
48. *Festschrift der Gasmotorenfabrik Deutz zum* 25 *jährigen Jubilaum*, Köln, 1889, p. 33.
49. Compare particularly the critical attitude (in part contrary to Norbert Wiener) of J. Diebold: *Automation. The Advent of the Automatic Factory*, New York, 1952, p. 154 ff.
50. GOETHE, J. W. V.: *Wilhelm Meisters Wanderjahre*, Book 3, Chap. 14, *Werke*, Sophien-Edn., Vol. 25, Part 1, Weimar, 1895, p. 272 f.

REFERENCES TO SOURCES

Acknowledgement is gratefully made to the copyright owners indicated for permission to reproduce copyright material.

Ref. No.

[1] Plato, *Dialogues* (*Georgias*), trans. Jowett; O.U.P. 1892.

[2] Plato, *Laws*; same edition as above.

[3] Plutarch, *Lives*, trans. Langhorne; Chatto, 1876.

[4] Seneca, *Letters to Lucilius*, No. LXXVIII, trans. Barker; Clarendon Press, 1932.

[5] Seneca, *Letters to Lucilius*, No. XC (as above).

[6–8] Pseudo-Aristotle, *Mechanical Problems*, trans. Hett; London and Harvard University Press, Loeb Classical Lib., William Heinemann Ltd., 1936.

[9] Xenophon, *Instruction in Child-rearing*, trans. Miller; Loeb Classical Lib., 1914.

[10] Pliny, *Natural History*, trans. Rackham and Jones; Loeb Classical Lib., 1938.

[11] Philo of Byzantium, *Mechanics* (Book IV).

[12], [13] Vitruvius, *Architecture*; Loeb Classical Lib., 1914.

[14–17] Hero of Alexandria, *Concerning the building of Missiles*.

[18–21] Hero of Alexandria, *Pressure Machines and Automatatheatres*.

[22] Pappus Alexandrinus, *Collections*, from German ed. Hultsch, 1900.

[23–26] Vitruvius, *On Architecture*, trans. Granger; Loeb Classical Lib., 1944–56.

[27], [28] Pliny, *Natural History*; see ref. no. 10.

[29] Gregory of Nyssa, *The Catathetical Oration*, trans. Strawley; S.P.C.K.

[30], [31] Gregory of Nyssa, *On the Making of Man*, from *Select* Writings and Letters, trans. Wilson, 1893.

[32] Taoist Author, from *Progress and Religion* by Christopher Dawson; Sheed and Ward, 1929.

[33], [34] St Augustine, *The City of God*, trans. Healy; Everyman Lib., J. M. Dent & Sons.

[35] *Manumission of a Slave* (*unfree*) *Woman in the year* 837 *A.D.*, from German version by Günther Franz in *Deutsches Bauerntum*, Weimar, 1940.

[36–41] Theophilus Presbyter (also called, but rarely, Rugerus) *Scheme of Various Arts*, trans. Hendric; 1847.

[42] Thomas Aquinas, *Summa Theologica*.

[43] Hugo de Sancto Victore, *Didascalicon* (*For Instruction*) from the German version by Joseph Freundgen; Paderborn, 1896.

[44] Benū Mūsā, *Instrument to Raise Articles out of Water*, from the German translation of the ninth century *Kitab al Hijal*, by Friedrich Hauser, Erlangen, 1922.

[45] Benū Mūsā, *Fi'l Hijal*, from the German translation by Wiedemann and Hauser, in *Islam*, 1918.

[46–48] *Windmills among the Arabians, according to Extracts from al-Mas'udi, al-Qazwini and al-Dimaschqi* from Wiedemann's *Zur Mechanik und Technik bei Arabern*, Erlangen, 1906.

Ref. No.

[49] Wiedemann, Eilhard, *Zur Technik bei den Arabern*, Erlangen, 1906.

[50–53] Villard de Honnecourt: (Plates 2, 9, 17, 44) from R. Willis: *Facsimile of the Sketch-book of Wilars de Honecort*, London, 1859.

[54–55] Pierre de Maricourt, *De Magnete*; cf. Peter Peregrinus, *De Magnete*; cf. also *The Letter of Peregrinus on the Magnet*, 1269 in *Electrical World*, Vol. 43, 1904.

[56], [57] Bacon, Roger, *Opus Maius*, trans. Robert Belle Burke, Univ. of Pennsylvania, 1928.

[58] Bacon, Roger, *Epistola de secretis operibus artis et naturae.*

[59] *Inventory of Nuremberg Handworkers in the Year* 1363 from *Chroniken der deutschen Städte*, Vol. II, ed. Hegel; Leipzig, 1864.

[60] Kyeser von Eichstätt, Konrad, *Bellifortis*, from the German translation of Graf Carl von Klinckowstroem, 1928.

[61] Kyeser von Eichstätt, Konrad, from same manuscript as 60 trans. by A. Becker-Freyseng, from the German version.

[62] *Letter of Introduction for the Gunmaster Merckln Gast*, of about the year 1390, Municipal Archives of Frankfurt-am-Main, from a German version in B. Rathgen: *Das Geschütz im Mittelalter*, Berlin, 1928.

[63] *Das Feuerwerksbuch*, translated from the German version edited by W. Hassenstein, Munich, 1943.

[64–67] Alberti, Leon Battista, *Architectura*, trans. by James Leoni, London, 1755.

[68] Francesco di Giorgio Martini. *Trattato di architettura.* (Torino 1841).

[69–72] Leonardo da Vinci, *The Literary Works of Leonardo da Vinci compiled and edited from the original Manuscripts* by J. P. Richter, O.U.P., 1939.

[73] Tartaglia, Niccolo, *Quesiti et inventioni diverse*, Venice, 1546, in *Three Books of Colloquies concerning the Arts of Shooting*, Cyprian Lucas, 1588.

[74] Filarete, Antonio Averlino, *Trattato dell' Architettura*, 1464, from the German version by W. von Oettingen.

[75] Bourbon, Nicolas, *Ferraria*, from the German version by Ludwig Harald Schütz, Göttingen, 1895.

[76] Biringuccio, V., *Pirotecnia*, Venice, 1540 trans. by Prof. Cyril Smith Chicago.

[77–78] Bombastus von Hohenheim (Paracelsus), *Das Buch Paragranum*, from *Paracelsus Selected Works*, trans. Guterman, Routledge & Kegan Paul, 1951. (Partly fresh translation.)

[79] Bombastus von Hohenheim (Paracelsus), *Die Bücher von den unsichtbaren Krankheiten*, 1531–2; from Sudhoff's *Paracelsus Sämtliche Werke*, Vol. 9, Munich, 1924.

[80] Bombastus von Hohenheim (Paracelsus), *Labyrinthus Medicorum errantium*, 1537–8, trans. A. E. Waite: *The Hermetical and Alchemical Writings of Paracelsus*, 1894.

[81] Bombastus von Hohenheim (Paracelsus), *De Tempore laboris et requiei*, trans. from Goldammer's *Paracelsus*: *Sozialethische und sozialpolitische Schriften*, Tübingen, 1952.

[82] Agricola, Georgius (1556), *De re metallica*, partly from the translation by H. C. and L. H. Hoover. *The Mining Magazine.*

[83] Hampe, *Nürnberger Ratsverlässe*, trans. from *Quellen-Schriften für*

Ref. No. *Kunstgeschichte und Kunsttechnik*, Vols. 11–13, Vienna and Leipzig, 1904.

[84] Del Monte, Guidobaldo, *Mechanicorum*, Pesaro, 1577; from Daniel Mögling's *Mechanische Kunstkammer*, Frankfurt, 1679.

[85], [86] Lorini, Buonaiuto, *Delle fortificationi*, Venice, 1597 from the German version.

[87], [88] Fuller, Advocate, Part of Speech delivered in 1602 in the Royal Law Court, London, concerning a monopoly of the right to import, manufacture and sell playing cards alleged to be based on the grant of a Royal Privilege, Quarterly Papers on Engineering, Vol. 5. London, 1846.

[89] Bacon, Francis, *Novum Organum*; copied from Devey ed. of *Works of Francis Bacon*.

[90] Bacon, Francis, *New Atlantis*; see 89.

[91], [92] Galilei, Galileo, *Discorsi e dimostrazioni matematiche*; 1638. Trans. Crew and de Salvio, Northwestern University Press.

[93] Descartes, René, *Letter to Mersenne of Oct.* 11, 1638 concerning Galileo's *Discorsi*, Descartes *Œuvres*, ed. Adam and Tannery, 1898.

[94] Descartes, René, *Discourse on Method*; trans. Rawlings.

[95] Pascal, Blaise, *Lettre dédicatoire à Monseigneur le chancelier (Pierre Séguier) sur le sujet de la machine nouvellement inventée par le Sieur B.P.*, Paris, 1643.

[96], [97] Borelli, Giovanni Alfonso, *De motu animalium*, Rome, 1680.

[98] Glauber, J. R., *On Metals*; Frankfurt, 1651, trans. Packe, 1689.

[99] Glauber, J. R., *Germany's Welfare*; Amsterdam, 1660, trans. Packe.

[100] Glauber, J. R., *The Miracle of the World*; Amsterdam, 1660, trans. Packe.

[101] Böckler, Georg Andreas, *Theatrum Machinarum Novarum*; Nuremberg, 1661.

[102] Wilkins, John, *Mathematical Magic*, London, 1648.

[103] Boyle, Robert, *Some considerations touching the usefulness of Experimental Natural Philosophy*.

[104] Boyle, Robert, *Testament*.

[105–108] Baxter, Richard, *Christian Directory*; from the 2nd ed., London, 1678.

[109] Barclay, Robert, *Theologiae verae Christianae Apologia*; London, 1678.

[110] Edwards, John, *A complete history or survey of all the dispensations and methods of religion*, London, 1694.

[111] Fontana, Domenico, Removal of the Vatican Obelisk, Rome, 1590.

[112] Bélidor, B. Forest de, *Architecture hydraulique*, Paris, 1739.

[113] Leupold, Jakob, *Description of the great machine at Marly*.

[114] Calvör, Henning, *Historisch-Chronologische Nachricht . . . des Maschinenwesens . . . auf dem Oberharze*; Brunswick, 1763.

[115] Leibniz, G. W., *Wasserhebung mittels der Kraft des Windes*.

[116–118] Huygens, Christian, from *Œuvres Complètes*, Vols. vi, xxii, vii, The Hague, 1895–1950.

[119] *Ad majorem Dei gloriam*; published in *Nouvelles de la République des Lettres*, Amsterdam, 1687.

[120] Papin, Denis, *Fasciculus dissertationum de novis uibusdam machinis*; Marburg, 1695.

[121–124] *Leibniz und Huygens' Briefwechsel mit Papin*; ed. Gerland, Berlin, 1881.

Ref. No.

[125] Leupold, Jakob, *Theatrum Machinarum oder Schauplatz der Hebezeuge*; Leipzig, 1725.

[126–128] Leupold, Jakob, *Theatrum machinarum hydraulicarum*, Leipzig, 1725.

[129] Leupold, Jakob, *Kurzer Entwurff, auf was Arth die Verbesserung des Machinen-Wesens bey denen Bergwercken zu veranstalten*, Leipzig, 1725.

[130] Fairbairn, William, *Treatise on Mills and Mill-work*, London, 1861.

[131] Polhem, Christopher, *Patriotic Testament*, 1746.

[132] Réaumur, *Introduction* to *Description des Arts et Métiers*, published by the Académie Royale of Paris, 1761.

[133] Justi, J. H. G. von, *Schauplatz der Künste und Handwerke*, Berlin, 1762.

[134] Perronet, J. R., *Remarques à l'art de l'Epinglier*, Paris 1762.

[135–138] Smith, Adam, *Inquiry into the nature and causes of the Wealth of Nations*, London, 1776.

[139] Franklin, Benjamin, *Good counsel to a young handicraftsman*,1748. (*Complete Works*, London, 1806, Vol III.)

[140] Darby, Abiah, *Letter written in* 1775 *concerning the discovery by her father-in-law how to extract iron from ore by the use only of coke obtained from coal*, (in the *Economic Series* of the University of Manchester, No. 111, 1924).

[141] Watt, James, *Steam Engines*; *Specification number* 913, 5 *Jan.* 1769. (From J. P. Muirhead *Specifications of James Watt's Patents*, London 1854.)

[142] Reichenbach, Georg von, *Diary of a Journey to England in* 1791. MS. in Deutsches Museum, Munich.

[143] Poleni, Giovanni, *Historical memoir concerning the great dome of the Vatican* (with passages in inverted commas inserted quoted from an anonymous work *Sentimento d'un filosofo* published in Venice, 1744) Padua, 1748.

[144] Emperor Frederick the Great, *Letter written to Voltaire in* 1778 from *Briefwechsel Friedrichs des Grossen und Voltaire*, ed. Koser & Droysen, Royal Prussian State Archives, Vol. 86.

[145] Coulomb, Charles Augustin de, *Théorie des machines simples*, Paris 1821.

[146] Coulomb, Charles Augustin de, *Essai sur une application des règles de maximis et minimis à quelques problèmes de statique relatif à l'Architecture* in *Mathematical Memoires* presented to the Académie des Sciences, Tom. 7 Paris, 1776.

[147] Pinet, G., *Histoire de l'École Polytechnique*, Paris 1887.

[148] Dinnendahl, Franz, *Autobiography*, published in *Beiträge zur Geschichte von Stadt und Stift Essen*, Vol. 28, 1905.

[149] Dinnendahl, Franz, *Estimate of cost of a* 15-*inch double-acting steam winch* (1807) from Dinnendahl's manuscript sketch book in the Deutsches Museum, Munich.

[150] Carnot, Sadi, *Réflections sur la puissance motrice du feu et sur les Machines*, Paris 1824.

[151] Harkort, Friedrich, *Eisenbahnen*, first published in *Hermann*, No. 26 of March 30, 1825.

[152] Harkort, Friedrich, *Abhandlung über Schienenwege*, 1826. A memorial

Ref. No. volume of 1826 published in the volume *Friedrich Harkort in his unknown poems and unpublished letters and documents*, ed. Hilgenstock & Bacmeister, Essen, 1937.

[153] Nasmyth, James, *Autobiography*, ed. Samuel Smiles, London, 1883.

[154] Babbage, Charles, *On the Economy of Machinery and Manufactures*, London, 1832.

[155–159] Baines, Edward, *History of the Cotton Manufacture in Great Britain*, London, 1835.

[160–163] Ure, Andrew, *Philosophy of Manufactures*, London, 1835.

[164–166] Fischer, Johann Conrad, *Tagebücher*, 1794–1851, from the new edition pubd. in Schaffhausen in 1951.

[167] Schinkel, Karl Friedrich, *Nachlass: Reisetagebücher, Briefe und Aphorismen*, ed. von Wolzogen, Berlin 1863.

[168] Beuth, Christian P. W., *Letter to Schinkel written in* 1823. See 167.

[169] Ludlow, John M. and Lloyd Jones, *Progress of the Working Class*, 1832–1867, London, 1867.

[170–175] Goethe, Johann Wolfgang von, *Wilhelm Meisters Wanderjahre* 1829, trans. R. O. Moon, 1947. G. T. Foulis & Co. Ltd.

[176] Harkort, Friedrich, *Bemerkungen über die Hindernisse der Zivilisation und Emanzipation der untern Klassen*, (ed. Julius Ziehen, Frankfurt, 1919.

[177–184] Marx, Karl, *Das Kapital*, Hamburg, 1867, trans. Eden & Cedar Paul, London 1930.

[185] Redtenbacher, Ferdinand, *Resultate für den Maschinenbau*, Mannheim 1848.

[186] Redtenbacher, Ferdinand, *Der Maschinenbau*, Mannheim 1862.

[187–189] Redtenbacher, Ferdinand, *Prinzipien der Mechanik und des Maschinenbaues*, Mannheim, 1852.

[190–191] Reuleaux, Franz, *Briefe aus Philadelphia*, Brunswick, 1877.

[192–193] Göldy, A., *Bericht . . . über die Maschinen-Industrie . . . auf der Internationalen Ausstellung in Philadelphia*, Winterthur, 1877.

[194–195] Goldschmidt, Friedrich, *Die Weltausstellung in Philadelphia und die Deutsche Industrie*, Berlin 1877.

[196] *The Westinghouse Foundry, near Pittsburg, Pennsylvania*, from *The Scientific American*, Vol. 62, No. 24, 1890.

[197] Ford, Henry, *My Life and Work*, New York: Doubleday & Co. Inc., 1922.

[198] Weber, Max Maria von, *Aus der Welt der Arbeit*, Berlin 1867.

[199] Reuleaux, Franz, *Theoretische Kinematik*, Brunswick, 1875, trans. Kennedy, London, 1876.

[200–201] Diesel, Rudolf, *Die Entstehung des Dieselmotors*, Berlin, 1913.

[202] Reuleaux, Franz, Manuscript 4967 of the Deutsches Museum, Munich.

[203–205] Vorsselmann de Heer, P.O.C., *Über den Elektro-magnetismus als bewegende Kraft*, in *Annalen der Physik und der Chemie*, Vol. 123, 1839.

[206] Hamel, J., *Colossale magneto-elektrische Maschine zum Versilbern und Vergolden*, from *Journal für praktische Chemie*, Vol. 41, 1847.

[207–209] Siemens, Werner, *Ein kurzgefasstes Lebensbild nebst einer Auswahl seiner Briefe*, ed. Matschoss, Berlin, 1916.

[210] Wilke, Arthur, *Die Elektrizität*, Leipzig, 1893, (in the 8th edition of *Buch der Erfindungen*).

Ref. No.

[211] Lilienthal, Otto, *Weshalb ist es so schwierig das Fliegen zu erfinden*, in *Prometheus*, No. 261, Berlin, 1895.

[212] Braun, Ferdinand, *Drahtlose Telegraphie durch Wasser und Luft*. Leipzig, 1901.

[213–214] Braun, Ferdinand, *Elektrische Schwingungen und drahtlose Telegraphie*, Stockholm, 1910.

[215] Arco, Count Georg von, *Mögliches und Unmögliches in der drahtlosen Telegraphie*, *Berliner Tageblatt*, May 18, 1904.

[216] Le Rossignol, Robert, *Zur Geschichte der Herstellung des synthetischen Ammoniaks*, in *Die Naturwissenschaften*, Berlin, 1928.

[217] Bosch, Carl, *Über die Entwicklung der chemischen Hochdrucktechnik bei dem Aufbau der neuen Ammoniakindustrie*, Nobel address of 1932.

[218] Smyth, Henry de Wolf, *Atomic energy for Military Purposes, Official report of the development of the Atomic bomb*. (Report of the U.S.A. War Ministry on the test in New Mexico on July 16, 1945.) U.S. Government Printing Office.

[219] Wiener, Norbert, *The human use of human beings* (Cybernetics and Society) Boston: The Houghton Mifflin Company. London: Eyre & Spottiswoode (Publishers) Ltd., 1949.

SELECT BIBLIOGRAPHY

General

BECKMANN, JOHANN: *A History of Inventions*, London, 1797–1814.

DAVISON, C. ST C. B.: *Historic Books on Machines*, London, 1953.

DERRY, T. K. and WILLIAMS, T. I.: *A Short History of Technology*, Oxford, 1960.

FINCH, JAMES KIP.: *Engineering and Western Civilisation*, New York, 1951.

FINCH, J. K.: *The Story of Engineering*, New York, 1960.

FLEMING, A. P. M. and BROCKLEHURST, H. J.: *A History of Engineering*, London, 1925.

FORBES, R. J.: *Man the Maker*. New York, 1950.

KIRBY, RICHARD S. and LAURSON, P. G.: *The Early Years of Modern Civil Engineering*, New Haven 1932.

MUMFORD, LEWIS: *Technics and Civilization*, New York, 1934.

MUMFORD, LEWIS: *Art and Technics*, London, 1952.

OLIVER, J. W.: *History of American Technic*, New York, 1956.

RICKARD, THOMAS A.: *Man and Metals*: *A History of Mining*, Vol. 1, 2, New York, 1932.

SARTON, G.: *Introduction to the History of Science*. Vol. 1–3, Baltimore, 1927–47.

SINGER, CHARLES, HOLMYARD, E. J., HALL, A. R., and WILLIAMS, TREVOR I.: *A History of Technology*. Vol. 1–5, London, 1954–1958. *Technology and Culture*. Vol. 1 sqs. Detroit, 1959–1960 sqs.

Transactions of the Newcomen Society. Vol. I ff. London, 1920/21 ff.

USHER, ABBOTT PAYSON: *A History of Mechanical Inventions*, New York, 1929.

I. Graeco-Roman Antiquity

ASHBY, TH.: *The Aqueducts of Ancient Rome*, Oxford, 1935.

DAVIES, O.: *Roman Mines In Europe*, Oxford, 1935.

DRACHMANN, A. G.: *Ancient Oil Mills and Presses*, Copenhagen, 1932.

DRACHMANN, A. G.: *The Mechanical Technology of Greek and Roman Antiquity*, Copenhagen, 1963.

FORBES, R. J.: *Studies in Ancient Technology*, Vol. 1–8, London, 1955–64.

The Legacy of Egypt, ed. S. R. K. Glanville, Oxford, 1947.

The Legacy of Greece, ed. R. W. Livingstone, Oxford, 1951.

The Legacy of Persia, ed. A. J. Arberry, Oxford, 1953.

The Legacy of Rome, ed. C. Bailey, Oxford, 1951.

The Legacy of Israel, ed. E. R. Bevan, Oxford, 1948.

PARTINGTON, J. R.: *Origins and Development of Applied Chemistry*, London, 1935.

ROSTOVTZEFF, MICHAEL: *Social and Economic History of the Hellenistic World*, Oxford, 1953.

TOUTAIN, J.: *The Economic Life of the Ancient World*, 2nd ed. London, 1951.

II. The Middle Ages

BRØGGER, A. W. and SHETELIG, H.: *The Viking Ships*. Oslo, 1953.

CROMBIE, A. C.: *Augustine to Galilei. The History of Science* A.D.400–1650. London, 1952.

CROMBIE, A. C.: *Robert Grosseteste and the origins of Experimental Science*, Oxford, 1953.

DURAND, DANA B.: 'Nicole Oresme and the Mediaeval Origins of Modern Science,' *Speculum*, Vol. 16. 1941.

EASTON, ST C.: *Roger Bacon and his Search for a Universal Science*, Oxford, 1952.

FORBES, R. J.: *Short History of Distillation*, Leiden, 1948.

HASKINS, C. H.: *Studies in the History of Mediaeval Science*, 2nd. ed. Cambridge (Mass) 1927.

HOLMES, U. T.: *Daily Living in the 12th Century*, Madison, 1952.

JOHNSON, R. P.: 'The Manuscripts of the Schedula of Theophilus Presbyter.' *Speculum*, Vol. 13, 1938.

The Legacy of the Middle Ages, ed. E. G. Crump and E. F. Jacob, London, 1951.

The Legacy of Islam, ed. T. Arnold and A. Guillaume, London, 1949.

NEEDHAM, JOS.: *Science and Civilisation in China*, Vol. 1 ff. Cambridge, 1954 ff.

O'LEARY, DE LACY: *How Greek Science passed to the Arabs*, 2nd imp. London, 1951.

PEVSNER, N.:'The Term "Architect" in the Middle Ages,' *Speculum*, Vol. 17, 1942.

POSTAN, M. and E. E. RICH: 'Trade and Industry in the Middle Ages,' (*The Cambridge Economic History of Europe*, Vol. 2) Cambridge, 1952.

SALZMANN, L. F.: *Building in England down to* 1540, Oxford, 1952.

THORNDIKE, LYNN: *A History of Magic and Experimental Science*, Vol. 1–8, New York, 1923–58.

THORNDIKE, LYNN: 'Invention of the Mechanical Clock about 1271 A.D.' *Speculum* Vol. 16, 1941.

WHITE, LYNN: 'Technology and Invention in the Middle Ages,' *Speculum*, Vol. 15, 1940.

WHITE, LYNN: 'Natural Science and Naturalistic Art in the Middle Ages,' *Amer. Hist. Review*, Vol. 52, 1947, No. 3, pp. 421–35.

WHITE, LYNN: *Medieval Technology and Social Change*, Oxford, 1962.

III. The Renaissance

HALL, A. R.: *Ballistics in the 17th Century*, Cambridge, 1952.

MCCURDY, EDWARD: *Leonardo da Vinci's Notebooks*. New York, 1923.

MASON, S. F.: 'The Scientific Revolution and the Protestant Reformation,' *Annals of Science*, Vol. 9, 1953.

PARSONS, W. B.: *Engineers and Engineering in the Renaissance*, Baltimore, 1939.

TAYLOR, R. E.: *Luca Pacioli and his Times*, Chapel Hill, 1942.

IV. The Baroque Period

BAXTER, RICHARD: *Autobiography* (*Reliquiae Baxterianae*, 1696) ed. J. M. Lloyd Thomas, London, 1925.

CUNNINGHAM, W.: *Christianity and Economic Science*, London, 1911.

CUNNINGHAM, W.: *The Growth of English Industry and Commerce in Modern Times*, London, 1915–19.

DAVIES, D. S.: 'Early History of the Patent Specification,' *Law Quarterly Review*, London.

DIRCKS, HENRY: *The Life, Times and Scientific Labours of the Second Marquis of Worcester*, London, 1865.

FARRINGTON, BENJAMIN: *Francis Bacon. Philosopher of Industrial Science*, New York, 1949.

FRUMKIN, M.: 'Early History of Patents for Invention,' *Transact. Newcomen Soc.* Vol. 26, for 1947–49 (London, 1953).

GOMME, A. A.: 'Patents of Invention.' *Origin and Growth of the Patent System in Britain*, London, 1948.

JENKINS, RHYS: 'The Protection of Inventions during the Commonwealth and Protectorate.' In: *Jenkins Collected Papers*, Cambridge, 1936.

MERTON, ROBERT K.: 'Science, Technology and Society in 17th Century England,' *Osiris*, Vol. 4, 1938.

MORE, LOUIS TRENCHARD: *The Life and Works of Robert Boyle*, London, 1944.

'The Progress of Machinery and Manufactures in Great Britain,' *Quarterly Papers of Engineering*, Vol. 5, London, 1846.

RAISTRICK, ARTHUR: *Quakers in Science and Industry*, London, 1950.

SCOVILLE, W. C.: 'The Huguenots and the Diffusion of Technology.' *Journal of Political Economy*, Vol. 60, 1952.

TAWNEY, R. H.: *Religion and the Rise of Capitalism*, New York, 1926.

WOLF, A.: *History of Science, Technology and Philosophy in the 16th and 17th Centuries*, Vol. 1, 2, London, 1935, 2nd ed., 1950.

V. The Age of Rationalism

ASHTON, THOMAS SOUTHCLIFFE: *Iron and Steel in the Industrial Revolution*, 2nd ed. Manchester, 1951.

CLOW, A. and CLOW, N. L.: *The Chemical Revolution*, London, 1952.

COHEN, I. BERNARD: 'Prejudice against the Introduction of Lightning Rods.' *Journal Franklin Inst.* Vol. 253.

COHEN, I. BERNARD: *Benjamin Franklin*, Indianapolis, 1953.

DICKINSON, H. W. and JENKINS, R.: *James Watt and the Steam Engine*, Oxford, 1927.

FAIRBAIRN, WILLIAM: *Treatise on Mills and Millwork*, London, 1861–63

NEF, J. U.: *The Rise of the British Coal Industry*, Vol. 1, 2, London, 1932.

RAISTRICK, A.: *Dynasty of Ironfounders. The Darby and Coalbrookdale*, London, 1953.

TOKSVIG, S.: *Swedenborg*, London, 1949.

VI. The Industrial Era

Charles Babbage's Calculating Engines. A collection of papers relating to them. ed. H. P. Babbage, London, 1889.

GIDDENS, P. H.: *Early days on Oil*, Princeton (N.J.) 1948.

GLOAG, J. and BRIDGEWATER, D.: *A History of Cast Iron in Architecture*, London, 1948.

HABAKKUK, H. J.: *American and British Technology in the 19th Century*, Cambridge, 1962.

HENDERSON, W. O.: *Britain and Industrial Europe* 1750–1870, Liverpool, 1954.

OSBORN, F. M.: *The Story of the Mushets*, London, 1952.

POLE, WILLIAM: *Life of Fairbairn*, London, 1877. Abridged Edition, 1878.

ROE, JOSEPH W.: *English and American Tool Builders*, 2nd ed., New York, 1926.

SPRATT, H. P.: *Outline History of Transatlantic Steam Navigation*, London, 1950.

VII. Technology becomes a World Power

ALLEN, G. C.: *A Short Economic History of Modern Japan*, 1869–1937, London, 1946.

BERKELEY, EDMUND: *Giant Brains*, New York, 1950.

BORTH, CHRISTY: *Masters of Mass Production*, New York, 1945.

COLE, G. H. D.: *Introduction to Economic History*, 1750–1950, London, 1953.

DIEBOLD, JOHN: *Automation. The Advent of the Automatic Factory*, New York, 1952.

EVANS, A. F.: *The History of the Oil Engine*, London, 1930.

GIEDION, S.: *Mechanization Takes Command*, New York, 1948.

GLOAG, J. and BRIDGEWATER, D.: *A History of Cast Iron in Architecture*, London, 1948.

GLOVER, J. G. and CORNELL, W. B.: *The development of American Industries*, 3rd ed. New York, 1951.

LOCKWOOD, W. W.: *The Economic Development of Japan*, 1868–1938, Princeton, (N.J.) 1954.

MOTTELAY, P. F.: *History of Electricity. Engineering*. Vol. 51, (1891) to Vol. 53, (1892).

MOTTELAY, P. F.: *Bibliographical History of Electricity and Magnetism*, London, 1922.

PASSER, H. C.: *The Electrical Manufacturers*, 1875–1900, Cambridge, U.S.A., 1953.

TAYLOR, FREDERIC W.: *The Principles of Scientific Management*, New York, 1911.

WALKER, J. B.: *The Epic American Industry*, New York, 1949.

WIENER, NORBERT: *Cybernetics*, New York, 1948.

INDEX OF NAMES

GENERAL INDEX